SYMPOSIA OF THE
INTERNATIONAL SOCIETY
FOR CELL BIOLOGY

VOLUME 6

# Formation and Fate of Cell Organelles

*Edited by*

KATHERINE BREHME WARREN

*National Institutes of Health, Cell Biology Study Section*
*Division of Research Grants*
*Bethesda, Maryland*

1967

ACADEMIC PRESS

New York and London

ACADEMIC PRESS INC.
111 Fifth Avenue, New York, New York 10003

*United Kingdom Edition published by*
ACADEMIC PRESS INC. (LONDON) LTD.
Berkeley Square House, London W.1

Library of Congress Catalog Card Number: 62–13095

PRINTED IN THE UNITED STATES OF AMERICA

# LIST OF CONTRIBUTORS

Numbers in parentheses indicate the pages on which the authors' contributions begin.


MARCEL BESSIS, *School of Medicine, University of Paris, Paris, France* (233)

B. R. BRINKLEY, *Section of Cytology, Department of Biology, University of Texas M. D. Anderson Hospital and Tumor Institute, Houston, Texas* (175)

J. F. DANIELLI, *Unit for Theoretical Biology, School of Pharmacy, and Center for Theoretical Biology, State University of New York at Buffalo, Buffalo, New York* (239)

T. L. DORMANDY, *Department of Chemical Pathology, Whittington Hospital, London, England* (275)

I. R. GIBBONS,[1] *The Biological Laboratories, Harvard University, Cambridge, Massachusetts* (99)

AHARON GIBOR,[2] *Rockefeller University, New York, New York* (305)

AUDREY M. GLAUERT, *Strangeways Research Laboratory, Cambridge, England* (19)

A. V. GRIMSTONE, *Department of Zoology, University of Cambridge, Cambridge, England* (219)

MARTIN HAGOPIAN, *Department of Pathology, College of Physicians and Surgeons, New York, New York* (71)

D. E. HOOKES,[3] *Department of Biophysics, University of London, Kings College, London, England* (115)

J. M. HOPKINS, *Medical Research Council, Biophysics Research Unit, London, England* (115)

A. KLUG, *Medical Research Council, Laboratory of Molecular Biology, Cambridge, England* (1)

MYRON C. LEDBETTER, *Biology Department, Brookhaven National Laboratory, Upton, New York* (55)

J. A. LUCY,[4] *Strangeways Research Laboratory, Cambridge, England* (19)

[1] Present address: Pacific Biomedical Research Center, University of Hawaii, Honolulu, Hawaii.

[2] Present address: Department of Biological Sciences, University of California, Santa Barbara, California.

[3] Present address: Medical Research Council, Biophysics Research Unit, London, England.

[4] Present address: Department of Biochemistry, Royal Free Hospital School of Medicine, University of London, London, England.

A. H. MADDY, *Chemical Biology Unit, Department of Zoology, University of Edinburgh, Edinburgh, Scotland* (255)

DANIEL MAZIA, *Department of Zoology, University of California, Berkeley, California* (39)

ERNEST C. POLLARD, *Biophysics Department, The Pennsylvania State University, University Park, Pennsylvania* (291)

SIR JOHN RANDALL, *Department of Biophysics, University of London, King's College, London, England* (115)

RUTH SAGER, *Department of Biological Sciences, Hunter College of the City University of New York, New York, New York* (317)

DAVID SPIRO, *Department of Pathology, College of Physicians and Surgeons, New York, New York* (71)

ELTON STUBBLEFIELD, *Section of Cytology, Department of Biology, University of Texas M. D. Anderson Hospital and Tumor Institute, Houston, Texas* (175)

# PREFACE

This symposium is concerned with the manner in which cellular structures arise from the component molecules. From the study of the behavior of relatively simple viruses it has been shown that some simple structures may arise by spontaneous aggregation. We must therefore ask the question: Can all cellular organelles and cells themselves arise by spontaneous assembly, or is some regulation involved? If regulation is involved: What are the mechanisms of regulation?

In some cases, regulation may be very simple in nature. For example, the protein of tobacco mosaic virus will assemble spontaneously to give linear helical aggregates which are not determinate in length. But the RNA which is characteristic of this virus has a defined length, so that in the presence of both RNA and protein regulation of *size* appears. Another simple mechanism is available through nucleation—the presence of a large number of identical protein units may not, in itself, be a sufficient condition for assembly, i.e., the presence of a second molecular species which can nucleate may be necessary. Regulation of the *number* of structures formed may thus depend on nucleation. A third type of regulation may involve different conformations of a given species of macromolecule. In one conformation the sites required for intermolecular interaction leading to assembly may be concealed, and in a second conformation suitable sites may be exposed. The distribution of a macromolecular species among its possible conformations can often be controlled by the presence or absence of a species of small molecule. Regulation may then be effected by controlling the concentration of the small molecular species.

If we wish to do more than observe spontaneous assembly and regulation, we must be able to calculate the differences in free energy of the molecules in their various states. The lower the free energy of a particular array of molecules, the more stable will be that array, so that knowing the free energies we can not only observe, but also predict, what arrays will form. Successful prediction is the hallmark of true understanding in this as in most other branches of science. To make these predictions we need to know accurately the magnitude of intermolecular forces.

In a structure which has resulted from spontaneous assembly, stability of forms depends upon the detailed nature of the bonding. Thus, in tobacco mosaic virus the free energy of addition of protein subunits to the helical nucleoprotein structure is of the order of 7000 calories. In a phospholipid bilayer the free energy of addition of phospholipid molecules to the bilayer is about 20,000 calories. Consequently, both structures are stable in the sense that they are not readily destroyed by thermal agitation at physiological temperatures. However, the morphology of the virus is stable, whereas that of the bilayer is not. This arises from the fact that the bonding in the virus has stereochemical specificity, whereas in the bilayer this is not so.

In this report, the contributors present their views on a variety of aspects of the problem of spontaneous assembly; the volume is concluded by a review of some recent work on cytoplasmic inheritance. This focuses our attention on some presently unsolved problems; for example, why is it that, whereas most of the enzymes of mitochondria are coded for in the nuclear DNA, nevertheless there is DNA in the mitochondria themselves. The problem of regulation of assembly is scarcely touched in this volume, but the papers contributed make it abundantly clear that it must soon receive concentrated attention.

The International Society for Cell Biology wishes to record its thanks to the International Union of Biological Sciences for partial support of the symposium.

JAMES F. DANIELLI
*Center for Theoretical Biology*
*State University of New York at Buffalo*

# CONTENTS

# CONTENTS OF PREVIOUS VOLUMES

## Volume 1—The Interpretation of Ultrastructure

## Volume 2—Cell Growth and Cell Division

## Volume 3—Cytogenetics of Cells in Culture

## Volume 4—The Use of Radioautography in Investigating Protein Synthesis

## Volume 5—Intracellular Transport

# THE DESIGN OF SELF-ASSEMBLING SYSTEMS OF EQUAL UNITS

A. KLUG

*Medical Research Council, Laboratory of Molecular Biology*
*Cambridge, England*

## INTRODUCTION

Many large, ordered biological structures, such as the coats of virus particles, muscle filaments, microtubules, and bacterial flagella, consist of a large number of protein subunits, but the number of different types of subunits is in general small. In fact, in many protein structures there may be only one type of subunit. Examples of highly ordered biological structures made up of subunits which are, or at least appear to be, identical are the protein coats of tobacco mosaic and turnip yellow mosaic viruses, the actin and myosin filaments of muscle, and the microtubules of cilia and of the mitotic spindle. Some of these structures can be dissociated into their components, and the intact components can in turn be reassembled under appropriate conditions *in vitro* to produce a structure which appears to be the same as, or at least very similar to, the original organized structure assembled *in vivo*.

These large, organized structures are formed by making use of the specificity of the relatively weak noncovalent interactions that are possible between macromolecules, particularly between protein molecules (for a review, see Caspar, [4, 5]). Proteins play an important role in the assembly of such structures, possibly the most important role, because it appears possible to make them in a wide variety of shapes and to confer on them bonding properties required for any particular job. Moreover, many different protein molecules, such as the subunits of oligomeric enzymes (to use the terminology of Monod *et al.* [17]), can reversibly associate into definite structures held together by noncovalent bonds between the units. These bonds may be activated under certain physiological conditions, for example by interaction of the subunits with the substrate, which may thus also have a regulatory function.

The most significant features of organized structures built in this way is that their design and stability can be determined completely by the bonding properties of their constituent units. Thus, once the component parts are made they may "assemble themselves" [7] without a template or other specific external control, although, in the case of the certain enzymes and in the sheath of the T4 bacteriophage tails, a controlling function may be exerted by the binding of some other small molecule(s); this changes the configuration of the protein molecule so that it has the appropriate bonding properties.

A biological advantage of a self-assembly design for any large structure is that it can be completely specified by the genetic information required to direct a synthesis of the component molecules. Economical use of the genetic information carried by the nucleic acid of a gene will require that identical copies of some basic molecule or group of molecules will be used to build any large structure. These large structures built up of subunits can also be built efficiently and with great accuracy because of the possibility of checking; that is to say, any bad or wrong copy of the basic molecule will be rejected during the assembly process.

## ORDERED STRUCTURES—SYMMETRY AND SUBUNITS

The simplest way of making a large organized structure is to pack identical chemical units so that the same kind of contacts between units is used over and over again. In the final structure, each subunit must be bonded to its neighbors in exactly the same way; thus all subunits are in identical environments and are *equivalent* to each other in the crystallographic or mathematical sense. In other words, the final structure must necessarily be a symmetric one. Note that a structure of this kind necessarily possesses the property that fixing the bond distances and angles between a subunit[1] and its neighbors is sufficient to determine the whole structure. A structure which possesses crystallographic symmetry can therefore be built by self-assembly of its subunits.

If, therefore, we wish to find all ways of assembling identical units so that the final structure is determined by merely fixing a set of bonds, we can treat the problem abstractly by considering all the symmetries possible for a group of identical units arranged in identical environments. This is a classical mathematical problem which was finally solved in the

[1] The unit which builds the structure in this way is sometimes composed of more than one protein molecule or chemical subunit. It has therefore been defined as a *structure unit* [7] or *protomer* [17], but we shall not find it necessary to make this distinction in the general treatment given in this paper.

We have used the term "equal" rather than "identical" in the title of this paper, to cover the possibility that more than one chemical species of protein molecule (as defined by the amino acid sequence) may be present, but all the molecules are functionally equivalent in building the structure.

nineteenth century, but its relevance to the problem of constructing biological structures was first clearly stated by Crick and Watson [8] with reference to the structure of simple viruses.

In discussing this problem we shall make much use of the concept of a "bond site" and a "bond" [7]. By "bond site" we mean a particular region on the surface of a protein molecule which has the property of combining, under appropriate conditions, with a complementary region on another protein molecule, usually of the same kind. The coming together of the two bond sites creates a bond. Naturally, the attachment of two protein molecules to each other may involve a number of different patches of contact, but they may usually all be subsumed under the name of a single geometrical bond lying in a direction normal to the surface of contact, which is itself abstractly represented by a plane. The use of the terms is illustrated in Figs. 2 and 5.

For the case of an aggregate of protein molecules, we may think of each protein molecule as being equipped with a set of bond sites pointing in different directions, which are built into the very structure of the folded protein chain itself. Under the appropriate physiological conditions a number of such molecules may come together so that complementary sites on different molecules interact to form the bonds. The overall free energy of a pool or assembly of such molecules is reduced by making the bonds, and the minimum energy will be obtained when the randomly arranged pool of isolated molecules is converted to a smaller number of ordered aggregates.

## SPATIAL SYMMETRY AND THE DESIGN OF AGGREGATES

The possible types of spatial symmetry are restricted by the geometry of space. Since biological structures are built of molecules which are different from their mirror image, mirror or inversion symmetry is not possible at the molecular level of organization. The only kinds of spatial symmetry operations possible are rotations and translations. All the possible combinations of these two operations have been enumerated mathematically and are represented by the enantiomorphic point, line, plane, and space groups (see, for example, the book by Weyl [20], which provides a lucid discussion and many examples of the mathematics of spatial relations).

### *Line Groups*

Helical structures, such as tobacco mosaic virus (TMV, Fig. 1a), and more generally all linearly periodic assemblies, have some kind of *line group* symmetry. Line group symmetry results from a generalized screw operation (rotation plus translation parallel to a line) possibly combined

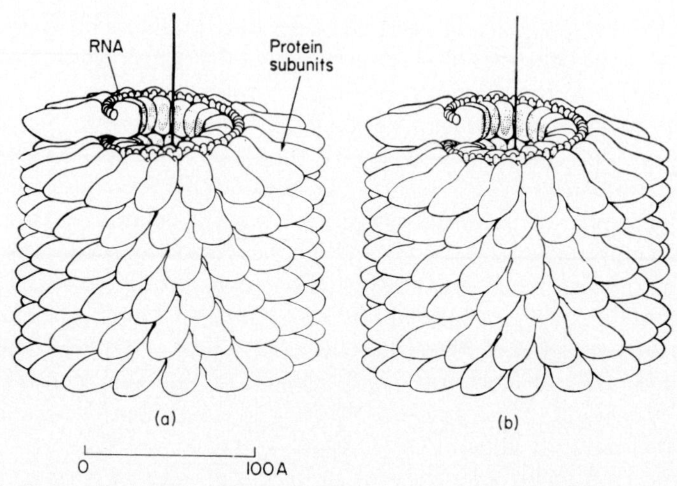


FIG. 1. Comparison of the structures of (a) the common strain of TMV and (b) the Dahlemense strain (from Caspar [3]). The segments shown correspond to about one-twentieth the length of the intact virus. The nucleic acid is represented as a chain coiled between the turns of the protein helix. (a) The diagram illustrates how each protein subunit of the common strain is equivalently related to its neighbors (cf. Fig. 2). (b) The packing of the subunits of the Dahlemense strain is very similar to that of the common strain, but there is a periodic perturbation of the positions of the ends of the subunits which brings turns of the helix near the outside surface alternately closer together and farther apart. The net result is that identical subunits in the Dahlemense strain are packed in 98 symmetrically distinct, but quasi-equivalent, environments.

with cyclic or dihedral point group symmetry. In the simplest type of helix, each unit is related to the next by a rotation plus a translation. Since the angle between subunits along a helix can take any value, there is no geometrical restriction on the number that can be placed in one turn, nor need this number even be integral. The length of a helical structure such as TMV is not determined by symmetry since the structure can repeat indefinitely along a line (the helix axis). A helical arrangement of subunits is therefore theoretically infinite in extent, but in the case of TMV the length of the particle is determined by the length of the RNA which interacts with the protein subunits and so provides an additional energy factor in the system. This method of fixing the length of the structure by the cocrystallization of a second component may be of more widespread occurrence. Thus, for example, synthetic filaments of actin appear to vary in length, but in the striated muscles the length of the actin filament is fixed, suggesting the presence of a second component. There are, however, other ways of terminating a linearly periodic structure, and these have been discussed by Caspar [4].

The bonding pattern at a particular radius in the TMV particle is

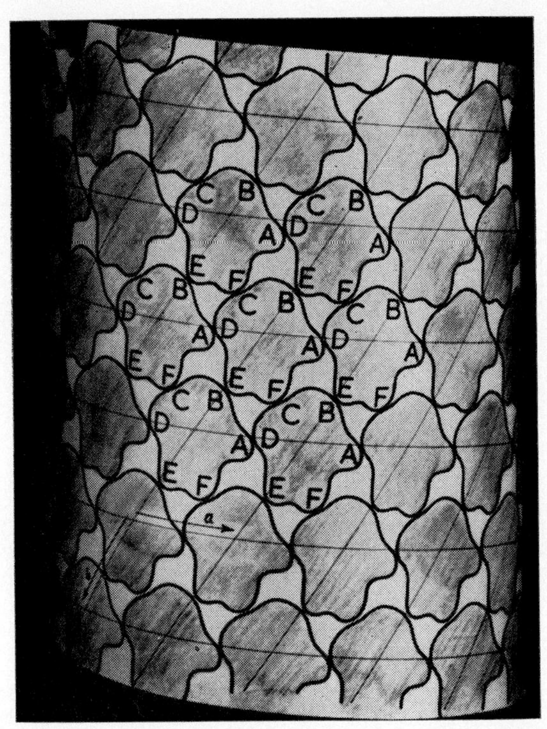


FIG. 2. Arrangement of two-dimensional units on a cylindrical surface illustrating equivalence in an array with helical symmetry. The surface lattice is defined by two of the families of helical lines that can be drawn through the lattice points. Note that each unit makes an identical set of "bonds" with its neighbors. The "bond sites" are A, B, C, D, E, and F, and the "bonds" are AD, BE, and CF.

shown diagrammatically in Fig. 2. If we were to look at another section of the structure at a different radius, the pattern would look different in detail because the physical contacts are different, but the *geometrical* relations between the subunits would be unchanged. We may therefore represent the structure by a two-dimensional rolled-up net, or cylindrical surface lattice. The rolled-up net is of course, a formal concept: in the real structure the rolling up will occur automatically in the assembly process because of the inherent curvature provided by the out-of-plane angles of the bonds, or, in other words, by the shape of the molecule, i.e., by the way that the combining regions at different radii on the molecule are fixed relative to one another.

### *Point Groups*

For a particle of finite extent, built of subunits regularly packed about a central point rather than about a line, the only symmetry possible is that of one of the *point groups* which contain a set of rota-

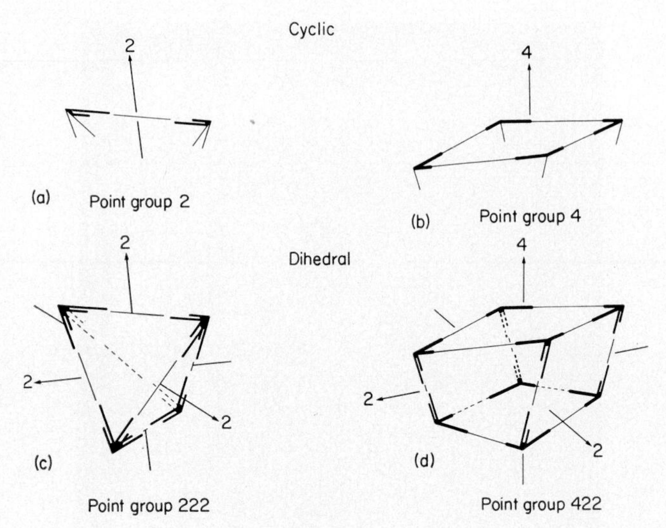


FIG. 3

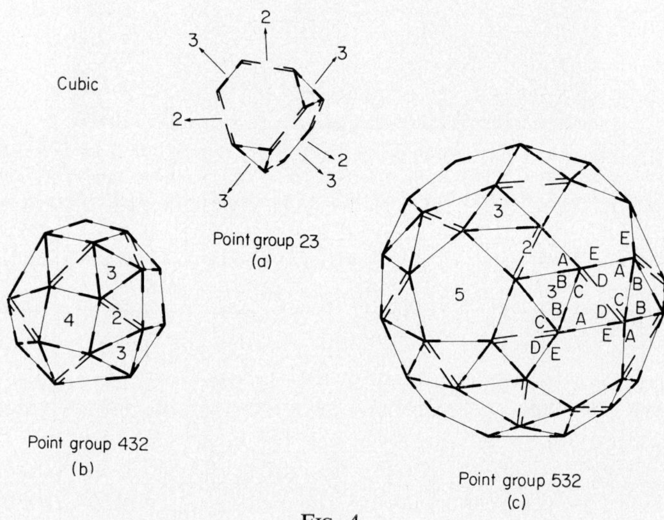


FIG. 4

FIGS. 3 and 4. Bonding patterns for asymmetric units ("structure units") arranged with point group symmetry (a structure unit may consist of more than one chemical subunit). The structure units are represented by the spiky objects at the vertices of a polygon or polyhedron, the edges of which represent the bonds between the units. The spikes represent the bond sites on a structure unit.

The pair of bonds formed about a 2-fold rotation axis forms a 2-sided polygon, which degenerates into a single edge in the diagrams. It should, however, still be thought of as a *double* bond composed of two identical single bonds; this is indicated by a pair of spikes emanating from the structure units.

The rotation axes are indicated in Fig. 3 by the arrows passing through the center,

tional symmetry axes passing through a point. The number of subunits must be a certain definite integer, depending on the particular combination of rotation axes present. This type of point group design is found for small protein aggregates, such as oligomeric enzymes, and for protein shells, such as occur in ferritin and the small, regular viruses. There are three types of point groups.

1. The simplest arrangement is that of *cyclic* symmetry, i.e., the arrangement which results when an integral number of subunits, $n$, are regularly arranged, head to tail, in a circle. A structure of this kind contains a single $n$-fold rotation axis, where $n$ is not restricted. Examples with $n = 2$ and $n = 4$ are shown in Figs. 3a and 3b, respectively.

Symmetry of a higher order results when there are a number (or "group") of rotation axes intersecting at a point. The number of ways in which various $n$-fold axes can be added is very limited.

2. Twofold axes can be combined at right angles with any single $n$-fold axis: this is the case of *dihedral* symmetry, in which the number of subunits is $2n$, $n$ not being restricted. Examples with $n = 2$ and $n = 4$ are shown in Figs. 3c and 3d, respectively.

3. The only other possible combinations of rotation axes are described by the three *cubic* point groups (Fig. 4a,b,c), which take their name from the fact that each of these point groups has at least one set of four 3-fold axes arranged as the four body diagonals of a cube, or, what amounts to the same thing, as the four axes connecting the center of a regular tetrahedron to its vertices. The number and relative arrangements of the rotation axes for these three types of cubic symmetry can be represented by the regular tetrahedron, a cube and the icosahedron ([8]; see Fig. 11 in [12]). The number of asymmetric units or subunits is, respectively, 12, 24, or 60.

In the case of cyclic symmetry the subunits lie on a circle. In the case of dihedral and cubic symmetry the subunits are arranged on a sphere, although this may not look very obvious. To put it more precisely, if a reference point is taken in any one subunit then the corresponding reference points in the other equivalent subunits of the

---

the numbers giving the order of the axes. In Figs. 4b and c, the directions of the rotation axes are omitted for clarity and replaced by the appropriate numbers on representative faces of the polyhedron.

The lettering in Fig. 4c refers to bond sites and is further explained in the legend to Fig. 5.

Note that the cyclic groups possess only a single class of bond. The dihedral and cubic groups formally possess three distinct classes of bond, but only two of these classes need to be formed to produce a complete, cohering structure. For example in Fig. 3c, the pair of bonds on two opposite edges of the polyhedron could be cut without destroying the formal integrity of the structure.

point group all lie on the same spherical surface. If one draws lines or "bonds" joining a reference point to its neighbors, then a polyhedron will result which can be used to symbolize the bonding pattern in the structure (Figs. 3 and 4). The vertices of the polyhedron fall on a sphere and can be thought of as forming the lattice points of a closed surface lattice. However, because of the variety of shapes of subunits, the appearances

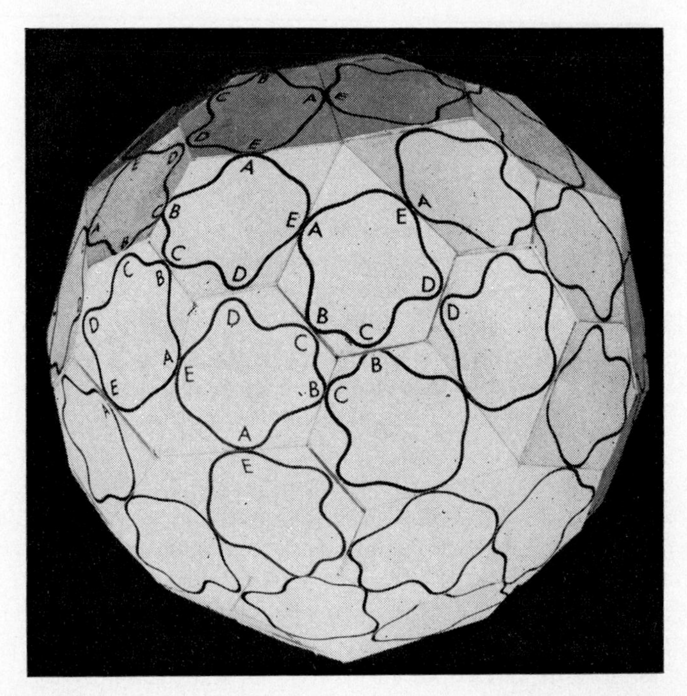


FIG. 5. Arrangement of two-dimensional units on a spherical (or strictly, polyhedral) surface illustrating equivalence in a structure with icosahedral symmetry. There are 60 identical units, and each unit makes an identical set of bonds with its neighbors. In this particular case, each unit is equipped with five "bond sites" *A, B, C, D,* and *E*. These form three different classes of "bond," namely a "pentamer" bond *AE*, a "trimer" bond *BC,* and a "dimer" bond *DD*. (The bond *DD* is strictly a double bond and the bond site *D* is composed of two identical parts related by a 2-fold axis.) Note that only two of these classes of bond are absolutely essential for the coherence of the structure.

of structures having the same type of symmetry can vary widely. Identification of structures of these kinds is also made difficult by the fact that the particular arrangement adopted in any one case of a real structure may be such as to disguise the inherent symmetry. Thus, for example, in the case of the tetrahedral symmetry (Fig. 4a), the 12 subunits may be arranged in the form of four, rather tight, clusters of 3 subunits each, and if the individual subunits are not resolved by the method of investigation used (e.g., the electron microscope or low-angle X-ray diffraction), then the structure may look as though it has only four gross or—in the terminology of Klug and Caspar [12]—*morphological,* units, and thus look as though it has the point group symmetry 222 (Fig. 3c). A fuller discussion of the connection between symmetry and morphology is given by Klug and Caspar [12].

### *Plane and Space Groups*

The *plane groups* are defined by two nonparallel translation vectors and may include axes of 2-, 3-, 4-, or 6-fold symmetry; examples of their occurrence in nature are the bacterial cell walls which seem to be two-dimensional layer structures built out of subunits. *Space groups* are the combination of the three-unit translations in three directions with the crystallographic point groups. All true regular three-dimensional structures, whatever their nature, have to have the microscopic symmetry of one of the space groups. We shall not here be concerned with plane and space groups since these theoretically lead to structures of indefinite extent, although, of course, in any real case they are terminated by various mechanisms.

### *Closed Designs and Intermediate Aggregates*

The important point about the point groups is that they lead to structures of finite extent, that is, they are the closed type of design that would be used for an aggregate of well-determined size, such as an oligomeric enzyme. In such a structure the design is necessarily one in which the specific bonding potential of the parts is saturated in some closed configuration. That is, a fixed number of molecules come together, and no more can be added using the same kind of bonds.

It is also worth stressing, among all the implications of such structures which we have not the space to deal with here, that the point groups with dihedral and higher symmetry can be factorized into subgroups. That is to say, they contain more than one class of bond, as discussed in the legends to Figs. 3, 4, and 5. Bonds of the same class form closed cycles centered on the rotational symmetry axes of the same class. This idea of classes of bond is of importance in analyzing association-dissociation phenomena. Thus, under any particular physiological state or under certain conditions *in vitro,* it is not necessary that the different classes of bonds be formed simultaneously. We thus have the possibility of forming stable intermediate aggregates in which only one class of bonds is made. For example, in a structure possessing tetrahedral point group symmetry 23, which we have already used as an example, it may be that under certain conditions in solution only the "trimer" bonds, i.e., those linking the subunits into groups of three, are formed, so that trimers will be found as intermediates aggregates. If the conditions are now changed and the additional classes of bond activated, for example, by the binding to the aggregate of a small molecule with a regulatory function, the trimers will come together to form the complete and final structure to which no more subunits can be added.

*Point-group Structures with More Than One Kind of Subunit*

Many multi-enzyme complexes are known, in which different protein molecules carrying out different enzymic functions are associated together to form a particle which is functionally significant and which appears to possess point-group symmetry [19.] If the symmetry is exact then the ultimate subunits corresponding to each enzyme function type must be combined in equal proportions. Crystallographically, the asymmetric unit of the structure would then consists of a set of molecules in which each kind of enzyme is represented once. If this set of molecules—an abstract notion—actually forms a *physical* grouping in the whole structure, the possibility arises that the whole particle can be "factorized" in terms of function into these separate groups, with of course, the additional possibility of cooperative (?) interactions between these groups.

## QUASI-EQUIVALENT BONDING

Because of the complex nature of protein molecules and the fact that the interactions between them may involve several distinct sets of combining regions (or bond sets, in our terminology), many ordered structures built of equal subunits with specific bonding properties cannot be represented by using the abstraction of strict equivalence. It is often possible to arrive at a structure of lower free energy by systematically deforming a set of bonds in a number of slightly different ways. This could come about, for instance, by the action of another component in the system which interacts with the protein molecules, or more generally, by the formation of another set of bonds which is not completely compatible with the principal set of bonds. However, if each subunit in the final structure still forms the same types or sets of bonds with its neighbors, then, although the units are no longer exactly equivalently related, they may still be said [7] to be *quasi-equivalently* related. Quasi-equivalence in ordered structures can be defined [4] as any small nonrandom variation in a regular bonding pattern that leads to a more stable structure than does strictly equivalent bonding.

A clear illustration of quasi-equivalent bonding is provided by the case of the Dahlemense strain of tobacco mosaic virus, whose structure [3, 6] is illustrated in Fig. 1b. In the common strain of the virus each subunit is equivalently bonded to its neighbors (Fig. 1a), but in the

Dahlemense strain the outer end of a subunit can bend up or down into slightly different positions so that chemically identical parts of different molecules are packed in slightly different but quasi-equivalent environments. The maximum up or down displacement from the position of equivalence is only about 0.5 Å, and the displacements are coordinated in such a way that there is a regular perturbation of the simple original helical structure which brings turns of the helix near the outside surface alternately closer together and farther apart. The net result is that the chemically identical subunits are packed, in this case, in 98 symmetrically distinct, but quasi-equivalent, environments. The origin of this perturbation probably lies in a weak interaction between the outside ends of the subunits which is not present in the normal strain of the virus. In order to form these bonds, chemically identical parts of neighboring molecules must be brought closer together than the pitch of the helix, which is determined by the main set of bonds controlling the helical packing. There is no strictly regular way in which all these new possible bonds can be formed, but, by periodically deforming the packing in the way shown, it is possible to bring some of the outer ends of the subunits closer together, that is, to make some of the new bonds at the outer radius, at the expense of straining the internal configuration of the protein molecules and moving the ends of other subunits further apart. There is evidently a decrease in the total free energy in this process, but the particular form of the displacements and deformation in the general case will depend upon the particular form of the energy curves describing the various interactions between the subunits and the internal configuration parameters.

## QUASI-EQUIVALENCE: ICOSAHEDRAL VIRUSES

Possible designs for ordered structures built of quasi-equivalently related units can be analyzed by considering the energetically plausible systematic modifications of strictly regular designs. Since quasi-equivalence cannot be expressed in absolute mathematical terms, the decision as to what constitutes quasi-equivalence depends on the *physical* analysis of the energetically allowed variation in the bonding of particular systems. Thus, a comprehensive description of all possible designs for structures built of quasi-equivalently related units cannot be obtained by the rigorous enumeration that has been applied to strict symmetry groups. Nevertheless the plausible minimum-energy designs for particular type of structures can be systematically described. Thus, for example, the possible designs for surface lattices have been analyzed by Caspar and Klug [7] by considering the ways in which regular plane lattices can be connected to form cylinders or closed shells in which each structure unit forms the same pattern of contacts with its neighbors. The description of the design of helical or polyhedral aggregates in terms of surface lattices has the advantage that the local bonding relations determining the overall symmetry can be recognized easily. It turns out that quasi-equivalent bonding is a topological necessity in the design of closed shells which are to be constructed from a large number of equal units.

The optimal design for a closed container built of a large number of equal subunits is the icosahedral surface lattice, which is derived from the plane equilateral triangular net (i.e., the hexagonal lattice) by turning some of the 6-fold axes into 5-fold axes; this process has the effect of folding the plane into a closed surface. This type of design is found in the construction of the protein coats of many isometric virus particles, for example, in that of turnip yellow mosaic virus (Fig. 6). The protein coat of this virus is made up of 180 protein subunits arranged in one of the icosahedral surface lattices predicted by the analysis. This particular structure provides a clear illus-

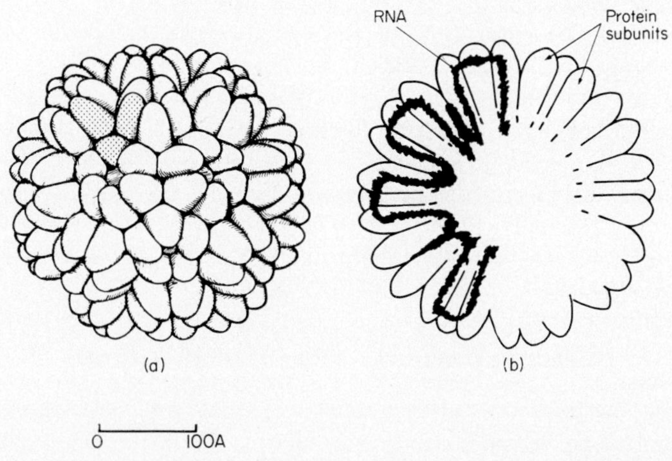


FIG. 6. (a) A drawing of the outer surface of the particle of turnip yellow mosaic virus [9]. The protein shell of the virus is made up of 180 protein subunits arranged with quasi-equivalent symmetry. The strict overall symmetry is icosahedral (Fig. 4c), and from this point of view the shell is made up of 60 strictly equivalent asymmetric units, one of which is shown stippled. This strict crystallographic asymmetric unit is composed of three subunits, which are related to each other by a local or quasi-3-fold axis. The subunits are clustered at an outer radius to form 32 gross morphological units, 20 of which are hexamers and 12 pentamers. In some other viruses with the same number of subunits, other modes of clustering occur, giving an extremely different appearance [13].

(b) A section through a diametral plane of the virus particle showing schematically the relation of the RNA distribution in the virus to the arrangement of protein subunits [14]. The single RNA chain penetrates deep into the protein shell and is bunched in the form of packets associated with the 32 morphological units.

tration of the fact that the overall symmetry of the particle (namely 532 or icosahedral symmetry) no longer defines all the structurally significant bonding relations between the subunits. Thus from the point of view of strict crystallography the asymmetric unit in the coat of turnip yellow mosaic virus is a group of three subunits (shown stippled in Fig. 6a), but an inspection of the structure shows that the bonds between the members of a group of three are of the same class and the group possesses a local or quasi-3-fold rotation axis. To describe all the bonding relations between all the subunits it is thus necessary to include the quasi-symmetry relations as well as those that relate to the strict icosahedral symmetry. In the terminology of Caspar and Klug [7], this structure is one based on the $T = 3$ icosahedral surface lattice, which in addition to the strict elements of icosahedral symmetry also possesses local or quasi- 6-fold, 3-fold, and 2-fold axes. We have already seen that the local 3-fold axes relate to each group of three subunits. The rotation axes passing through the center of the hexamers in the structure are strictly 3-fold axes of the whole structure, but are locally 6-fold axes.

## QUASI-EQUIVALENCE AND HELICAL WAVES ON BACTERIAL FLAGELLA

As an illustration of the application of the ideas outlined above to an unsolved problem in organelle formation, we shall consider the case of bacterial flagella. A new hypothesis can be proposed (Klug, in preparation) on the origin of the helical shape of a flagellum and the mode of propagation of the undulatory motion that it can transmit.

Flagella consist largely of only one type of protein molecule (flagellin) and apparently lack ATPase activity. Moreover flagella which have been detached from the cell body seem incapable of movement and it seems very likely that the source of movement is to be found in the basal body or root attaching the base of a flagellum to the cell. This basal body provides an energy source and contains the driving mechanism, the flagellum then being merely a passive means for transmitting the helical wave. This view is not universally held, but it will be taken as one of the basic assumptions of the hypothesis to be outlined here. Indeed the mechanism proposed is one which can transmit such waves without the need for energy production along its length.

Recent electron microscope work ([11, 15] and Finch, Klug and Berger, in preparation) has shown that the bacterial flagella are built up of approximately globular subunits about 50 Å in diameter arranged in a helical surface lattice of the type shown in Fig. 7a. These globular subunits can with some confidence be identified with the flagellin molecules. There is some uncertainty about the exact helical parameters in any particular species, but this number and a knowl-

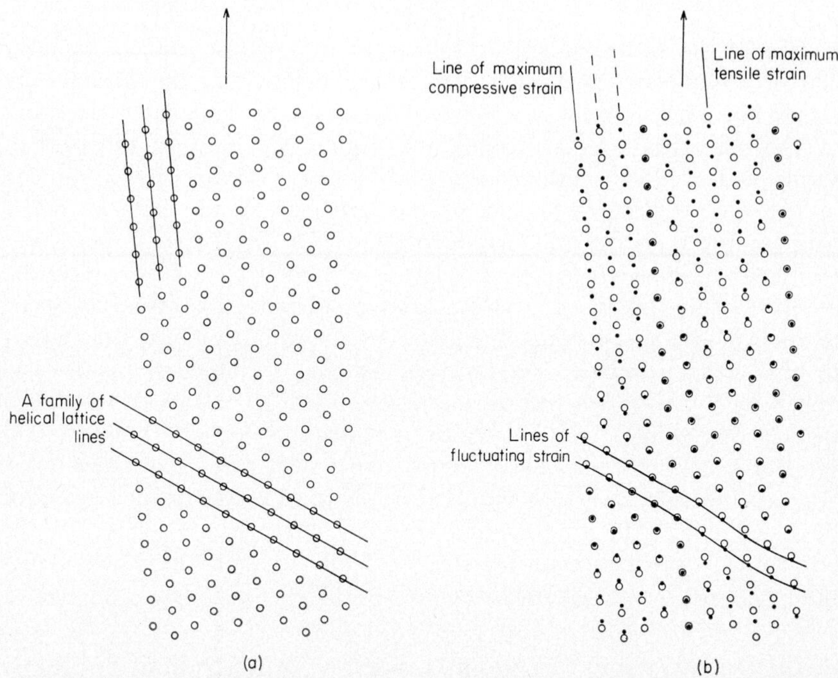


FIG. 7. (a) Part of the surface lattice of a helical structure with a straight axis, i.e., giving rise to a straight rod-shaped particle. Each lattice point represents a structure unit and the diagram shows the cylindrical surface at a particular radius, when it is unrolled flat. A family of lattice lines on the diagram corresponds to a family of helical lines on the cylindrical surface.

(b) The surface lattice of a coiled-coil structure. The black dots mark the new positions of the lattice points when the axis of the structure in (a) is deformed into a helix of large pitch and small radius. The diagram is very schematic since the twist of the axis and the radial perturbations in the subunit positions cannot be shown in a two-dimensional picture of this kind. In fact, helical lattice lines become deformed into coiled coils.

edge of the detailed structure is not necessary for our argument. The flagella do not appear to possess an appreciable central hole—or, if there is one, it is impenetrable to the negative stain used for the electron microscopy. This observation, supported by other types of evidence, has been taken by some authors [16] to mean that there is a central core formed by some additional component, whose presence can be invoked to explain the motion of flagella by periodic interaction of the two components. However, there is no need to postulate such a core to explain the motion, and the brunt of the following argument is to show that it can be produced with one kind of protein subunit only.

### *The Shape of a Flagellum*

The first fact to be explained is the helical shape of a flagellum at rest. (A detached flagellum in the electron microscope has the appearance of a sine curve, corresponding to a helix that has been flattened on drying.) The results described above on the Dahlemense strain of TMV suggest an explanation of how a helical shape could be conferred to the whole rod, even when the rod is made up of identical chemical subunits. In the common strain of TMV (Fig. 1a) the subunits pack into exactly equivalent environments, leading to a simple helical rod with a straight axis: this type of structure can be represented by an idealized subunit equipped with a single set of bonds at one radius (Fig. 2). In the case of the Dahlemense virus (Fig. 1b), there are additional interactions incompatible with a description by a single set of bonds, and it is necessary to invoke the operation of an additional weaker set of bonds operating at a different radius, in this particular case at an outer radius of the particle. This second set of bonds cannot be made for all subunits simultaneously because it is not exactly compatible with the packing determined by the dominant set, and it gives rise to a regular perturbation of the surface lattice (see Fig. 7). The overall shape of the particle is still determined by the dominant set of bonds and is that of a straight cylindrical rod.

Now imagine that the second, not quite compatible, set of bonds at an outer radius is made stronger, equal in strength to the set at the inner radius. Each set of bonds operating on its own would lead to a simple, unstrained helical structure with a straight axis, and the two helical structures would have slightly different parameters. When both sets of bonds are operating, a simple helical structure with a straight axis is incompatible with the requirement that both sets of bonds be made in their unstrained form throughout the entire particle. It can be shown that a compromise structure of the form of a coiled coil will result in which the axis of the particle will itself be deformed into a helix, so that the strain energy over both sets of bonds is a minimum. This helix will be of very long pitch since we have assumed that the individual helical parameters determined by the two sets of bonds acting on their own are only slightly different, so that the pitch of the coiled coil will be something like a beat between those of the two contributing sets.

This beat will involve a perturbation in the radial positions of the subunits as well as in the axial and azimuthal directions, and the whole particle will assume a helical shape. In the terminology used earlier in this paper, the subunits are in quasi-equivalent environments, since all subunits physically make the same sets of bonds with each other. The strain in the bonds is, however, different over a very large number of crystallographically distinct subunits, and the strain distribution is periodic with a very large repeating period. The helical lattice lines which join the subunits in the undeformed helical structure (Fig. 7a) become coiled coils in the twisted helical surface lattice (Fig. 7b). It will be possible to pick out a helical line which corresponds to the direction of maximum contraction, that is to say maximum compressive strain, and at the same time another helical line of the same family lying at an azimuth of 180 degrees with respect to the first on which there is maximal tensile strain. There is no need to postulate that the direction of the line of contraction be in any particular direction with respect to the surface lattice, although, if the helical undulation is to be of long wavelength, it would clearly be advantageous if this line lay approximately in a longitudinal direction.

If this picture of flagellar structure is correct, then the long wavelength period of the flagellum is a property of the three-dimensional structure of the flagellin molecule and ultimately of its amino-acid sequence. A mutation leading to a change in sequence could lead to a different packing and to a changed morphology as in the case of "curly mutants," which have a shorter wavelength than normal flagella.

### *Propagation of Helical Waves*

A quasi-equivalent structure of this kind possesses the basic formal requirement for helical undulations, pointed out by Reichert in 1909 and quoted by Astbury *et al.* [2], namely, that there shall be a helical line of contraction present which can be continuously displaced along the length of the flagellum. The transfer of such a line of contraction from one longitudinal helical line to the adjacent member of the same family produces the overall helical undulation. In this type of wave there is no bodily rotation of the flagellum relative to the body of the bacterium.

A consequence of the type of model postulated is that the amplitude of the helical wave, in other words the radius of the major helix of the coiled coil, is a function of the bonding properties of the subunits.[2] It corresponds to a state of minimum free energy of the whole assembly of protein subunits. The possibility of changes in the internal configuration of the subunits ("strain"), and the small shifts in position that have to be made to propagate a wave, are thus built in to the subunits themselves. The amplitude of the perturbation of the helical wave will therefore be independent of the resistance of the medium. The driving machinery transfers the base of the line of maximum contraction from the base of one member of the helical family

[2] This is not to say that other factors may not contribute in the case of an intact flagellum attached to the cell. Thus, for instance, the mode of connection of the flagellum with the basal body may influence all the parameters of the coiled coil, including amplitude, if the bonding between subunits is not too stringent.

to the next, a process which in this type of model requires no expenditure of energy in the flagellum. The more distal subunits will rearrange themselves accordingly, and it is this progressive rearrangement of the pattern of bond strain that produces the overall helical wave. The *rate* of propagation of the wave will of course, be dependent upon many different factors such as the rate of energy input from the driving machinery, the activation energy for the transfer of bond strain between the subunits, and the viscosity of the surrounding medium. However, it is a strong prediction of the model that the amplitude is independent of the viscosity of the medium. Moreover—as has been observed—the waves will not be damped out as they pass along the flagellum, as would be expected if the flagellum behaved like a simple passive, elastic element.

When one builds a model of a structure like this in the isolated state, there is a certain degree of ambiguity about the construction, since one has arbitrarily to fix one of the helical lines of a family as possessing the highest degree of compression, that is, containing the highest compressive strain energy in the bonds in that direction. This constraint would have to be inserted arbitrarily in the construction of the model. The self-assembly of a structure of this kind would require a very high activation energy, and if indeed bacterial flagella do assemble out of their subunits in this way, it would explain the necessity, in carrying out the reconstitution of flagella *in vitro,* for seeding with small fragments of flagella that act as nuclei [1]. Experiments of this kind give information on the stringency of the bonds between the flagellin molecules, since the seeds apparently select one out of a number of various possible structures which have nearly equal thermodynamic stability [18].

In conclusion, it could be said, whether or not a bacterial flagellum is a passive element of the type postulated, that a rod with a built-in helical twist could assemble itself out of a set of identical subunits, provided the bonding between them is of the quasi-equivalent type, and that such a structure could propagate helical waves without the necessity for supplying energy continuously along its length. The prescription we have given for a model involves a minimum of two slightly incompatible sets of bonds. In a real structure, the use of only two sets of bonds to describe the detailed configuration would be an idealization, and the strain would presumably be distributed over the whole bonding surface of the subunit and would involve very small deformations indeed, but the essential idea of different strain distributions at different radii would still hold. Simple calculation in the case of bacterial flagella with a wavelength of about 2 $\mu$ and an intersubunit distance of about 50 Å shows that deformations only of the order of 1 Å need occur. This certainly appears to be within the competence of protein subunits, as judged by the quasi-equivalent deformations that have been observed in the Dahlemense virus and in the spherical viruses referred to above.

## CONCLUDING REMARKS

There are many topics that have not been mentioned or have been only touched on in this paper—for example, the problem of metastable intermediate aggregates (see Caspar's analysis [3] of the association of TMV protein); the selection, by interaction with a second component, of a unique structure out of a number of similar variant structures; and the related problem of nucleation in self-assembly [3, 13].

More general considerations on the design and assembly of organized structures may be found in the article by Caspar [4], on which certain parts of this paper have drawn heavily. The morphopoiesis of bacteriophage, which represents the most complex system as yet tackled in depth, is discussed by Kellenberger [10].

### References


1. Asakura, S., Eguchi, G., and Iino, T., *J. Mol. Biol.* **10,** 42 (1964).
2. Astbury, W. T., Beighton, E., and Weibull, C., *Symp. Soc. Exptl. Biol.* **9,** 282 (1955).
3. Caspar, D. L. D., *Advan. Protein Chem.* **18,** 37 (1963).
4. Caspar, D. L. D., *in* "Molecular Architecture in Cell Physiology," Symp. Soc. Gen. Physiologists, p. 191. Prentice-Hall, Englewood Cliffs, New Jersey, 1966.
5. Caspar, D. L. D., *in* "Principles of Biomolecular Organisation," Ciba Found. Symp., p. 7. Churchill, London, 1966.
6. Caspar, D. L. D., and Holmes, K. C., Submitted for publication, 1968.
7. Caspar, D. L. D., and Klug, A., *Cold Spring Harbor Symp. Quant. Biol.* **27,** 1 (1962).
8. Crick, F. H. C., and Watson, J. D., *Ciba Found. Symp. Nature Viruses* p. 5 (1957).
9. Finch, J. T., and Klug, A., *J. Mol. Biol.* **15,** 344 (1966).
10. Kellenberger, E., *in* "Principles of Biomolecular Organisation," Ciba Found. Symp., p. 192. Churchill, London, 1966.
11. Kerridge, D., Horne, R. W., and Glauert, A. M., *J. Mol. Biol.* **4,** 227 (1962).
12. Klug, A., and Caspar, D. L. D., *Advan. Virus Res.* **7,** 225 (1960).
13. Klug, A., Finch, J. T., Leberman, R., and Longley, W., *in* "Principles of Biomolecular Organisation," Ciba Found. Symp., p. 158. Churchill, London, 1966.
14. Klug, A., Longley, W., and Leberman, R., *J. Mol. Biol.* **15,** 315 (1966).
15. Lowy, J., and Hanson, J., *J. Mol. Biol.* **11,** 293 (1965).
16. Lowy, J., Hanson, J., Elliott, G. F., Millman, B. M., and McDonough, M. W., *in* "Principles of Biomolecular Organisation," Ciba Found. Symp., p. 229. Churchill, London, 1966.
17. Monod, J., Wyman, J., and Changeux, J.-P., *J. Mol. Biol.* **12,** 88 (1965).
18. Oosawa, F., Kasai, M., Hatano, S., and Asakura, S., *in* "Principles of Biomolecular Organisation," Ciba Found. Symp., p. 273. Churchill, London, 1966.
19. Reed, L. J., and Cox, D. J., *Ann. Rev. Biochem.* **35,** 57 (1966).
20. Weyl, H., "Symmetry." Princeton Univ. Press, Princeton, New Jersey, 1952.

# ASSEMBLY OF MACROMOLECULAR LIPID STRUCTURES *IN VITRO*

J. A. LUCY[1] AND AUDREY M. GLAUERT[2]

*Strangeways Research Laboratory, Cambridge, England*

## INTRODUCTION

The formation and fate of many intracellular organelles is intimately connected with the mechanism of the formation and the fate of the lipoprotein membranes that delineate a number of the different organelles of living cells. Leaving aside the complexities of the biochemical syntheses of the varied substances which comprise lipoprotein membranes, we are faced with the question of how these components become appropriately organized in the macromolecular assemblies that constitute membranes, and furthermore what is the nature of this organization. This paper is concerned with some of the ways in which lipids, particularly the phospholipid lecithin, assemble spontaneously into complex structures when the lipid molecules are dispersed in an aqueous environment *in vitro*.

For many years, it has been commonly supposed that phospholipids can be organized in only one way in aqueous systems, namely in bimolecular leaflets, and the term "membrane" has often been equated with the term "bimolecular leaflet." Other configurations of lipid have generally been regarded either as artifacts, or as occurring only as the products of the lytic actions of substances like saponin that destroy the integrity of membranes. There is no doubt that when saponin interacts with lipoprotein membranes it enables structures to form that are not apparently explicable in terms of bimolecular leaflets. This phenomenon seems to reflect the stability of globular aggregates of saponin molecules and, for the first part of this paper, it is proposed to discuss the self-assembling features of lipid dispersions containing saponin. Subsequently, we shall discuss structures that contain lecithin but no saponin and it will become apparent that, even in the absence of saponin, structures are formed that seemingly are not constructed from bimolecular leaflets. Thus even a phospholipid such as lecithin, which forms stable bimolecular leaflets very readily, nevertheless still

[1] Present address: Department of Biochemistry, Royal Free Hospital School of Medicine, University of London, London, England.

[2] Sir Halley Stewart Research Fellow.

has a tendency to form globular micelles in appropriate circumstances. Finally, the possible significance of globular micelles of phospholipid will be considered in relation to the formation and properties of membranes, and also in relation to the fate of certain subcellular particles.

## ASSEMBLIES CONTAINING SAPONIN

### *Saponin Alone*

Electron micrographs of negatively stained preparations of saponin, in which the main aglycon is the triterpene gypsogenine, have the rather unimpressive appearance of randomly distributed particles which vary in diameter from approximately 40 to 75 Å. These subunits occasionally form short chains and small areas of laterally fused rings (Fig. 1). Variation in size of the subunits may perhaps result from the difficulty of forming uniformly small aggregates with the bulky and rigid triterpene moieties of the saponin molecules; alternatively, this variability may stem from the heterogeneity of the saponin.

### *Cholesterol and Saponin*

While saponin alone shows little tendency, at low concentrations, to form organized macromolecular assemblies, the addition of cholesterol yields a system that is, in contrast, capable of much more order. Specimens obtained from monolayers of cholesterol treated with saponin are seen, in negatively stained preparations, to contain many laterally fused rings that are present in a specific, hexagonal array. Each ring is apparently composed of globular subunits, 30–35 Å in diameter (Fig. 2, arrow). A similar hexagonal array is observed in electron micrographs of negatively stained preparations of erythrocytes treated with saponin (Fig. 3) [5]. In this instance, the cholesterol necessary for the formation of the hexagonal array is presumably provided by the cell membrane, and it is possible that the hexagonal array forms in the plane of the original membrane.

### *Lecithin, Cholesterol, and Saponin*

Negatively stained preparations of cells treated with saponin more commonly reveal the structure that was originally observed by Dourmashkin *et al.* [4], rather than the assembly illustrated in Fig. 3. In the absence of membranes, structures that are similar to those observed by Dourmashkin *et al.* assemble spontaneously in aqueous dis-

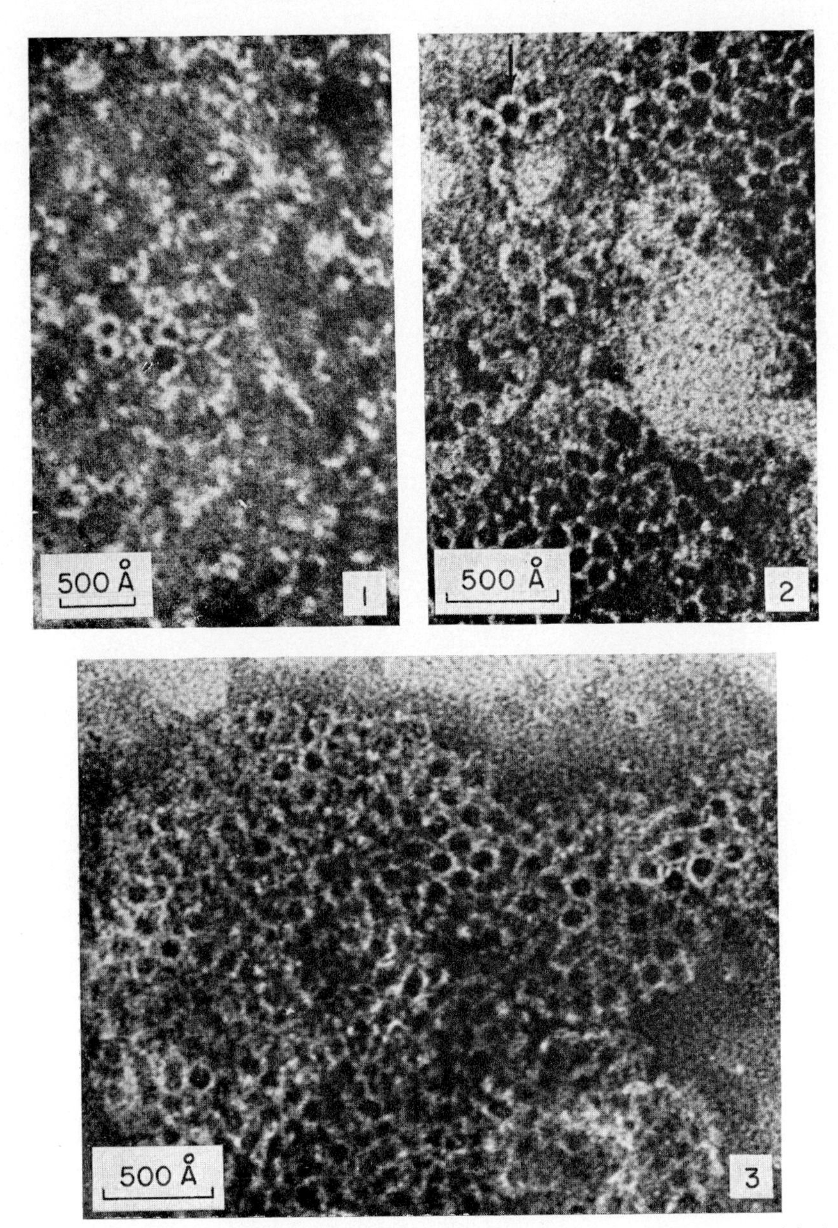


FIG. 1. A negatively stained preparation of 0.05% saponin contains many globular subunits which form chains and some rings. × 200,000. From Lucy and Glauert [12].

FIG. 2. A negatively stained preparation of saponin and cholesterol contains many laterally fused rings which form hexagonal arrays. Chains and rings are apparently composed of globular subunits (arrow). × 300,000. From Lucy and Glauert [12].

FIG. 3. A negatively stained preparation of rabbit erythrocytes after hemolysis by saponin contains an array of fused rings similar to those in Fig. 2. × 300,000.

persions of lipid if lecithin is present in addition to cholesterol and saponin [1, 12], as illustrated in Fig. 4, which is a micrograph of a negatively stained preparation of a dispersion of lecithin, cholesterol, and saponin. This type of structure, which can achieve a remarkably high degree of regularity, is seen to be composed of a hexagonal array of rings of lighter density that surround dark areas that are about 80 Å

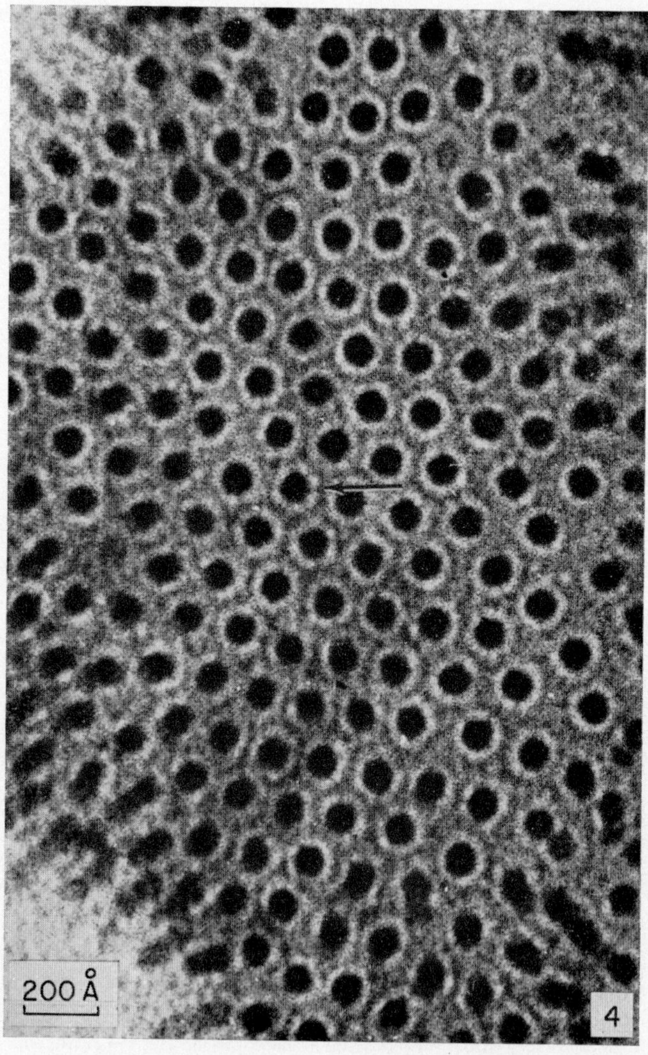


Fig. 4. A negatively stained preparation of lecithin, cholesterol and saponin. The rings in the hexagonal array appear to be composed of globular subunits (arrow). × 500,000.

in diameter; it is noticeable that each of the light rings is separate from its neighbors. It can be seen that each of the rings of light density is again, as in the saponin-cholesterol complex, made up of globular subunits about 35 Å in diameter (Fig. 4, arrow). For reasons that are explained in detail elsewhere [12], it is thought that an aqueous dispersion of lecithin, cholesterol, and saponin contains two species of globular micelles, one composed mainly of lecithin and cholesterol and the other containing primarily saponin and cholesterol. These pairs of components are known to form stable complexes with each other. Any stable arrangement of these two kinds of micelle will have aqueous areas that are sufficiently large to accommodate the carbohydrate moieties of the saponin molecules, and the arrangement that is adopted spontaneously by these micelles is thought to be that illustrated diagrammatically in Fig. 5 [13]. In the proposed structure, micelles of saponin-cholesterol are arranged in rings around aqueous areas, 80 Å in diameter, and each ring is separated from its neighbor in the hexagonal array by interstitial micelles of lecithin-cholesterol. This arrangement gives a hexagonal structure in which the center-to-center distance between the aqueous areas is approximately 170 Å, a value

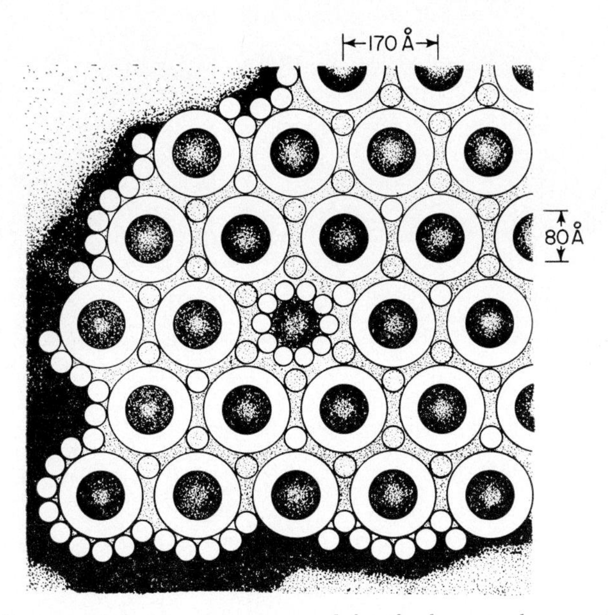


FIG. 5. A diagram of the structure proposed for the hexagonal array observed in preparations of lecithin, cholesterol, and saponin. The white rings are composed of saponin-cholesterol micelles, as illustrated in detail for the ring in the center of the diagram. The rings are separated from one another by interstitial micelles of lecithin containing cholesterol. From Lucy and Glauert [13].

similar to the dimensions actually observed. It is thought that the hexagonal array of rings is, in fact, probably a three-dimensional arrangement of tubes or spheres which have a pseudo-two-dimensional appearance in the negatively stained preparations.

Electron micrographs of negatively stained preparations of lecithin, cholesterol, and saponin dispersed in an aqueous environment reveal a number of other interesting, but nonhexagonal, assemblies. Thus, isolated rings and short piles of rings lying on their sides are also observed (Fig. 6, arrows). On closer inspection (Fig. 7, arrows), it may

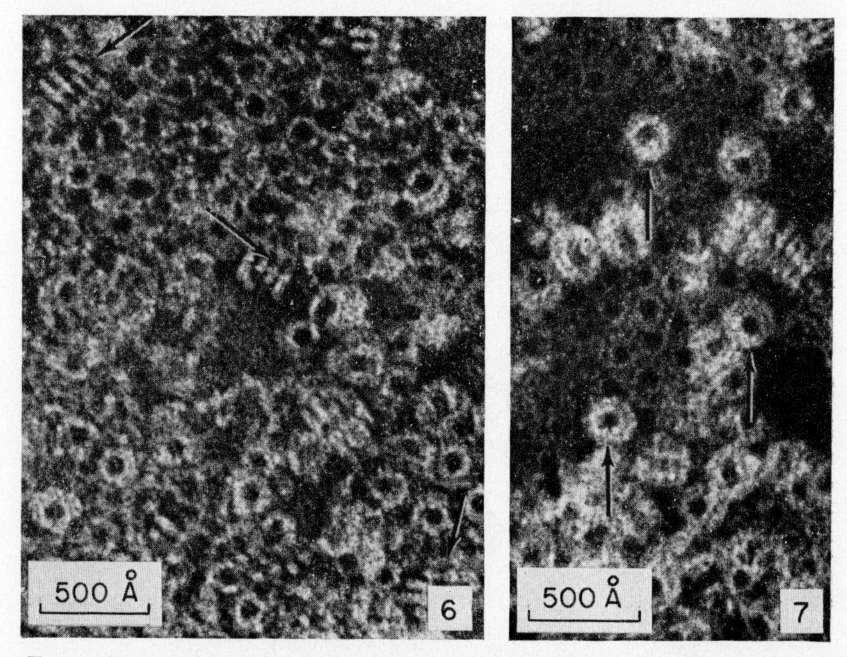


FIG. 6. Short piles of rings (arrows) in a mixture of lecithin, cholesterol, and saponin. × 300,000. From Lucy and Glauert [12].

FIG. 7. Isolated rings in a mixture of lecithin, cholesterol, and saponin appear to be composed of two concentric rings of subunits (arrows). × 300,000. From Lucy and Glauert [13].

be seen that each of the isolated rings is composed of two concentric rings of subunits, and these rings therefore differ in structure from those present in the hexagonal array of Fig. 4. It has been proposed that in the isolated "double rings," the units comprising the outer of the two concentric rings of subunits, are lecithin-cholesterol micelles, and that the inner ring is composed of saponin-cholesterol micelles [13]. If the aqueous dispersion is allowed to stand for a sufficiently

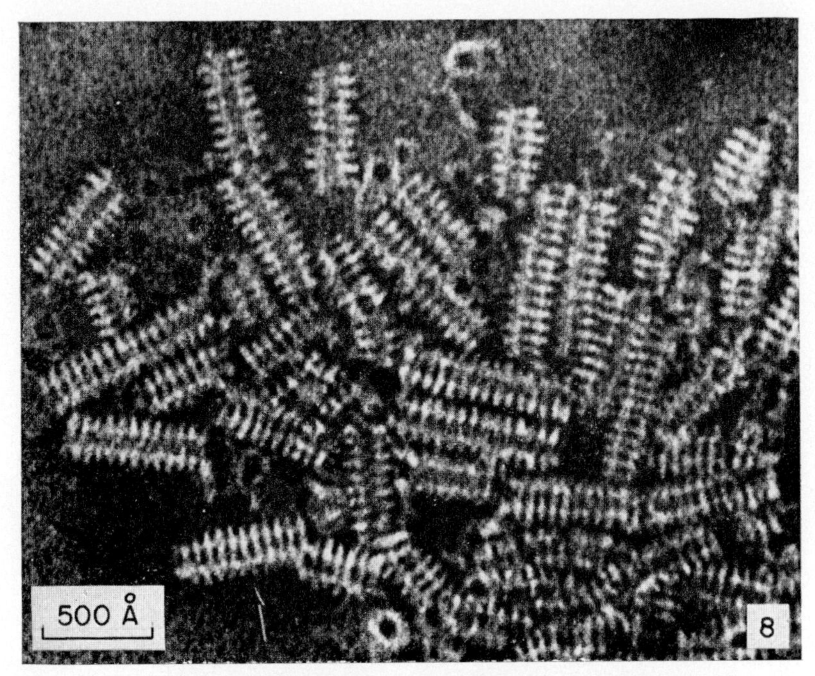


FIG. 8. Long piles of rings (arrow) and helices in a preparation of lecithin, cholesterol, and saponin. × 300,000. From Lucy and Glauert [13].

long period (several hours at room temperature) before samples are removed for electron microscopy, it is found that specific aggregates of the double rings are present in which the rings are associated in quite long piles (Fig. 8, arrow). In addition, the piles of rings are accompanied by helical structures (Figs. 8 and 9), and it would seem that a "stacked disk-helix" transition, similar to that observed in tobacco mosaic virus, may occur once the pile of rings exceeds a certain length. The overall similarity of the helices formed in these dispersions of lipids to helical viruses can be appreciated by comparing the structure shown in Fig. 9 with the internal component of Sendai virus in Fig. 10.

### *Effects of pH and Different Negative Stains*

We have made a number of experiments on the effects of varying the pH, and the cationic and anionic environment, on the lecithin-cholesterol-saponin helix. Variations in these different parameters can affect the rate of assembly of the helix from its constituent subunits and can also modify both the stability and the precise morphology of the completed helical structures. These experiments are of interest because they

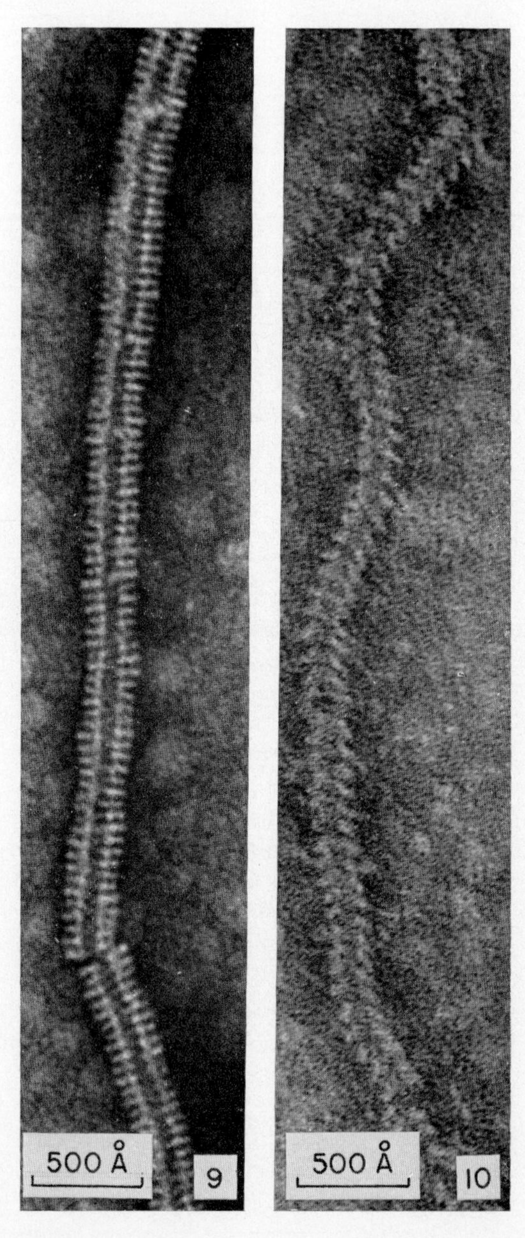


FIG. 9. A long helix in a preparation of lecithin, cholesterol, and saponin. Calcium phosphotungstate. × 300,000.

FIG. 10. The internal component of Sendai virus. × 300,000. (Unpublished micrograph by Mrs. June D. Almeida.)

demonstrate that quite extensive changes in macromolecular structure can occur when the environment of the lipid molecules is only slightly altered. The assemblies shown in Fig. 8 were formed by lipids dispersed in a solution of potassium phosphotungstate at pH 7.0. Essentially similar structures are observed if the lipids are dispersed for some hours in water free from potassium phosphotungstate and the negative stain (at pH 7) is added immediately prior to electron microscopy, although it is noticeable that the helical structures assemble relatively rapidly in these circumstances. If, however, the potassium phosphotungstate that is added immediately before electron microscopy is at pH 4.4, extensively disrupted helices are observed as is shown in Fig. 11.

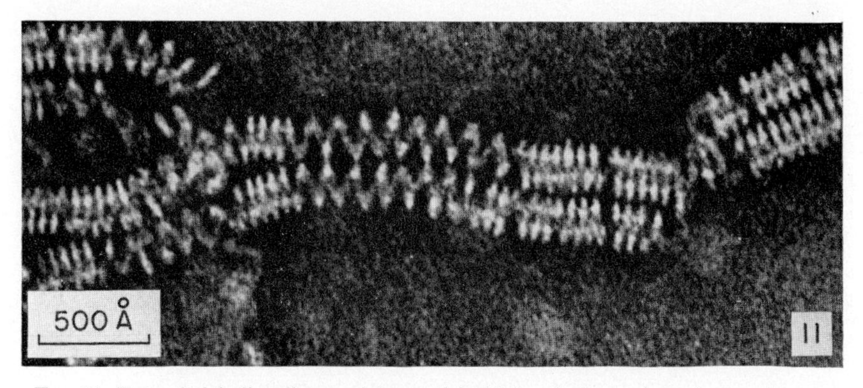


FIG. 11. Extended helices in a mixture of lecithin, cholesterol, and saponin negatively stained with potassium phosphotungstate at pH 4.4 $\times$ 300,000.

Replacing the potassium of the negative stain by calcium (at pH 7) has a different effect. In the presence of calcium phosphotungstate, the rate of assembly of the macromolecular structures is very slow. If, however, the dispersions containing calcium are allowed to stand at 21°C for several days, or alternatively if calcium phosphotungstate is added to lipids previously dispersed in water, one observes structures that show two interesting features. First, the stacked disk configuration is much more prevalent than in preparations containing the potassium ion. Second, additional subunits are present around the perimeters of most of the stacked disk structures, and it appears that the rings of which the disks are assembled contain more than two concentric rings of subunits (Fig. 12). These additional subunits may be extra lecithin-cholesterol micelles that are held in position by divalent calcium ions.

If the lipids are dispersed in a solution of ammonium molybdate having a pH of 5.2, relatively few of the isolated "double rings" are

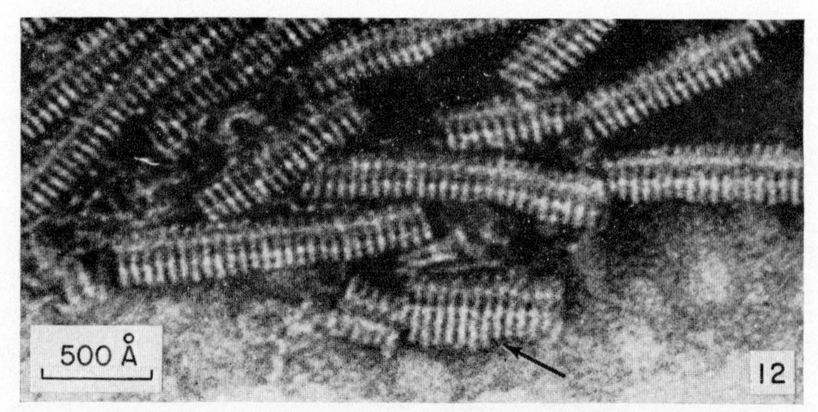


FIG. 12. Helices negatively stained with calcium phosphotungstate appear broad with additional lateral subunits (arrow). × 300,000.

observed in the early stages of the assembly process. Instead, the structures shown in Fig. 13 are seen; these structures have not so far been observed with any other negative stain. It is thought that these assemblies, which resemble negatively stained preparations of small spherical viruses such as human wart virus in appearance [8], may be either flat disks containing many subunits or, alternatively, may be spherical objects like the wart virus. With time, lecithin, cholesterol, and saponin dispersed in solutions of ammonium molybdate yield

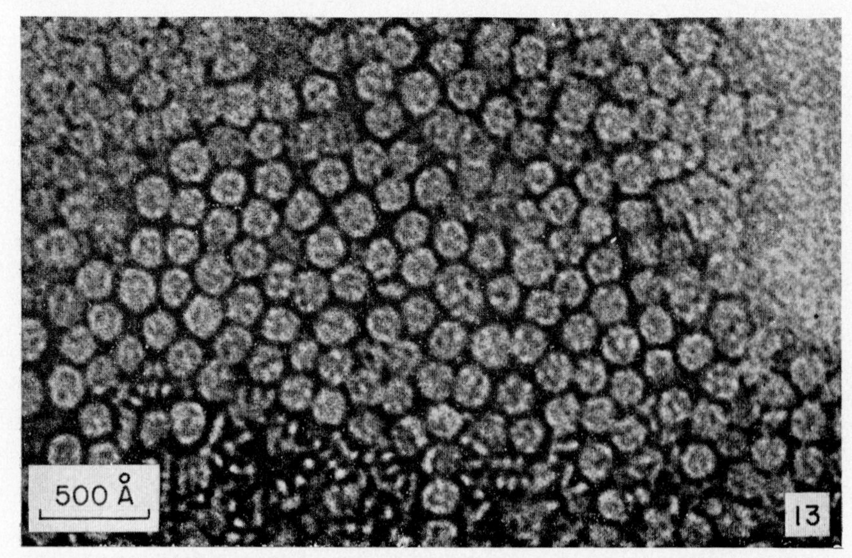


FIG. 13. Flat disks or spherical structures in a preparation of lecithin, cholesterol, and saponin negatively stained with ammonium molybdate. × 300,000.

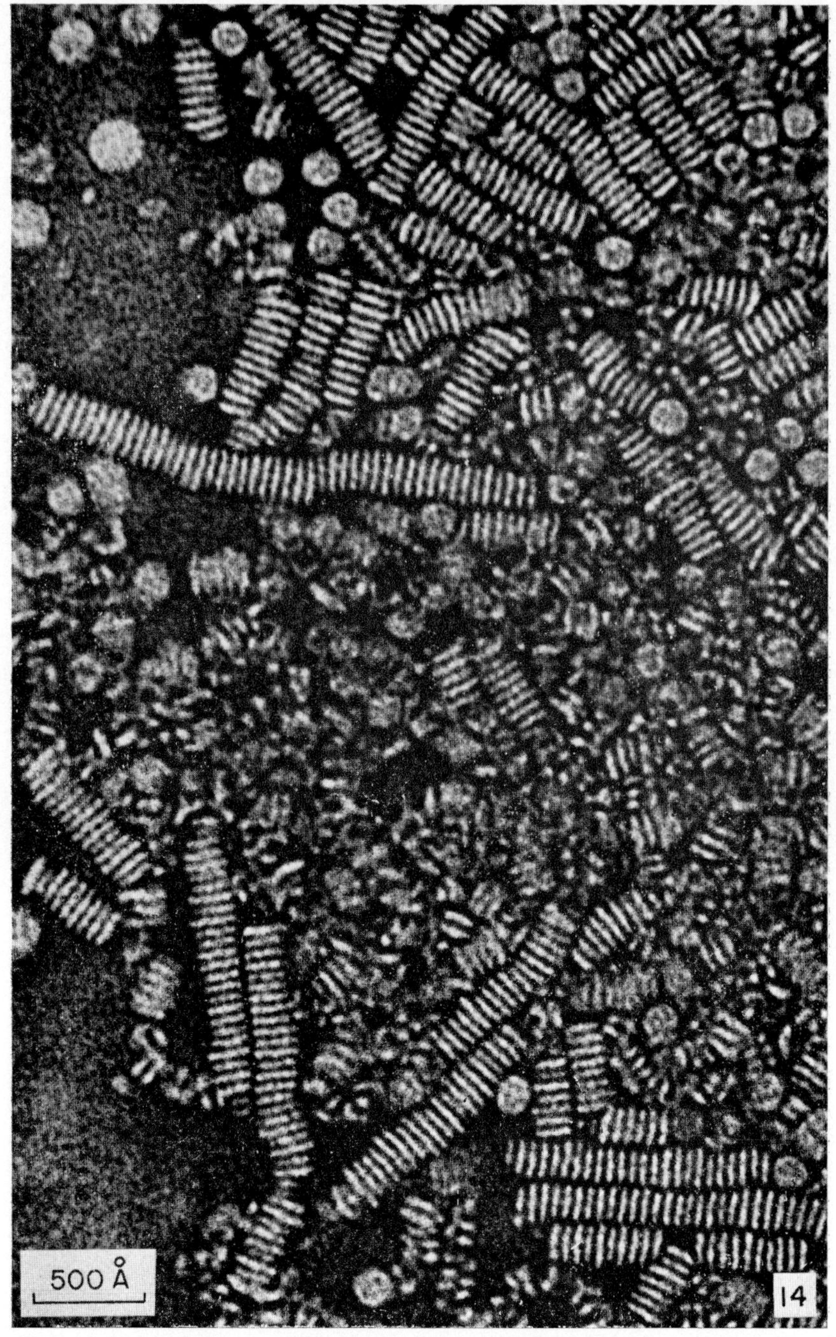


FIG. 14. Stacked disk assemblies in a preparation of lecithin, cholesterol, and saponin negatively stained with ammonium molybdate. × 300,000.

structures (Fig. 14) in which the stacked assemblies appear to be piles of relatively solid discs rather than the piles of rings seen in the presence of potassium phosphotungstate.

### *Problems of Interpretation*

It would seem that none of the structures that we have considered so far can be interpreted in terms of bimolecular leaflets, even when the assemblies contain lecithin. Furthermore, despite the variations that are apparent in the helical and stacked disk structures under differing conditions of pH, etc., the structures appear at all times to be composed of globular units. Perhaps the presence of saponin facilitates the association of lecithin molecules into globular micelles—rather than bimolecular leaflets—because the micelles fit into stable macromolecular assemblies that are thermodynamically favorable. Alternatively, as Bangham and Horne [1] have suggested, the charges on the ions of the negative stains may contribute toward the stability of lecithin in micellar form. We tend to think that the existence of a molecular association between lecithin and cholesterol is a major factor contributing to the stability of the micelles (cf. Haydon and Taylor [6]). Nevertheless, regardless of the precise reasons for the stability of globular micelles *in vitro,* it would appear that factors like electrostatic charge and chemical composition may well operate, not only in the model systems that we are discussing now, but also *in vivo,* to favor the presence of globular micelles of phospholipid within natural membranes under appropriate conditions.

## ASSEMBLIES FREE FROM SAPONIN

### *Lecithin Alone*

Assemblies formed by phospholipid dispersions in the absence of the hemolytic agent saponin are, naturally, of more direct relevance to the configuration of the lipids of biological membranes than the structures that we have considered so far. Figure 15 shows a dispersion of lecithin that has been negatively stained by potassium phosphotungstate. It may be expected that, in the absence of perturbing influences, lecithin most probably forms continuous bimolecular leaflets when dispersed in water [6]. In the aqueous dispersion, therefore, these phospholipid structures may be closed shells of concentric bimolecular layers of lecithin, although it is possible that the ions of the negative stain

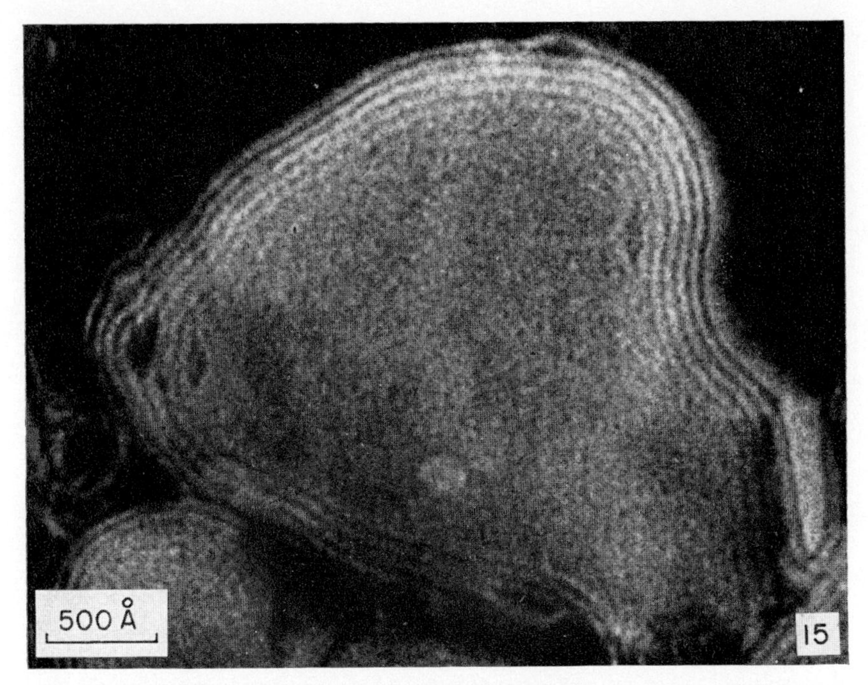


Fig. 15. Layered structures in a preparation of lecithin. Potassium phosphotungstate. × 300,000.

have modified the structures that are actually observed in the micrograph.

*Lecithin and Cholesterol*

While the configuration of purified lecithin in water may be the bimolecular layer, this need not necessarily be the structure of cell membranes since biological membranes contain considerable quantities of lipid that is not phospholipid. On the basis of general thermodynamic arguments, Haydon and Taylor [6] have suggested that, as the ratio of other lipids to phospholipids in the membrane is increased, the bimolecular leaflet structure will become unstable. In the light of these arguments, it might be expected that assemblies, of lecithin and cholesterol, for example, may be capable of forming structures that are not based on the bimolecular leaflet. The structures seen in Fig. 16 would appear to be an instance of this. Here, in a preparation of lecithin and cholesterol (originally dispersed in a 1 : 1 molar ratio) that has been negatively stained with calcium phosphotungstate, we see tubular structures that are about 110 Å in diameter. Figure 17 is a micrograph of a similar preparation stained with potassium phosphotungstate, in which hollow tubes filled with the negative stain are present both in profile and in end view. The tubes seem to be composed of globular units since, in those tubes which apparently do not contain the negative stain, a globular surface structure can be seen. This surface structure can be more clearly observed in Fig. 18, which indicates that these lipid assemblies have a morphological similarity to bacterial flagella [7]. Application of the photographic rotation technique of Markham

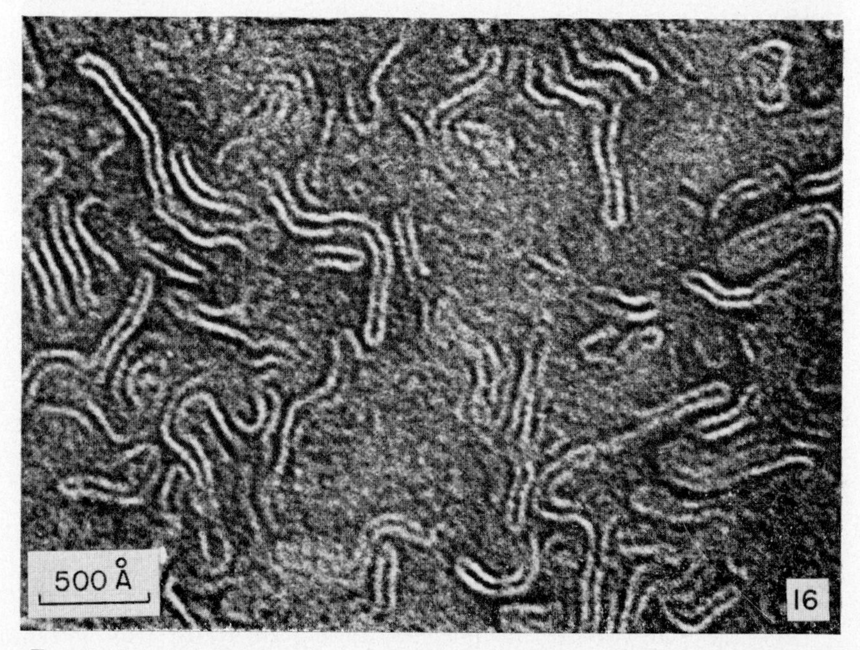


FIG. 16. Tubular structures in a preparation of lecithin and cholesterol. Calcium phosphotungstate. × 300,000.

*et al.* [14] to the end view of one of the tubes of Fig. 17 has indicated that the tubes have fivefold symmetry in cross section [12]. It has been suggested previously [12] that these tubular structures, which have also been observed by Bangham and Horne [1], represent specific assemblies of globular micelles that contain lecithin and cholesterol. Tubular structures of this kind have apparently been seen only in preparations of lecithin and cholesterol, and not in electron micrographs of lecithin alone. Cholesterol, rather than the negative stain, would therefore appear to be the perturbing influence that is responsible for the formation of the tubular assemblies and hence for the presence of the constituent, globular micelles of lipid.

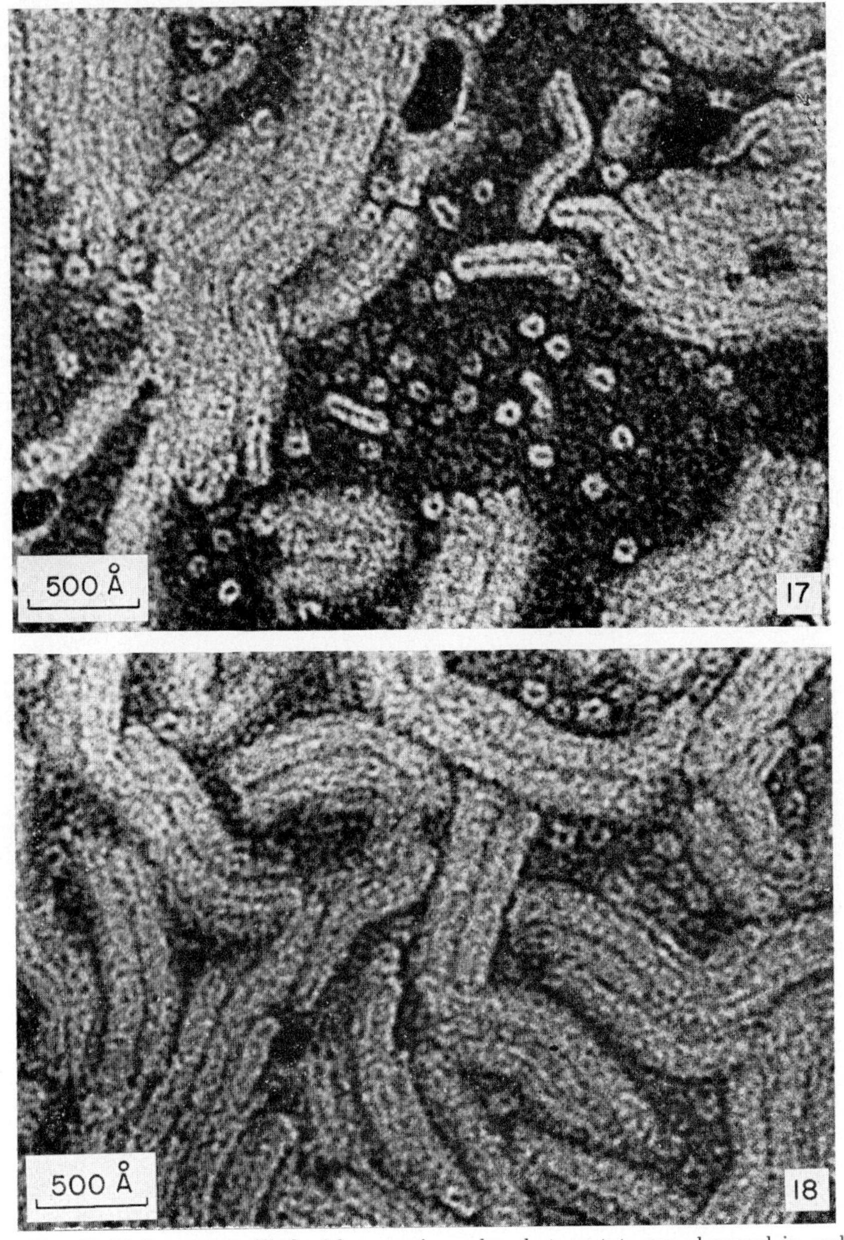


FIG. 17. Hollow tubes filled with potassium phosphotungstate are observed in end view and in profile in a preparation of lecithin and cholesterol. A globular surface structure can be seen in tubes which apparently do not contain the negative stain. × 300,000. From Lucy and Glauert [12].

FIG. 18. Surface structure of tubes composed of lecithin and cholesterol. Potassium phosphotungstate. × 300,000. From Lucy and Glauert [12].

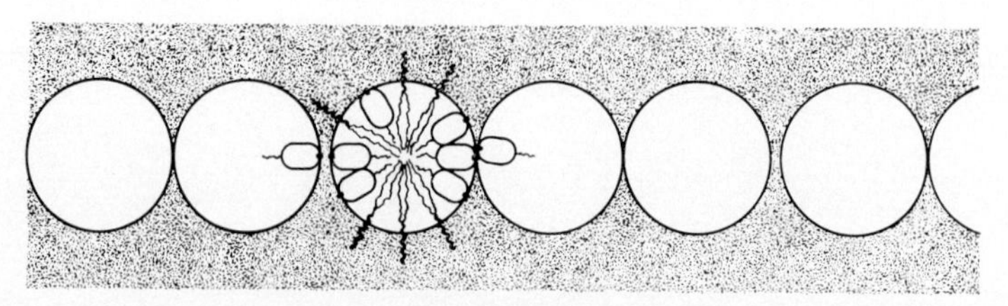

FIG. 19. A cross-sectional view of a model for a biological membrane in which the lipids are in small globular micelles. The micelles are arranged in a plane, and a layer of protein or glycoprotein is shown on each side of this plane of micelles. The organization of lipid molecules (phospholipid and cholesterol) within the micelles is illustrated diagrammatically, but not stoichiometrically, for one of the micelles in the row. From Lucy [9].

## GLOBULAR MICELLES AND BIOLOGICAL MEMBRANES

In this paper we have seen that globular micelles of phospholipid play an important role in the assembly of macromolecular lipid structures *in vitro*. The stability of globular micelles of lipid has led, in our experiments, to the rather surprising observation that assemblies of lipid molecules can have a close morphological resemblance to assemblies of protein subunits such as viruses and bacterial flagella which are structurally quite a remove from the bimolecular leaflet. We have therefore suggested that small globular micelles of lipid, as well as bimolecular leaflets, may function as building blocks in the formation of biological structures containing lipids [12].

One of us, in an extension of this approach, has also considered the possibility that the lipids of biological membranes may be present in the form of globular micelles in certain circumstances, and has put forward a theoretical micellar model for the lipids of biological membranes [9]. This micellar model is not intended as a mutually exclusive alternative structure to the bimolecular leaflet model [3], but as a complementary structure. It is envisaged that a micellar configuration for the lipids of membranes might be in dynamic equilibrium with the bimolecular leaflet structure, and which of the two configurations occurs at any particular time, at a given site, will probably depend on factors such as chemical composition, electrostatic charge, and pH. A cross-sectional view of the micellar model is illustrated diagrammatically in Fig. 19, and the surface view is illustrated in Fig. 20. The diagrammatic illustration of the fibroblast cell in Fig. 21 is intended to illustrate in graphic form the dynamic character of the plasma membrane and the concept that the membrane may simultaneously have areas of both micellar and leaflet configuration. From a consideration of factors that cannot be discussed in detail here, it would

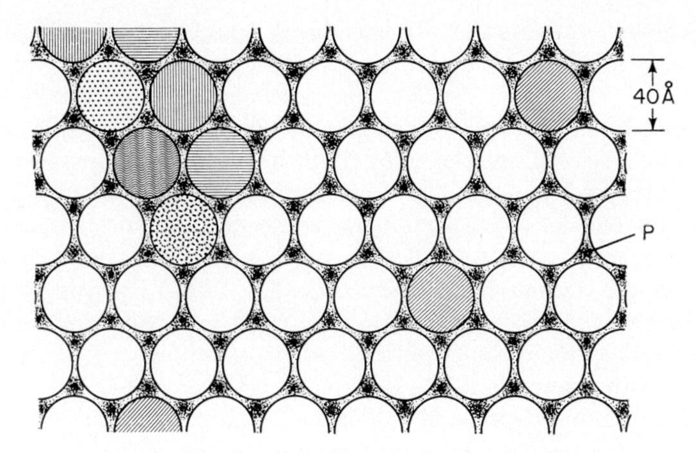


FIG. 20. A surface view of the lipid moiety in a micellar model for a biological membrane. The globular micelles of lipid (unshaded) are shown in hexagonal close packing, but it is thought that such a structure would be flexible and that individual micelles would be in continuous slight random movement. The lipid cores of these micelles would have diameters of about 40 Å. The shaded subunits are intended to represent globular proteins with enzymatic or hormonal properties that have replaced the lipid micelles at certain points in the plane of the flexible lattice. Aqueous pores (*P*), approximately 8 Å in diameter, are present. From Lucy [9].

seem probable that the total area of micellar configuration in most plasma membranes is quite small [9].

A number of the properties of the proposed micellar model for the structure of biological membranes are relevant to the formation and properties of the membranes of subcellular particles. Thus, it has been

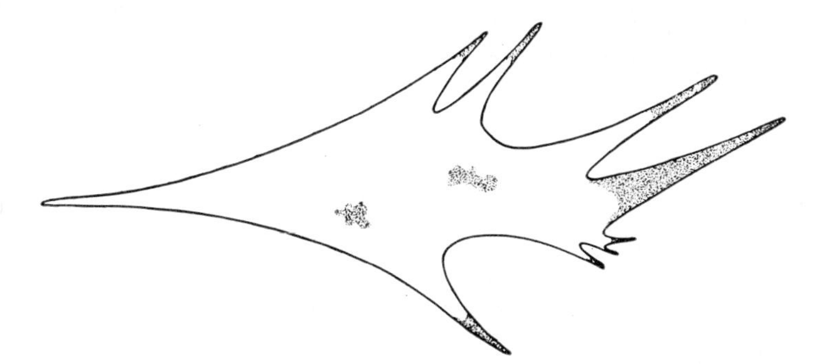

FIG. 21. A diagrammatic illustration of a fibroblast, intended to indicate possible equilibria between areas of surface membrane in which the lipids are arranged in globular micelles (shaded) and other areas in which the lipids are arranged in a bimolecular leaflet (unshaded). From Lucy [9].

suggested that formation of new membrane could well occur in areas in which the micellar configuration predominates since this structure would easily allow the insertion of additional lipid micelles into the membrane. It is conceivable that local conditions in the aqueous environment of the cell interior may initially favor the organization of newly synthesized molecules of membrane lipids into globular micelles rather than bimolecular leaflets. The newly formed micelles may then associate to form a sheet of micelles which may be in equilibrium with a more or less continuous bimolecular leaflet. It has recently been suggested [10] that a process of this kind may well be involved in the formation of the membrane-bound particles of phospholipid and dicetyl phosphoric acid that have been studied by Bangham and his colleagues [2] and which, in some respects, may be regarded as experimental models for subcellular particles like mitochondria and lysosomes.

During the functional lifetime of subcellular particles, the micellar organization offers a possible means of localizing particle-bound enzymes within the plane of the lipid layer of the limiting membrane. This idea is illustrated diagrammatically in Fig. 20, and it may perhaps apply to the membranes of the endoplasmic reticulum, as well as to lysosomes and mitochondria.

Finally, as far as destruction of subcellular particles is concerned, an increase in the proportion of micellar membrane, which would probably be a much more permeable structure than the bimolecular leaflet structure, would increase the permeability to water of the limiting membrane of the particle. It seems possible that under certain circumstances, as may occur for example in the action of excess of vitamin A on lysosomes [11], the increased flow of water into the particle resulting from an increased proportion of micellar membrane may burst open the lysosome, and thus destroy its integrity as a discrete, independent entity within the cell.

## CONCLUSIONS

In brief, our investigations on the assembly of macromolecular lipid structures in model systems have indicated that lipids possess a greater structural versatility than is usually attributed to them, and it is suggested that this versatility has been put to good use by some of the organelles on which we are focusing our attention at this symposium.

## SUMMARY

The existence of a subcellular particle may be regarded as being dependent on the presence of a limiting membrane system that maintains the stability and integrity of the particle. Information on the formation and fate of membranes and related structures, both *in vivo* and in model systems *in vitro*, is therefore directly relevant to the problem of the formation and fate of intracellular particles. The negative staining technique of electron microscopy has been used to study the structures of macromolecular assemblies that are formed spontaneously in aqueous dispersions of lipids *in vitro*. Assemblies formed by lecithin and cholesterol dispersed in water, in the presence and absence of saponin, have been investigated in detail. Few of the observed structures can be interpreted in terms of bimolecular leaflets. Most of the assemblies are apparently composed of specific arrangements of globular micelles of lipid, and a number of them have marked morphological similarities to structures that are composed of protein subunits, such as bacterial flagella, and helical and spherical virus particles. The possible significance of the occurrence of globular micelles of lipid in biological membranes is discussed in relation to the formation and stability of intracellular particles.

## Acknowledgments

We are very grateful to Mrs. June D. Almeida for providing the micrograph for Fig. 10.

Figs. 1, 2, 6, 17, and 18 are reproduced from Lucy and Glauert [12] and Figs. 19–21 from Lucy [9] by permission of Academic Press. Figs 5, 7 and 8 are reproduced from Lucy and Glauert [13] by permission of CNR, Rome.

## References

1. Bangham, A. D., and Horne, R. W., *J. Mol. Biol.* **8,** 660 (1964).
2. Bangham, A. D., Standish, M. M., and Watkins, J. C., *J. Mol. Biol.* **13,** 238 (1965).
3. Davson, H., and Danielli, J. F., "The Permeability of Natural Membranes," 2nd Ed., Cambridge Univ. Press, London and New York, 1952.
4. Dourmashkin, R. R., Dougherty, R. M., and Harris, R. J. C., *Nature* **194,** 1116 (1962).
5. Glauert, A. M., Dingle, J. T., and Lucy, J. A., *Nature* **196,** 953 (1962).
6. Haydon, D. A., and Taylor, J., *J. Theoret. Biol.* **4,** 281 (1963).
7. Kerridge, D., Horne, R. W., and Glauert, A. M., *J. Mol. Biol.* **4,** 227 (1962).
8. Klug, A., and Finch, J. T., *J. Mol. Biol.* **11,** 403 (1965).
9. Lucy, J. A., *J. Theoret. Biol.* **7,** 360 (1964).
10. Lucy, J. A., *in* "Biological Membranes; Fact and Function" (D. Chapman, ed.). In press. Academic Press.
11. Lucy, J. A., and Dingle, J. T., *Nature* **204,** 156 (1964).
12. Lucy, J. A., and Glauert, A. M., *J. Mol. Biol.* **8,** 727 (1964).
13. Lucy, J. A., and Glauert, A. M., *in* "Symposium on Electron Microscopy" (P. Buffa, ed.), p. 233. CNR, Rome, 1964.
14. Markham, R., Frey, S., and Hills, G. J., *Virology* **20,** 88 (1963).

# FIBRILLAR STRUCTURE IN THE MITOTIC APPARATUS[1]

**DANIEL MAZIA**

*Department of Zoology, University of California, Berkeley, California*

## INTRODUCTION

In this account, I shall summarize the progress that has been made toward the resolution of the spindle fibers of the mitotic apparatus. By resolution, I mean not only the identification of the molecular units, but also a description of the structure in terms of the units.

## CHEMICAL RESOLUTION

Let me date the chemical resolution of the spindle fibers from 1952, when Katsuma Dan and I isolated mitotic apparatus from sea urchin eggs and made some observations of the molecules obtained when the isolated mitotic apparatus was dissolved [19]. Two conclusions emerged from that work. First, there seemed to be strong indication that the assembly of the mitotic apparatus involved the participation of intermolecular S—S bonds. Second, despite the enormous structural complexity of the whole mitotic apparatus, a very large proportion of its mass consisted of a single "major protein" as judged by ultracentrifugal and electrophoretic criteria. In our 1952 paper, Dan and I described the major protein component of the mitotic apparatus isolated by our first and rather violent method as being a particle having a sedimentation constant of 4.0 S and a molecular weight of the order of magnitude of 45,000.

The idea of a "major protein of the mitotic apparatus" is one that has limited charm. It was useful, of course, as a guide to the problem of the provenance of the relatively huge apparatus, but was not very useful for thinking about the process of mitosis. One wants to know what part of the mitotic apparatus the protein belongs to, and of course one is particularly interested in the composition of the spindle fibers. Only recently have we had any good reason to believe that the major protein fraction is in fact a structural component of the spindle fibers, and that finding is the main point of the present account.

[1] This investigation was supported by U.S. Public Health Service Research Grant GM-13882 from the National Institutes of Health.

I will base my conclusions about this protein on the recent work of Sakai [26], from our laboratory. The main differences between Sakai's findings and those published earlier by ourselves and more recently by other workers seem to have their origin in the method of isolation of the mitotic apparatus. The problems of isolating the mitotic apparatus, as pointed out by all who have written about them, stem from problems of stability that are not well understood. The instability of the structure is a real phenomenon, expressed *in vivo* as well as in difficulties of isolation. The structure appears and disappears in the course of the cell cycle. Its stability *in vivo* depends on temperature and pressure [7, 30] and is affected by mild and reversible chemical modifications: e.g., the dismantling of all or part of the structure by the action of colchicine or its overstabilization by $D_2O$ [5]. When the cell is disrupted, the mitotic apparatus breaks down in the media commonly used for the isolation of mitochondria, nuclei, etc. At the present time, there exists a whole series of methods of isolating the mitotic apparatus, most of them tested on eggs of marine invertebrates. All are based on the provision of special conditions for stabilization, and none of the media used can be claimed to come close to reproducing the actual environment of the mitotic apparatus within the cell. But what medium of isolating any subcellular structure does that? Have we taken the fact that cells do not contain strong sucrose solutions seriously enough to ask what the counterpart conditions are inside the cell? I will not review the various procedures for isolating the mitotic apparatus; probably we shall be able to design better ones once we know why the existing ones provide stability. The problem is not even as simple as providing stability, for as soon as we have achieved that stability we are confronted with *overstability*. I mean by overstability something fairly precise in terms of actual laboratory operations: the overstabilized mitotic apparatus cannot be dissolved under the gentle conditions we deem desirable for obtaining its molecular constituents in solution. We who work in the field seem to be in agreement that ready solubility down to molecular dimension in 0.5 *M* KCl near neutral pH is a serviceable criterion. We may recall that in the first work by Dan and myself it was necessary to use 0.5 *N* NaOH to dissolve the isolated material.

The problem of stabilization is not, alas, fully encompassed by simple chemical considerations. The mitotic apparatus becomes more stable with time after isolation, even in the cold, and the empirical quality of the isolation techniques is aggravated by the fact that speed of working is a factor. Kane and Forer [10] have described some of the morphological changes which take place with time and have discussed them.

I would not take time to discuss the stabilization of the mitotic apparatus if it were merely a problem of laboratory manipulation. But it seems to me that the questions raised by experience with "overstabilization" are closely linked to the problems of the function of the spindle fibers and more generally of structures composed of microtubules. At this point, it is necessary only to insist that the structure of the mitotic apparatus seems to be poised in a state of dynamic and precarious stability between a dissociated state which is all too easy to achieve and a very stable state which is also too easy to achieve.

These experiences with stability have a strong bearing on our chemical results. Let us consider the findings of Sakai. He isolated the mitotic apparatus in a rather complex medium, not the easiest to work with, containing 1 *M* sucrose, 0.001–0.002 *M* EDTA, and 0.15 *M* dithiodipropanol, a modification of a medium developed by Mazia *et al.* [20] when the main purpose of a new isolation procedure was to achieve conditions under which ATPase would not be denatured. It is important to note that the pH was ca. 6.0–6.2; later work, especially that of Kane [9] has shown that a low pH is a major variable governing stability. Mitotic apparatus isolated in this medium are readily soluble in 0.53 *M* KCl at pH 8.5. By "soluble" I mean that the structure is completely dispersed and therefore that molecules essential for the structure go into solution. I do not, of course, mean that all the material of the mitotic apparatus goes into solution. On the contrary, 30% is sedimentable at 100,000 *g* (60 minutes), and it is the remaining 70% that we consider to be the soluble component. In view of the structural complexity revealed by the electron microscope, which I shall describe below, we had to ask whether the dissolved protein included the protein of the spindle fibers or not. If it did not, we would be as far as ever from a way of studying the chemistry of the spindle fibers. Fortunately, as I shall show, we found we were dissolving the spindle fibers and leaving behind some other components of the mitotic apparatus.

Let me summarize briefly those findings of Sakai that are relevant to our discussion of the resolution of the spindle fibers.

1. The ultracentrifugal pattern (Fig. 1) was qualitatively similar to that seen in earlier work [e.g., 29]. There was a major component, accounting for ca. 60% of all the dissolved material and several minor components.

2. The major component, the lightest, had a sedimentation constant of 3.5 S, and a molecular weight of ca. 68,000. A second component, 13 S, was constantly found in smaller amounts. A third component, 22 S, was found consistently.

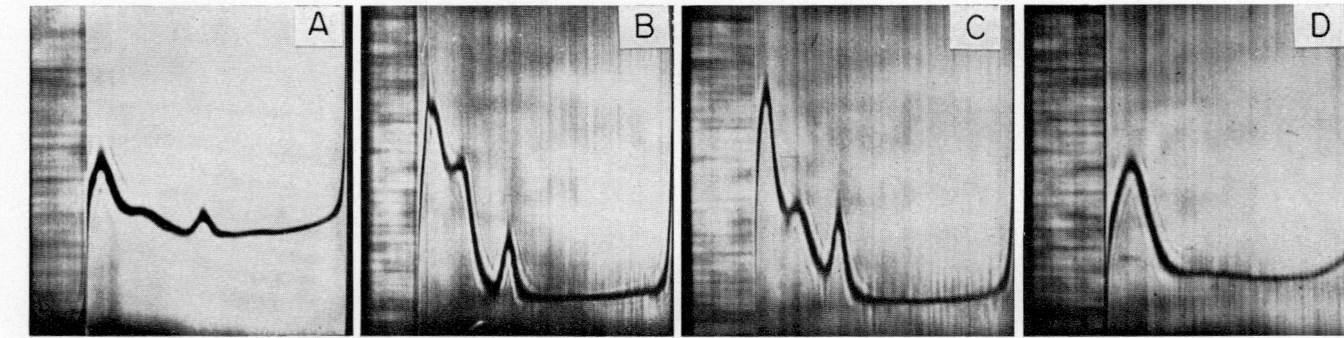


FIG. 1. The major soluble proteins of the isolated mitotic apparatus [from Sakai, 26]. (A) Ultracentrifugal pattern of soluble proteins from mitotic apparatus isolated by the method of Mazia *et al.* [20]. The major peaking is sedimenting at 3.5 S. (B) Effect of treatment of the material shown in (A) with 1 m$M$ sodium sulfite. Time and speed comparable to (A). Note that the major peak is splitting. (C) Same as (B) except that sodium sulfite concentration is 50 m$M$. The major peak is now sedimenting at 2.5 S. (D) Fraction of soluble protein of mitotic apparatus that is precipitated by 50 m$M$ $Ca^{++}$, sedimenting in 100 m$M$ sodium sulfite. Precipitation by $Ca^{++}$ yields a more homogeneous fraction.

3. The major component, 3.5 S, could be precipitated quantitatively by 50 m$M$ $Ca^{++}$ and could be redissolved in EDTA. Thus it is possible to obtain the 3.5 S unit, which accounted for so much of the dissolved material, in a relatively pure form.

4. The 3.5 S unit is split into two 2.5 S units (molecular weight ca. 34,000) by treatment with sodium sulfite or with dithiothreitol. There was also evidence that the 13 S component is split into 2.5 S units in the same way. Thus, it seems that the major component is a dimer of the 2.5 S units and the 13 S component may represent an aggregate of units which can be broken down into 2.5 S units.

5. The 3.5 S unit contains 6 free —SH groups. When it is split by sulfite, 7 —SH groups are recovered. Since splitting of an S—S bond by sulfite yields one —SH bond, it is concluded that the 3.5 S dimer is formed by uniting two 2.5 S monomers through one S—S bond. The conclusion is confirmed by reduction of the 3.5 S unit with dithiothreitol, which yields 2 additional —SH groups per mole of 3.5 S protein.

6. The major protein as isolated contains about 4.5% nucleotide. In earlier reports [e.g., 17, p. 247] the persistence of the association of nucleotide (according to Zimmerman a complex of nucleotides in ratios similar to those of the average RNA of the starting material) has been discussed. Obviously the hazard of spurious binding of the nucleotide (or RNA) in the course of isolation cannot be ignored. On the other hand, the fact that protein of the mitotic apparatus may be freed of the nucleotide [28] does not necessarily mean that the association was spurious.

In summary, the major protein fraction consists of a 3.5 S unit of molecular weight ca. 68,000, which is resolvable as a dimer of 2.5 S units of molecular weight ca. 34,000, the dimerization taking place through a single S—S bond.

Other workers, notably Stephens [28] and Miki-Noumura [21], have also obtained a unit sedimenting at about 2.5 S, but they have reported much larger units from the original dispersion of the isolated mitotic apparatus.

Having stated that all methods of isolation depend on the provision of conditions of stability, none of which has been proved to be the counterpart of the conditions of stability inside the cell when it contains a mitotic apparatus, one can hardly judge that one method is more "natural" than another. My conclusion that the 2.5 S or 3.5 S units are the actual subunits of structure will rest on their correspondence to the units actually seen with the electron microscope.

At this point, the conclusion from the efforts at chemical resolution of the mitotic apparatus is that the resolution, by one method or another, arrives finally at a unit of about 2.5 S to which Sakai has assigned a molecular weight of about 34,000.

## THE THIOL PARADOX

In the earliest work on the chemistry of the mitotic apparatus, the possible role of S—S bonds in the assembly of the structure was a prominent consideration. This line of thought dates back to the work of Rapkine [24] whose experiments dealt with a fall and rise of trichloroacetic acid-soluble —SH, thought by him to be glutathione, in the course of the mitotic cycle. His interpretation related the fluctuations to a reversible denaturation of proteins which was thought in turn to be related to the conversion of proteins to the fibrous state. In the work on the isolated mitotic apparatus by Dan and myself, we concluded that S—S bonds played an important role in the stability of the mitotic apparatus but presented evidence suggesting that the "essential morphology of the MA does not depend on S—S linkages alone."

There is no need to cite all the work that supports the original preconception that S—S bonds play a decisive part in the assembly of the mitotic apparatus. Much of the work has been done in my laboratory and in Dan's laboratory, and we can conclude that experiments asking whether S—S bonds were important in the structure always said that they were important. The evidence has been summarized in a number of review articles [13, 16, 17]. Some of the main lines of evidence are: the accompaniment of the mitotic cycle by the Rapkine cycle, which was confirmed by Sakai and Dan [27] who, however, found that the trichloroacetic acid-soluble component which underwent the cycle was a protein, not glutathione; the blocking of the formation of the mitotic apparatus by mercaptoethanol [15]; the cyclic changes, the cytochemical staining of the mitotic apparatus for —SH groups [11], and a considerable body of experience indicating that the stabilization of the mitotic apparatus involved S—S bonds [discussed by Mazia, 18]; the arrest of mitosis at high oxygen pressure [25].

On the other hand, the original assertion that S—S bonds alone were not responsible for the coherence of the mitotic apparatus was an understatement. If experiments designed to test the role of S—S bonds in the structure of the mitotic apparatus generally gave positive results, it is equally true that experiments asking whether its structure depended on weak (e.g., hydrogen) bonds generally say that such is the case! That is what I mean by the *thiol paradox*. First of all, one must be troubled by an inconsistency between the notorious instability of the mitotic apparatus and the notion of its being a covalent-bonded structure. This contradiction was discussed in a paper in which I proposed a hypothesis of a dynamic sulfur-bonding based on continuous mercaptide exchange [18]. But there is more direct evidence for an essential role of weak bonds in the structure. Such evidence comes from much of the work on birefringence [7], from which the stability of oriented structure in the mitotic apparatus seems to be extremely sensitive to temperature in a reversible way; from the profound effects of $D_2O$ in stabilizing the mitotic apparatus [5]; from a variety of studies on the effect of hydrostatic pressure [30]; from the study of the effects of agents such as colchicine which are difficult to relate to sulfur chemistry. In the discussion, I shall attempt to explain the thiol paradox.

## ELECTRON MICROSCOPIC RESOLUTION

Thus far, I have deliberately spoken only of "spindle fibers," using almost the oldest term for the structure in the mitotic spindle that expresses its polarization and the engagement of the chromosomes to the poles. One of the very earliest definitive descriptions [4] says: "From the studies of Bütschli, Strasburger, Hertwig, and Mayzel . . . we have known for some time of the existence, in different organisms, of those types of nuclear division figures which contain bundles of fine threads, usually arranged in the form of a spindle located between the poles." Thus the resolution of the spindle fiber begins with a structure recognized with the light microscope, usually after appropriate fixation procedures. The only really extensive observations of spindle fibers in the living cell are those made by means of advanced polarization microscopy, especially in the work of Inoué.

The earlier history of the electron microscopic investigation of the spindle seems peculiar in that the observers either failed to find the spindle fibers or could detect them only at a level of coarseness that did not much improve the image over what was seen with the light microscope. An account by Porter [23] is an early reference to the resolution of the spindle fibers into smaller filaments, seemingly tubular in character, and the conclusion that the spindle fibers are bundles of microtubules appeared in the work of DeHarven and Bernhard [3]. In my opinion, the whole body of further work, now amounting to a description of mitosis nearly as complete as we can expect from the present techniques of the electron microscopy of thin sections, has confirmed that conclusion.

The cytological aspects of the distribution and behavior of the microtubules in the dividing cell are discussed admirably well by Ledbetter and by Stubblefield in this symposium, and I will deal with the further steps of resolution. However, it seems important to recall that the mitotic apparatus, even apart from chromosomes and centers, is not just a mass of microtubules. It is a body, seemingly having considerable rigidity, within which the microtubules are embedded and within which they operate. It possesses a matrix that shows considerable differences from the surrounding cytoplasm, though I abstain from adding another Greek-rooted plasm to the lexicon of cell biology. The literature on the background structure of the mitotic apparatus is not extensive, and I will cite mainly the work of Patricia Harris [6] on the sea urchin egg. This material that has been observed after direct fixation of the cell and after isolation of the mitotic apparatus is thus far the only material on which we can correlate microscopic and biochemical data.

The matrix in which the spindle fibers lie is describable as a gel by virtue of its physical cohesion. The main component appears to be a dense mass of vesicles, and one imagines that the gel properties derive from the interaction of the vesicles, which is so strong that the larger particles such as mitochondria are excluded from the domain of the mitotic apparatus as it forms. The only particles one sees consistently within the mitotic apparatus are free ribosomes.

The existence of this massive matrix poses this question for our present discussion: Does the major protein fraction which is the most characteristic macromolecular component of the mitotic apparatus derive from the matrix or from the spindle fibers? This question has now been answered in a fairly direct way [12]. The major protein fraction is put into solution by treating isolated mitotic apparatus with 0.5 *M* KCl. We now observe the dispersion thus obtained and find that it contains no microtubules but still contains vesicles and small particles, and we conclude that the dissolved mitotic apparatus contains the substance of the microtubules in solution. To confirm the observation, we observe the pellet obtained after centrifugation at 100,000 *g* for 1 hour. It contains vesicles and some particles but does not contain microtubules. Thus, we conclude that the soluble molecules studied by Sakai include the molecules of the microtubules.

If the microtubules are composed of the major protein, the most direct evidence would come from the visual resolution of microtubules into units which are comparable to units that have been characterized in the chemical studies. The existing literature on microtubules—including some that appears in this symposium (Ledbetter, Hookes, Grimstone)—tells us that the microtubules are cylindrical arrays of filaments [termed protofilaments by André and Thiéry, 1] spaced at ca. 50 Å. From the work of Ledbetter and Porter, described by Dr.

Ledbetter in the present symposium, comes the conclusion that there are 13 such filaments in the microtubule. In the earliest work on the observation of microtubules at high resolution [1, 22] it was seen that the filaments consist of rows of globular subunits.

We have recently completed a study of the microtubules of the mitotic apparatus, using the technique of spreading the material on air-aqueous interfaces, followed by negative staining [12]. If the material was spread on uranyl acetate, we obtained images of the microtubules that agree in all respects with those obtained by others. When the material was spread on pure water, the microtubules were occasionally torn open and spread flat, so that the details could be seen a little better (Fig. 2). The pictures of the spread-out tubules show 12–13 filaments, and we think it likely that Ledbetter's idea that there are 13 is correct and general. The globular units of which the filaments are composed are seen quite clearly; they measure about 35 Å. Of course such a measurement made by the negative staining technique so close to the limit of resolution of our actual technique can only be approximate, but it does exclude some possibilities.

These observations are very similar to those made by Barnicot [2] on negative-stained spindle fibers from newt cells. He describes fibrils which ". . . like those described by Pease, had the appearance of rows of longitudinally connected granules about 35 Å in diameter."

The aim of this work was to resolve the spindle fibers as far as possible. Let us then summarize the results. By chemical approaches, it has been found that the mitotic apparatus can be dispersed into a number of components, the most conspicuous being the major protein having a molecular weight of ca. 68,000 which in turn can be resolved by the splitting of one S—S bond per molecule into two monomers having a molecular weight of ca. 34,000. The control of this fractionation by electron microscopy tells us that the dissolved proteins include the molecules of the microtubules; therefore it is possible to ask whether the structural units of the tubules are the molecules of the major protein.

If now we match the chemical and microscopic results, we are allowed the conclusion that molecules constituting the major protein are in fact the globular units which we see by electron microscopy. A particle showing a diameter of 35 Å in negative-stained preparations will not have a molecular weight greater than 68,000. Even if we make liberal allowances for error of measurement, the particles cannot be very much larger than the measured dimension, since the axis-to-axis distance between the filaments is given as ca. 50 Å. We can, for example, exclude the 22 S unit, which we see in small quantity and which is recovered

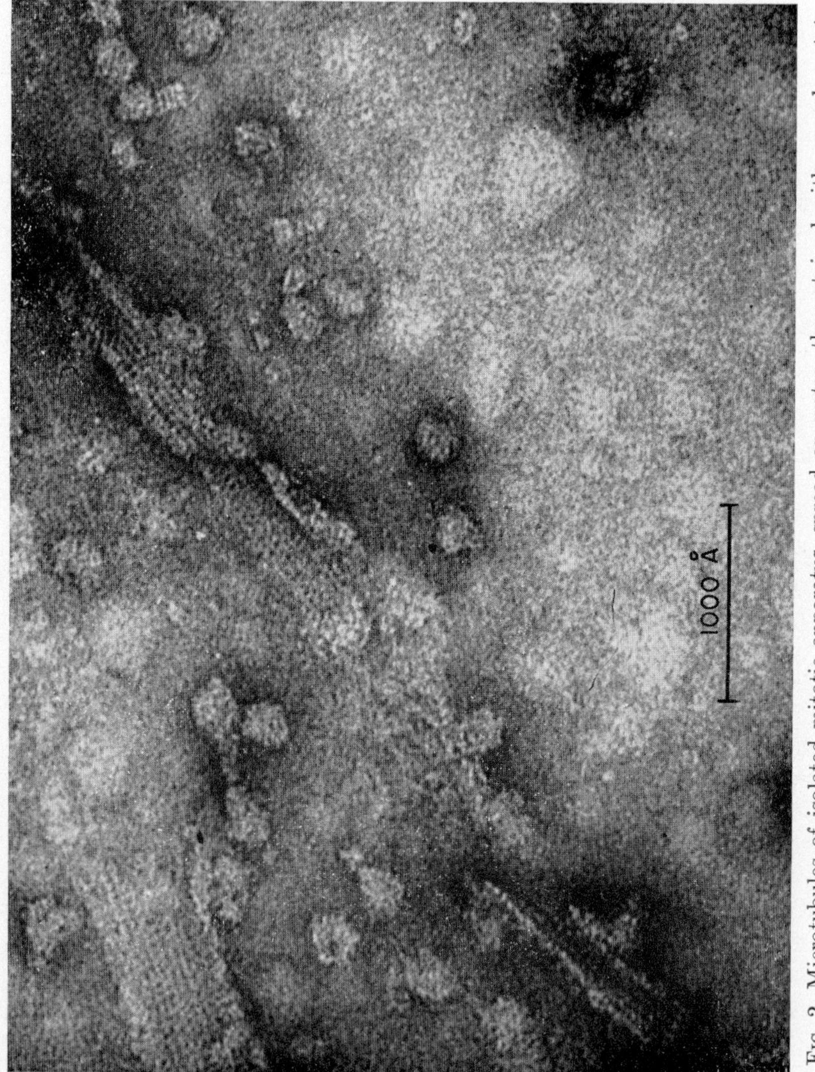


FIG. 2. Microtubules of isolated mitotic apparatus, spread on water, then stained with uranyl acetate. At lower left, a portion of a microtubule is well spread. Instrumental magnification × 40,000; total magnification × 263,000.

in large quantity by the techniques used by Kane [9]. That particle, appearing so consistently, may be an extremely important aggregate of the smaller ones and may play an important role at some stage of the assembly or disassembly of the microtubules, but it is far too large to be the unit we actually see as the unit of the filaments which make up the tubule.

We would like to know whether the unit we see, the ordering of which is discussed by Hookes and by Grimstone in this symposium, is the 3.5 S dimer or the 2.5 S monomer. In a published account [12] we gave reasons for thinking it is the 2.5 S monomer, but that argument can be no more rigorous than the measurement of the diameter of such particles with the electron microscope, which is not very rigorous. At least, our conclusion was an attempt to be objective in comparing the microscopic and the chemical findings. As will be seen in the following discussion, it would be convenient for theory building to make the 3.5 S dimer the structural unit of the microtubules.

This account has been based on experience with the microtubules of the mitotic apparatus. Of course we now view the microtubule in the generic sense—in the sense of a class of structures having a similar appearance—as being a widespread constituent of cells. That generalization has considerable basis in the comparison of two kinds of microtubules that have been studied in some detail: those of the mitotic apparatus and those comprising the nine outer fibers of cilia and flagella. Having cited the earlier work and the work reported elsewhere in this symposium, I can mention that Dr. Solari has made excellent pictures of the microtubules of the flagella of sea urchin spermatozoa by the spreading technique, and finds that the picture is similar in detail to that described here for the mitotic apparatus. From Dr. Gibbons' paper we learn of a major protein having a molecular weight of 60,000 which could be compared to our 3.5 S unit. Perhaps we are approaching the time when we know enough about similarities among the various kinds of microtubules to examine the differences in detail; after all, spindle fibers and flagella or cilia do not perform in exactly the same ways.

## DISCUSSION

It is superfluous to state that the objective of attempts at a molecular resolution of a part of the cell is to understand better how it works; but it is not at all superfluous to ask, once that resolution has progressed, whether that end has in fact been approached. It cannot be claimed that the progress in the study of the mitotic apparatus has provided such dazzling insights into the act of mitosis that the mysteries we contemplate in a time-lapse film are substantially diminished, but I think there is a good deal more to be said than we could say before. In this brief discussion, I will try to align what has been learned about the spindle fibers with the rudimentary facts about mitotic movement, serving the aims of our symposium to the extent that the problem of "formation and fate" of the spindle fibers is bound very closely to our present ways of thinking about the movements.

First, I would remind you that the separation of chromosomes is accompanied by a shortening of the chromosome-to-pole fibers and a lengthening of pole-to-pole fibers [17, pp. 272–280]. We are allowed to regard both these processes as active, as responsible for the actual separation of the chromosomes. A more cautious position would be that the shortening and lengthening of the spindle fibers is certainly the guidance mechanism in mitosis, but that it has not been proved to be the motor mechanism. However, we can surely regard the shortening and elongation of the microtubules as an intrinsic part of chromosome movement.

One of the consequences of the electron microscopic resolution of the spindle fibers into microtubules was the strong impression, perhaps more, that the microtubules actually and literally shorten and lengthen. That is to say that they do not become thinner or thicker nor do they curl or coil nor do they give any other evidence of contraction or expansion at constant mass. According to all observations so far, "contraction" leads to a shorter microtubule, which has changed only in length, and elongation appears as a growth of a microtubule.

Even before these were firm observations, there was some appreciation of the possibility that change in length of spindle fibers was a change in mass. As an example of a groping at that idea, I quote an old paper of mine [14] in which I discussed a possible mechanism of "contraction" of spindle fibers "by a statistical disaggregation in which some of the molecules are lost and those that remain are consolidated into a new and smaller association." Further, ". . . it [the model] merely describes the situation where some of the units, by contraction or by dissociation, lose their ability to interact with the rest and join the less oriented (or less associated) background material. It follows . . . that while the chromosomal fibers may be consumed by their contraction the spindle as a whole, which has the same protein composition, may be growing." Inoué [7, 8] has developed in a more quantitative form a conception which attributes the dimensional changes in the spindle to an equilibrium between the ordered (in fibers) and disordered (a background pool of molecules) phases of the molecular species making up the fibers. He has described the thermodynamic characteristics of the equilibrium and has cited a considerable variety of evidence that is consistent with the model.

Let us be conservative, avoiding for the moment the question how such a shortening or lengthening by dissociation and association could actually pull chromosomes to poles or push poles apart, and merely insist that these changes are important to the processes. We then find ourselves in the position of knowing something about the molecules involved and capable of speculating about how the molecules are added to or removed from the tubules.

We can propose that the overall assembly of the microtubules is analogous to the spontaneous assembly of viruses, etc. It depends on the weak bonding of unit molecules whose conformations allow close approach and close fit. Just as the TMV protein assembles spontaneously as a cylinder, so might the molecular units of the microtubules. One would view the overall inter-molecular bonding in the microtubules as being based on weak bonds, not covalent bonds.

The molecules would assemble spontaneously to form the microtubular array at the sites of lengthening, and would disassemble at the sites of shortening. I would like to speculate that the probability of assembly or disassembly is determined immediately by the conformation of the units. That is, there would be two states of the units, which I would call a *fit* state and a *non-fit* state. Speculating further, I would propose that the *fit* state is represented by the 3.5 S dimer, the *non-fit* state by the 2.5 S monomer. Thus, the old thiol paradox would be resolved. The microtubule, a weak-bonded structure in the sense that its characteristic morphology depends on weak bonds, would *appear* to be an S—S bonded structure in chemical experiments designed to test the importance of S—S bonds. It would appear so because the conformation of the molecular units that is essential for the assembly of the structure would be that of the dimers which are formed by S—S bonding. So far as I can see, all the old work on the importance of S—S bonds in the formation and structure of the mitotic apparatus could be reinterpreted in that way.

As was stated earlier, the precision of the measurements on negatively stained electron micrographs does not allow a clear decision whether the particles we see are the 3.5 S units or 2.5 S units. If they are the latter, we would have to say that they really entered the structure of the microtubules as dimers but that the evidence of the pairing is not preserved in the photographs. It is the dimers that we recover when we dissolve the mitotic apparatus. To argue confidently that the observed particles are the 3.5 S dimers, we would have to be sure that such particles would have the measured diameter of 35 Å when dried and negatively stained.

I would repeat that what I have given thus far is an interpretation in which I have added some facts about the unit molecules to an existing hypothesis concerning the shortening and lengthening of the microtubules. The rest of what I have to say is not additional hypothesis so much as a series of possible answers to inevitable questions about mitosis, the answers stating some implications of the hypothesis. How can we imagine that a shortening of fibers will result in an apparent "pulling" of chromosomes? If the dropping out of molecules from the microtubules is a localized, short-range affair, we can imagine that bonds will open, molecules will leave the structure, and new bonds will be formed, reducing the length of the tubules without "breaking" them. Mitotic movement is not a rapid process; the fibers shorten at the rate of about 0.1 mm per hour. How do we deal with the fact that some fibers—the chromosomal fibers—are shortening while others—the pole-to-pole fibers—are lengthening? Here there is no easy way out; in any hypothesis we have to invoke highly localized conditions making for lengthening or shortening. The general kinds of conditions are twofold. The conditions making for the formation of disulfide bonds are a little better known in the form of the changes of soluble —SH and of the interprotein electron transfers recently described by Sakai [26]. These can help us describe the overall growth of the mitotic apparatus as the cell enters division, but not the local events. The second set of conditions parallel those invoked in the cases of "self-assembly" that have been studied in other systems. Perhaps the term is a misnomer. The whole point is that the experimenter provides simple but precise conditions for assembly or disassembly of a virus or a collagen thread, perhaps something as simple as a change of pH or of the ionic medium. The problem is: How can the cell provide such conditions at a given point? It is attractive to recognize that we may have some landmarks. An interesting hypothesis that explains everything because we know so little is that the centrioles provide the sites of local assembly and kinetochores provide the sites of local disassembly. What are the implications for the explanation of antimitotic action? Here the view that the microtubules are assembled from molecules in the *fit* conformation can be useful in conjunction with Inoué's proposals about an equilibrium between the structured and unstructured states. We can imagine that antimitotic agents acting like colchicine do not necessarily react with the mitotic apparatus itself, but with the molecules in the free state, modifying their conformation so that they can not change to the *fit* state. In the case of another type of agent, mercaptoethanol, we can imagine that it impedes the dimerization. So far, no one has observed that either type of chemical does much to the isolated mitotic apparatus. It is also a fact that neither has much effect on the mitotic apparatus once it is in anaphase; the effects seem to be on the formation of the apparatus, perhaps at stages when the dimers are being prepared.[2]

## CONCLUDING STATEMENT

Having essayed speculative answers to questions that certainly will be asked about the consequences of some findings about the resolution of the microtubules of the mitotic apparatus, I would like to summarize the factual findings that are relevant to this symposium. These are:

(1) The long-studied "major protein" of the mitotic apparatus is a 3.5 S particle, to which a molecular weight of ca. 68,000 has been assigned.

(2) The 3.5 S particle is a dimer of 2.5 S particles of half that molecular weight. The dimer is made by forming one S—S bond.

(3) The microtubules are resolved into filaments which are linear arrays of the molecules of the major protein, though it is not certain whether the units visualized are the 3.5 S units or the 2.5 S units.

## References

1. André, J., and Thiéry, J. P., *J. Microscopie* **2,** 71 (1963).
2. Barnicot, N. A., *J. Cell Sci.* **1,** 217 (1966).
3. DeHarven, E., and Bernhard, W., *Z. Zellforsch. Mikroskop. Anat.* **45,** 378 (1956).
4. Flemming, W., *Arch. Mikroskop. Anat. Entwicklungsmech.* **18,** 151 (1880). [English transl. in *J. Cell Biol.* **25**(1), Pt. 2, 1 (1965).]
5. Gross, P. R., and Spindel, W., *Ann. N.Y. Acad. Sci.* **90,** 500 (1960).
6. Harris, P., *J. Cell Biol.* **14**(3), 475 (1962).
7. Inoué, S., *in* "Biophysical Science—A Study Program" (J. L. Oncley, ed.), pp. 402–408. Wiley, New York, 1959.
8. Inoué, S., *in* "Primitive Motile Systems in Cell Biology" (R. D. Allen and N. Kamiya, eds.), pp. 549–598. Academic Press, New York, 1964.
9. Kane, R. E., *J. Cell Biol.* **12,** 47 (1962).
10. Kane, R. E., and Forer, A., *J. Cell Biol.* **25**(3), Pt. 2, 31 (1965).
11. Kawamura, N., and Dan, K., *J. Biophys. Biochem. Cytol.* **4,** 615 (1958).

[2] Between the time of the symposium and the receipt of the proofs, the binding of colchicine to proteins thought to be of microtubular provenance has been reported by Borisey and Taylor (*J. Cell Biol.* **34,** 535, 1967) and by Wilson and Friedkin (*Biochemistry* **6,** 3126, 1967).

12. Kiefer, B., Sakai, H., Solari, A. J., and Mazia, D., *J. Mol. Biol.* **20,** 75 (1966).
13. Mazia, D., *in* "Glutathione" (S. Colowick *et al.,* eds.), pp. 209–222. Academic Press, New York, 1955.
14. Mazia, D., *in* "Advances in Biological and Medical Physics" (J. H. Lawrence and C. A. Tobias, eds.), Vol. IV, pp. 70–118. Academic Press, New York, 1956.
15. Mazia, D., *Exptl. Cell Res.* **14,** 486 (1958).
16. Mazia, D., *in* "Sulfur in Proteins" (R. Benesch *et al.,* eds.), pp. 367–390. Academic Press, New York, 1959.
17. Mazia, D., *in* "The Cell" (J. Brachet and A. E. Mirsky, eds.), Vol. III, pp. 78–412. Academic Press, New York, 1961.
18. Mazia, D., *in* "Biological Structure and Function" (T. W. Goodwin and O. Lindberg, eds.), Vol. II, pp. 475–496. Academic Press, New York, 1961.
19. Mazia, D., and Dan, K., *Proc. Natl. Acad. Sci. U.S.* **38,** 826 (1952).
20. Mazia, D., Mitchison, J. M., Medina, H., and Harris, P., *J. Biophys. Biochem. Cytol.* **10**(4), Pt. 1, 467 (1961).
21. Miki-Noumura, T., *Embryologia* (*Nagoya*) **9,** 98 (1965).
22. Pease, D., *J. Cell Biol.* **18,** 313 (1963).
23. Porter, K. R., *in* Symposium on the Fine Structure of Cells," pp. 236–250. Wiley (Interscience), New York, 1955.
24. Rapkine, L., *Ann. Physiol. Physiochim. Biol.* **7,** 382 (1931).
25. Rosenbaum, R. M., and Wittner, M., *Exptl. Cell Res.* **20,** 416 (1960).
26. Sakai, H., *Biochim. Biophys. Acta* **112,** 235 (1966).
27. Sakai, H., and Dan, K., *Exptl. Cell Res.* **16,** 24 (1959).
28. Stephens, R. E., *Biol. Bull.* **129,** 396 (1965).
28a. Stubblefield, E., and Brinkley, B. R., this symposium, p. 175.
29. Zimmerman, A. M., *Exptl. Cell Res.* **20,** 529 (1960).
30. Zimmerman, A. M., and Silberman, L., *Exptl. Cell Res.* **38,** 454 (1965).

# THE DISPOSITION OF MICROTUBULES IN PLANT CELLS DURING INTERPHASE AND MITOSIS[1]

**MYRON C. LEDBETTER**

*Biology Department, Brookhaven National Laboratory, Upton, New York*
With Drawings by John S. Defino, Yonkers, New York

## INTRODUCTION

Cytoplasmic tubules of the type to be discussed here may be seen in the light microscope only when they are present in relatively large numbers and oriented parallel to one another. Among the earliest observations of this sort were those of the mitotic spindle made almost a century ago at the time of Strasburger [27] and Flemming [9]. Fixed preparations of this time already suggested that the spindle was a gel composed of some kind of fibrous material. These workers actually saw groups of microtubules clumped together by the fixative; however, later experiments using micrurgy, centrifugation, and various optical methods, especially polarized light, conclusively established the fibrous nature of the spindle in the living state [12].

In the 1950's some studies with the electron microscope showed evidence of filaments in the mitotic spindle [8]; but a fuller understanding of its fine structure awaited improvements in the composition of the fixing solutions. The use in the fixative of divalent ions, sometimes with low pH, and especially the use of dialdehydes [26] sufficiently stabilizes the mitotic apparatus to enhance greatly the details discernible in our fixed and thin sectioned material. Further information on the cytotubules is available from negatively stained preparations [2, 5, 19] and from replicas of cells which have been freeze-fractured and etched [17].

With improved stabilization individual filaments of the mitotic apparatus appear as minute tubular structures about 240 Å in diameter. When Roth and Daniels [25] and shortly thereafter Harris [10] first made such observations on the mitotic spindle it was not clear whether the fibrous component was actually tubular, or whether this appearance was due to a high concentration of electron dense matter upon the surface of a solid rodlike structure. Evidence from negatively stained

[1] Research carried out at Brookhaven National Laboratory under the auspices of the U.S. Atomic Energy Commission.

material [2, 5, 19] and from particularly favorable cross sections of these structures [13] supports the tubular concept.

It is now recognized that microtubules are a constant feature of plant and animal cytoplasm though they vary over wide limits as to frequency and manner of distribution within the cell [21]. There is reason to believe that most of these elements are fundamentally similar in structure whether in the mitotic spindle, interphase cytoplasm, or in fibers of cilia and flagellae [14]. According to this assumption the mitotic spindle represents but one particular arrangement in a continuous display of tubules during the history of the cell.

## MICROTUBULES—GENERAL FEATURES

The outside diameter of these tubular structures is about 240 Å. This dimension is found both in dialdehyde-fixed material seen in thin section and in negatively stained preparations. The length of the tubules is more difficult to define. This is due partly to our dependence on thin sections for the bulk of information on the way the tubules course through the cytoplasm. In these preparations we are unable to discriminate between an end of a tubule and the point at which it passes out of the plane of sectioning. However, it is certain that tubules can extend for several microns, since they have been followed for that distance in single sections [7].

It can be inferred from the fact that the microtubules tend to follow straight lines or gentle curves that they possess more rigidity than the cytoplasm in which they are buried. There is, in fact, good evidence that they are sufficiently rigid to contribute significantly to the shape of cells [4, 7, 23, 29].

When viewed in transverse section the microtubules are seen to be surrounded by a "halo" of low density. Ribosomes are excluded from this zone and the tubules seem to be limited in the proximity to one another such that the center-to-center distance between tubules is not less than about 350 Å. Contact may be attained between the tubule wall and the dense surface of membranes, notably the plasmalemma, tonoplast, and pectin vesicles [13, 14].

## THE MITOTIC CYCLE

We will present a description of the microtubules in association with certain cytoplasmic elements as exhibited in cells of higher plants at interphase and during the events of mitosis. These features will be illustrated by three-dimensional reconstructions based on the study of numerous electron micrographs of thin-sectioned material fixed in glutaraldehyde and osmium tetroxide, and from the published literature. We have chosen to illustrate this sequence by a cell as it would appear lying in a root meristem cut longitudinally in such a way that one half of the cell is exposed to view. The principal axis of the root and the spindle axis will run vertically. The walls which encase the adjacent cells are shown as though cut near the cell of interest. Only those organelles are shown that we can relate directly to mitosis. Wide latitude has been taken in making these drawings to alter the relative size and number of structures in order to simplify and better illustrate the salient features. It should be understood, for instance, that the microtubules, nuclear pores, and plasmodesmata are relatively smaller and more numerous than depicted. The rather complicated forms in the nucleolus have been reduced to a simple cluster of granules. In presenting this series of illustrations our limited knowledge requires us to make certain assumptions some of which will undoubtedly have to be modified as more complete information becomes available.

After the description of mitosis and cell wall differentiation, a short discussion will be given of some details of the fine structure of microtubules.

## INTERPHASE

The organization of some of the components of a cell between mitotic events is portrayed in Fig. 1. The microtubules are confined largely to a peripheral layer of cytoplasm just within the plasma membrane to a depth of about 1000 Å. Furthermore, they are arranged in such a way that they correspond closely to the orientation of the cellulose microfibrils in the adjacent cell wall. That is, the tubules subjacent to the side wall run circumferentially about the cell axis, while those at the end walls are disposed at random, though sometimes found in bundles up to about six elements. Those tubules along the side walls have been illustrated in Fig. 1 as continuous about the cell circumference, but there is little evidence to help us decide whether they are truly continuous or possibly present in relatively short arcs about the cell wall, or even spirals of low pitch. The possible termination of the randomly arranged tubules at the end wall is equally open to speculation. While microtubules seem not to be necessary for randomly deposited cellulose, as found at the tips of root hairs [18], it appears that tubules are a constant accompaniment of the deposition of oriented cellulose microfibrils.

In these cells the chromatin is most prominent as loosely condensed masses distributed about the periphery of the spherical nucleus. The

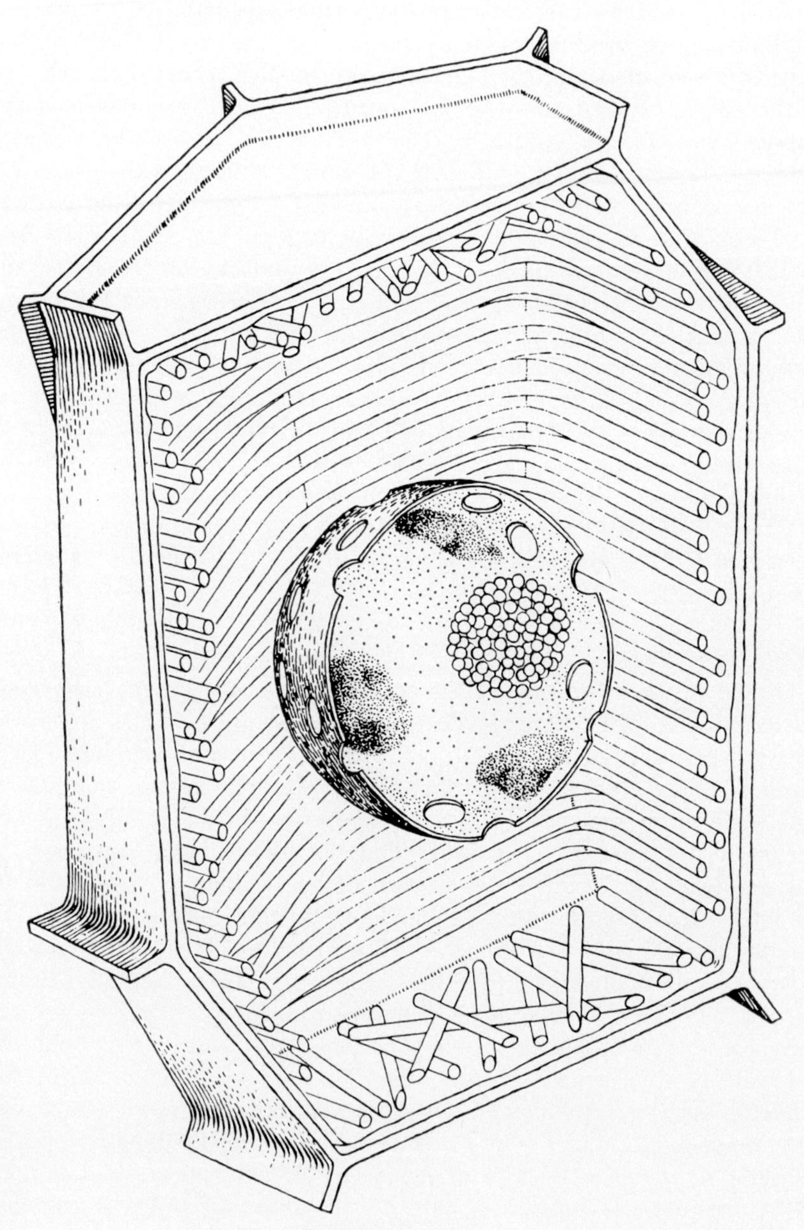

FIG. 1. Interphase.

envelope is perforated by numerous small pores. The nucleolus, often quite large in plant cells, is represented by a tightly packed cluster of granules.

The relationship of microtubules to secondary wall growth will be discussed in a later section.

## EARLY PROPHASE

The events described here were discovered by Pickett-Heaps and Northcote [20] in wheat meristems and in dividing complexes which give rise to the stomata in leaves.

Cells were examined at what they termed "preprophase" in an attempt to detect some early reorganization of materials at the polar zones about the nucleus. They found none. Instead, they detected an interesting change in the cortical cytoplasm. Here the microtubules of the peripheral layer of cytoplasm of the interphase cell, which the authors refer to as "wall tubules," are supplanted by 150 or more elements encircling the nucleus as shown in Fig. 2. They are restricted to an equatorial band which extends for over 2.5 $\mu$ along the wall and have a high density showing some evidence of close packing among the tubules. The band is composed of several layers of tubules which accumulate to over twice the depth into the cytoplasm (2400 Å) from the plasma membrane as the corticle tubules at interphase. Of special interest is the fact that this band predicts the location of the wall in the next division.

The chromatin begins to condense into tighter and more discrete clumps at early prophase (Fig. 2). This condensation was used as a marker of this stage by Pickett-Heaps and Northcote. The nucleus may undergo changes in shape at this time, sometimes becoming more flattened at the presumptive poles. The envelope remains intact and retains its pores. The granules of the nucleolus become less tightly packed and may begin to disperse into the nucleoplasm.

## PROPHASE

The equatorial band of tubules is not found in the cell which is entering full-fledged prophase (Fig. 3), the cytoplasm formerly occupied by the band now showing no obvious differences from the remainder of the cell cortex. Mitotic poles are established at opposite sides of the nucleus as delineated by sets of converging tubules. An axis through the poles would pass perpendicularly through the plane occupied by the previous equatorial band of tubules. The cap-shaped

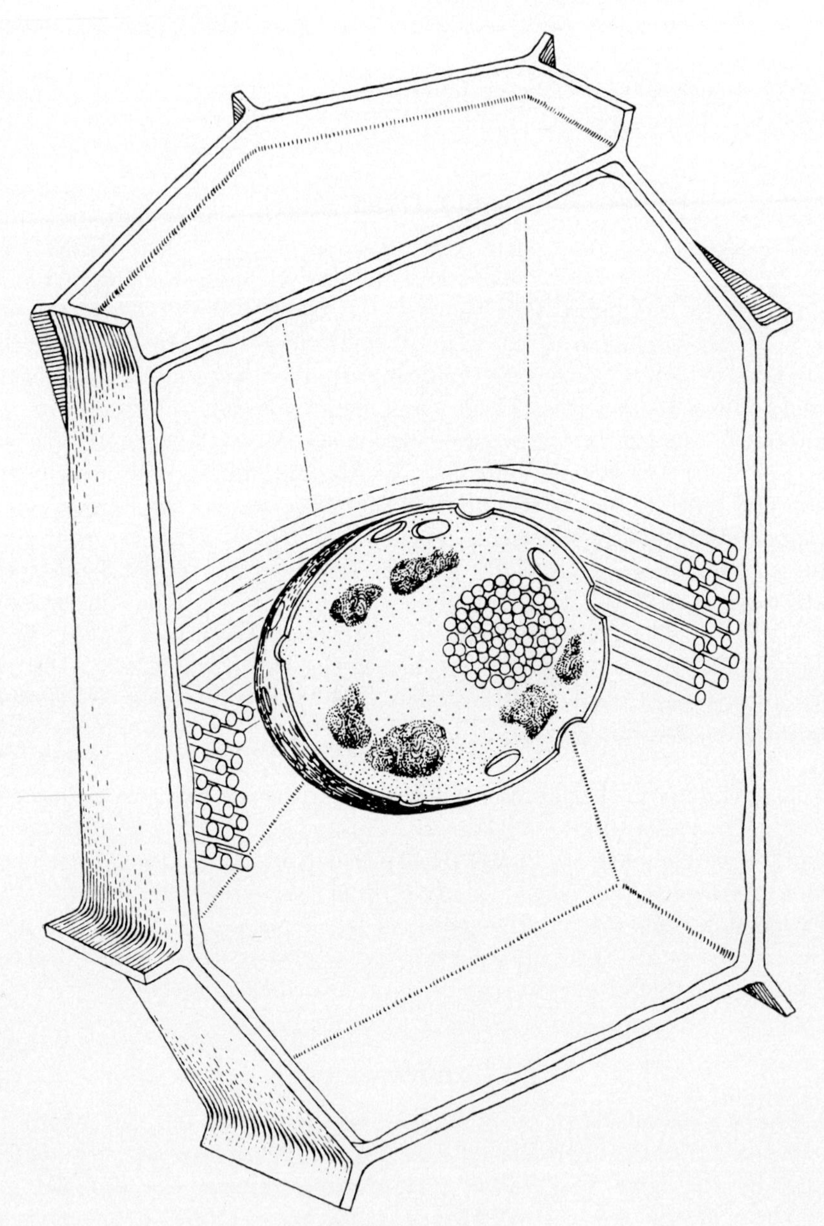

FIG. 2. Early prophase.

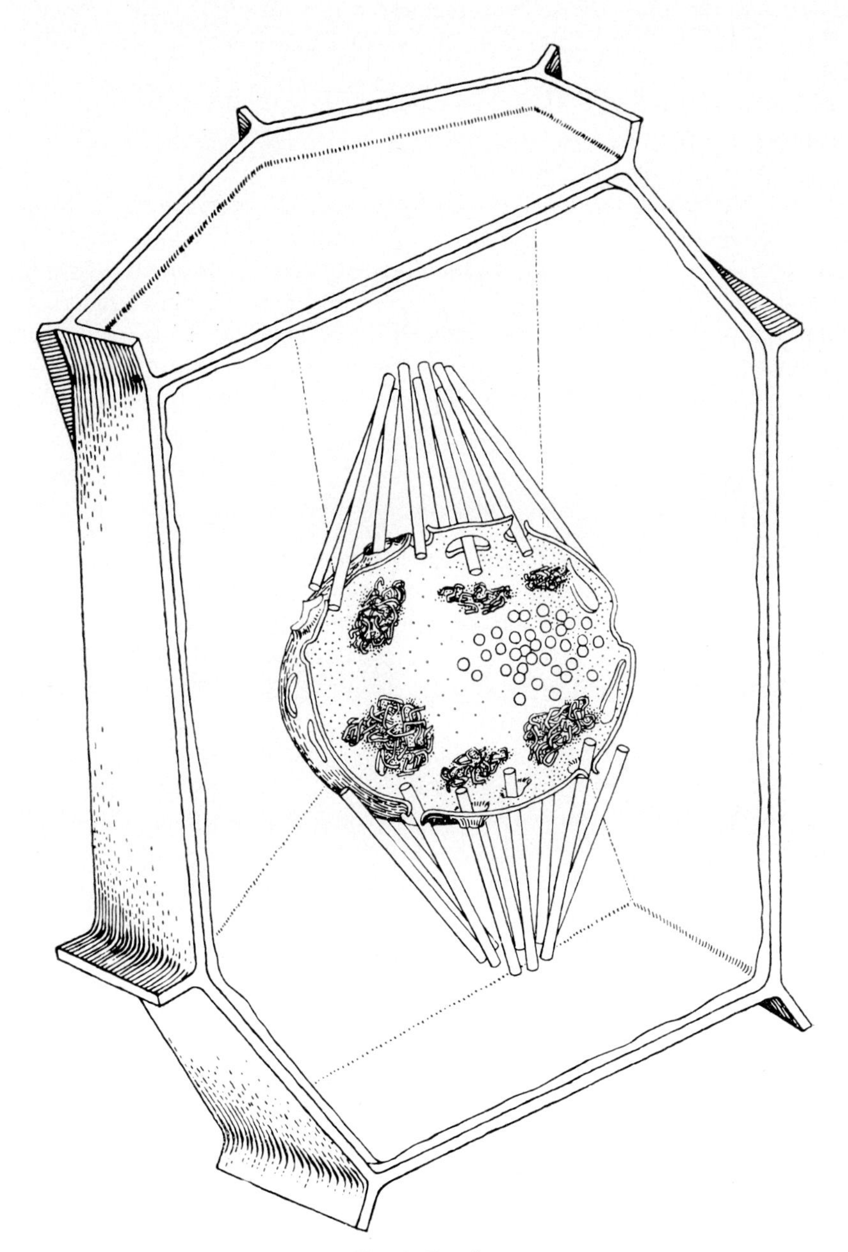

FIG. 3. Prophase.

clusters of tubules define the regions which are the "clear zones" [3] from which large organelles are excluded. Rigidity of the microtubules may provide a barrier against the entrance of organelles into these zones. The point of convergence of the tubules seems less well defined here than in cells that possess centrioles, and no astral rays are formed exterior to the pole.

Other changes are evident in the nucleus. The chromatin, now much more condensed, drifts away from the surface of the envelope. The nucleolus is usually lost as a discrete body as the particles disperse into the nucleoplasm. The envelope ruptures beneath the bundles of polar tubules. Microtubules are attached to the condensing chromatin through openings in the envelope. At the edges of the ruptures, which may result from enlargement of the nuclear pores, the envelope begins to extend up into the clear zone along the tubules.

## METAPHASE

At metaphase (Fig. 4) tubules extend from the polar regions to the chromosomes, which have now reached a maximum degree of condensation. The chromosomes are arranged with their kinetochores in a plane normal to the spindle axis. The point of attachment of the tubules to the chromatin in plant cells appears less complicated than the well-structured equivalents of animal cells [15, 24]. Plant cells lack the kinetochore plate characteristic of animal cells; the tubules instead pass directly into the chromatin (Fig. 5). It is not possible to say from the present evidence whether the chromatin occupies the "halo" of low density which characteristically surrounds the microtubules. The attachment sites of the tubules to the chromosomes in plant cells may be considerably more diffuse than those in animal cells. Tubules are found between the chromosomes, but the general impression gained is that there are fewer of these than similar tubules in animal cells. The length of individual microtubules which pass from pole-to-pole or chromosome-to-pole is not known although they certainly extend for several microns along the spindle. All the tubules appear to converge on the poles in like manner.

Near the poles there are numerous profiles of endoplasmic reticulum, presumably derived from the nuclear envelope. The degree of fragmentation suggests that the membrane surface of the endoplasmic reticulum may be proliferating in these zones. There must be a supply of endoplasmic reticulum sufficient to meet the needs of the two new cells, including that which will be required to envelop the daughter nuclei. This may be the time at which the necessary materials are

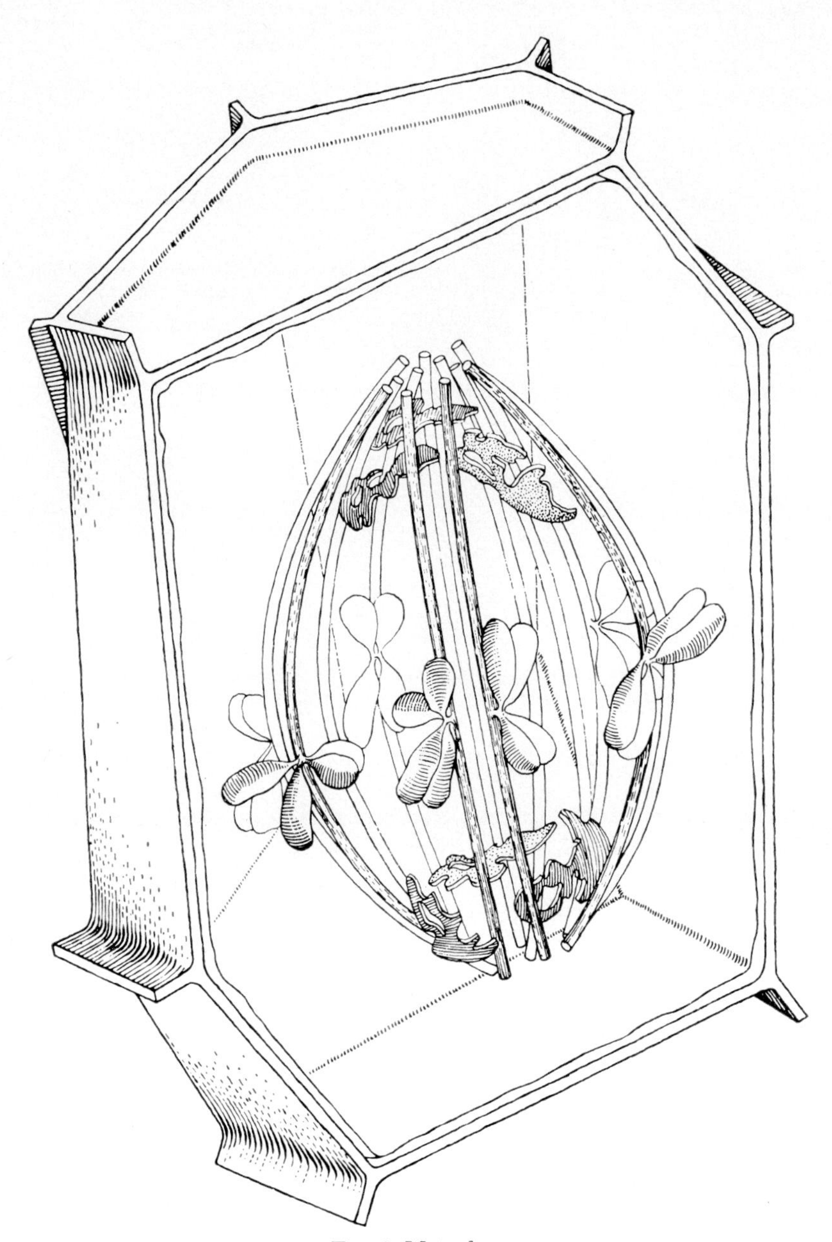

FIG. 4. Metaphase.

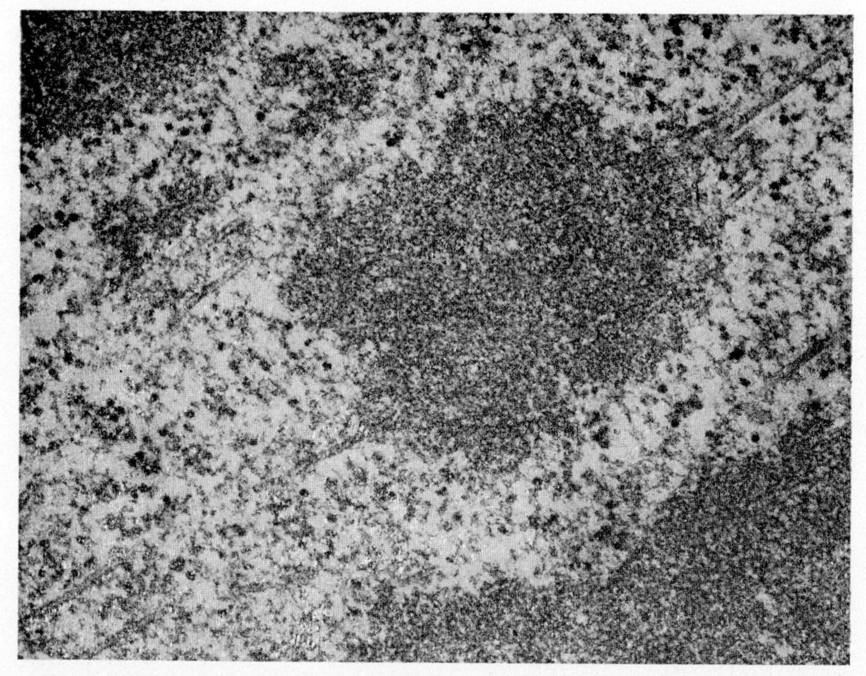

FIG. 5. Electron micrograph of portion of mitotic spindle at metaphase in anther of *Saintpaulia ionantha*. The spindle axis runs from lower left to upper right. Microtubules extend directly into the chromatin without first attaching to a kinetochore plate as in animal cells. $\times$ 52,000.

synthesized. Pickett-Heaps and Northcote have reported a close association between the cisternae of the endoplasmic reticulum and the spindle microtubules [20].

At metaphase the nucleolar material is not evident except in species which have persistent nucleoli.

## ANAPHASE

Anaphase is marked by an increased spindle length and the separation of the two sets of chromosomes (Fig. 6). The tubules of the polar regions are less well organized and have a tangled appearance. The chromosomes begin to uncoil and come together to form the daughter nuclei. At this time there is a tendency for the endoplasmic reticulum to adhere to the surface of the chromosomes [22].

Low magnification electron micrographs of the interzone between the two daughter sets of chromosomes appear structurally bland. Examination at higher magnification reveals that ribosomes predominate here,

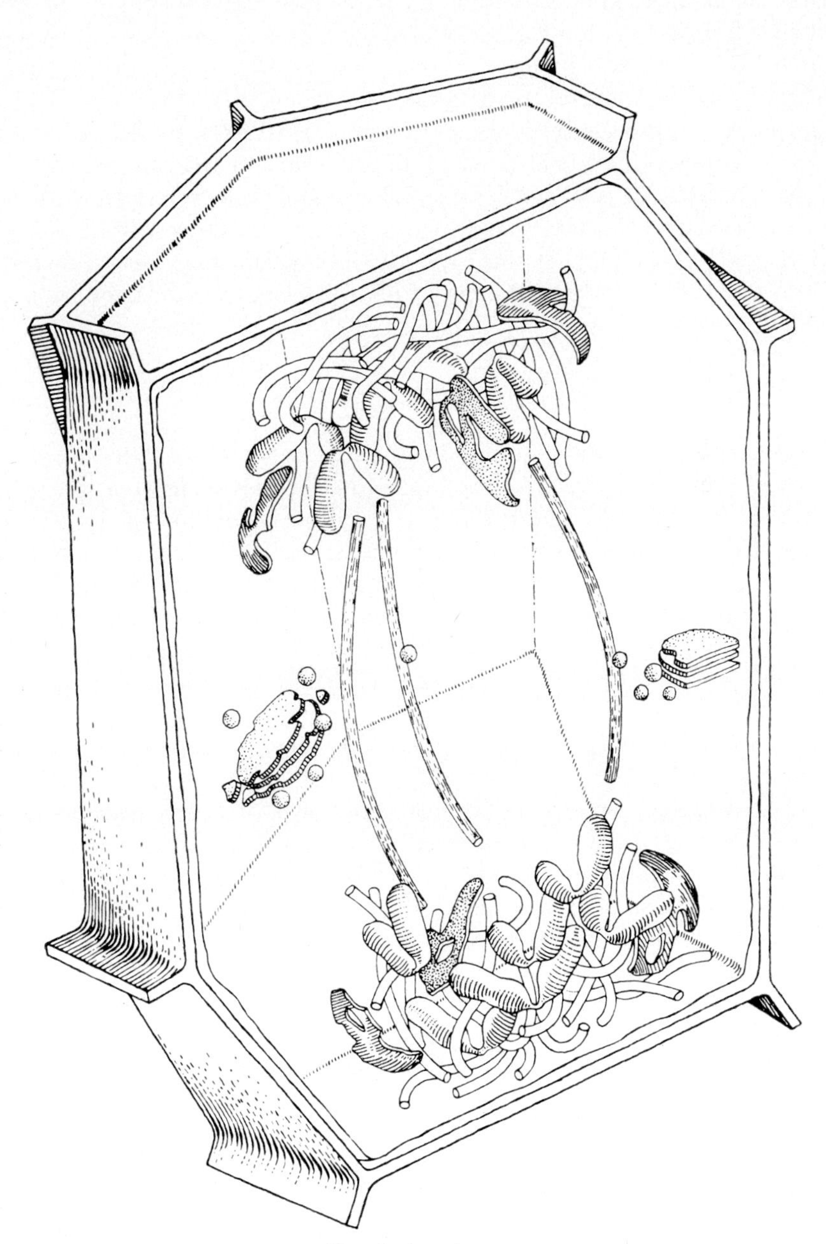

FIG. 6. Anaphase.

some in arrangements indicative of polysomes. Relatively few microtubules traverse this zone at early anaphase. Dictyosomes (which persist throughout the division cycle) are shown in Fig. 6 because of their relevance to the ensuing plate formation. They are found adjacent to the interzone, especially near its equator. Small vesicles derived from the dictyosomes, presumably containing pectin, are thought to play an important part in wall formation [32]. Some of these vesicles may be seen attached to the interzonal microtubules where they may be guided to the cell plate [13]. Besides the small vesicles some finely divided fibrous material (not shown here) is often associated with these tubules [13].

## TELOPHASE

The contacts between the endoplasmic reticulum and the chromosomes have now spread to encompass the entire surface of the two sets of chromosomes (Fig. 7). The resulting daughter nuclei retain somewhat the shape of the chromosome clusters of the anaphase cell. That is, they have a rather flattened shape with the flattened surfaces lying perpendicular to the previous spindle axis.

The chromosomes continue to uncoil, and granules appear along their surfaces in the rather scant nucleoplasm. These granules will condense to form the nucleoli. A few remnants of the spindle microtubules may be found about the nuclei, but they have lost any strong spindle orientation.

The interzone now displays a high population of microtubules organized into a phragmoplast [13]. Small vesicles collect along the surface of the phragmoplast microtubules and are found fusing at the equator to form a cell plate. These vesicles are derived from the dictyosomes which are clustered about the edge of the developing plate. The microtubules themselves seem to interdigitate at the region of the plate, though there is no evidence that they traverse this plane for any considerable distance. It has been suggested by Richard Allen [Allen and Bowen, 1] that the points at which the tubules penetrate the plate may later appear as plasmodesmata. Tubules are so placed that if extended they would converge approximately at a point previously occupied by the poles of the spindle. The phragmoplast begins as an assemblage of tubules between the daughter nuclei. As the plate grows, the phragmoplast of tubules expands into a torus as shown in Fig. 7 and eventually becomes disorganized as the plate contacts the parent cell wall. This circle of contact is that occupied by the previous equatorial band of early prophase [20]. As the phragmoplast begins to

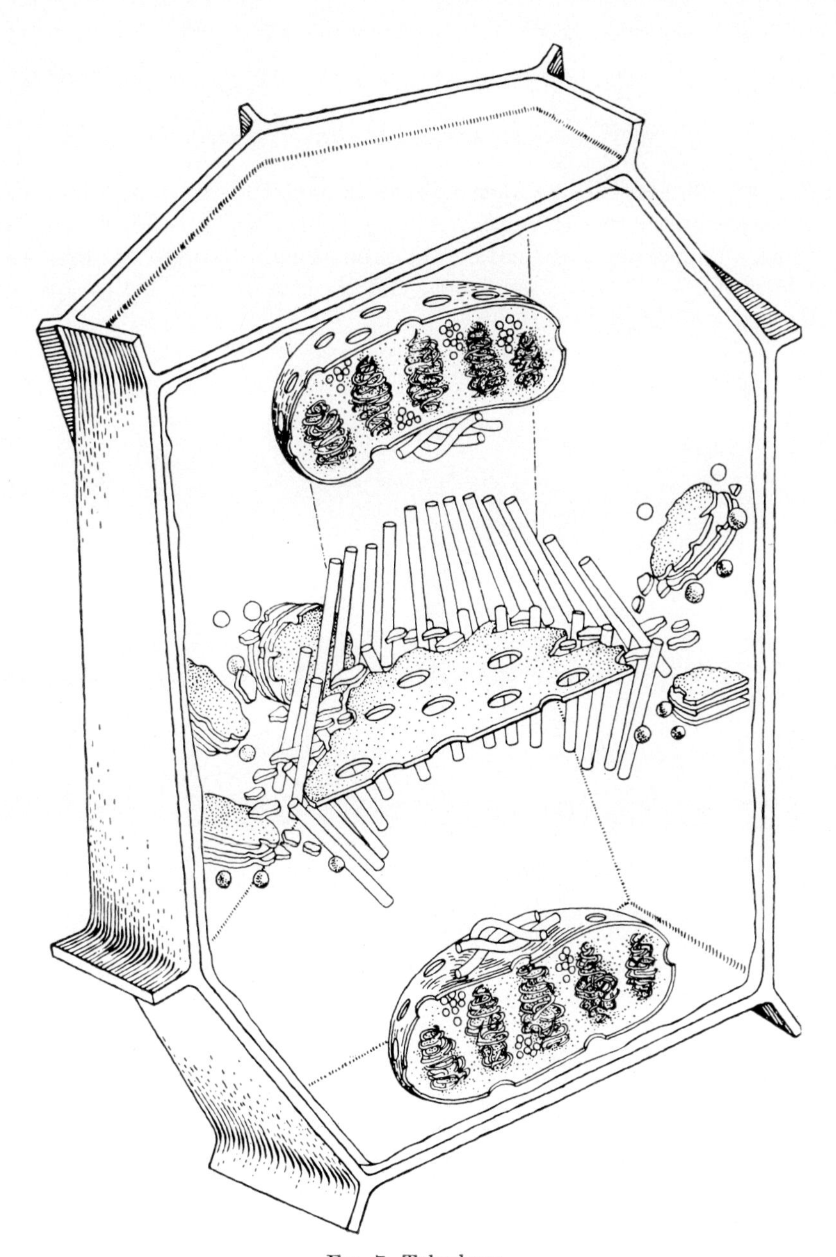

FIG. 7. Telophase.

disorganize, the cortical microtubules characteristic of the interphase cell again make their appearance, and the division cycle has been completed.

## MATURATION, WALL DIFFERENTIATION

The density of cortical microtubules at interphase is high in the young cells of the meristem (Fig. 1). As the cell enlarges and becomes vacuolate and mature, the density of tubules in the cortex declines. In the fully differentiated parenchymal cell the tubules may be so sparse that they are difficult to locate.

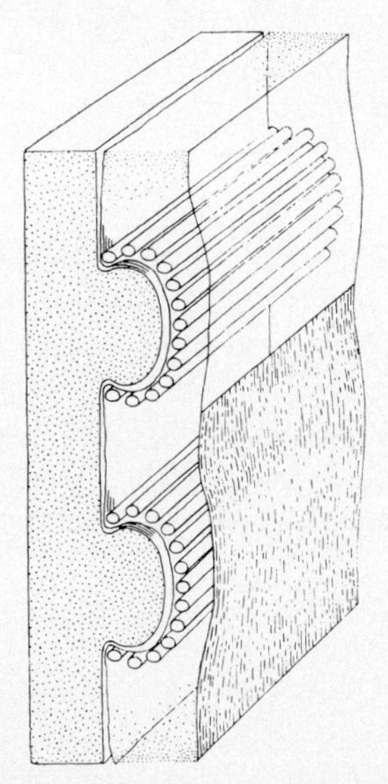

FIG. 8. Development of secondary wall thickenings. A portion of the cell wall with two banded thickenings is shown at the left. Immediately overlying the cell wall is a sheet representing the limiting membrane of the protoplast. At the right is the vacuolar membrane. Microtubules are confined to the areas of thickening and run parallel to their long axes.

Hepler and Newcomb [11] studied differentiating xylem cells which were developing secondary thickenings. They reported that microtubules are confined to the thickening areas, and run along the long axis of the thickenings (Fig. 8). We see again that the tubules mirror the orientation of the cellulose in the adjacent wall. It appears from a study of root hair development [18] that tubules are not essential for the deposition of cellulose; however, where the microtubules appear there is a close correspondence of their orientation and that of cellulose microfibrils being laid down in the wall. It is difficult to escape the notion that the microtubules in some way influence the oriented deposition of cellulose microfibrils.

## THE FINE STRUCTURE OF MICROTUBULES

The degree of homology among the various cytotubules found in plant and animal cells is at present unclear. Some features of substructure suggest a fundamentally similar arrangement of building units may exist. As stated earlier, the outside diameter is about 240 Å and the length may extend for several microns. Isolated and negatively stained preparations [2, 15, 19] from the mitotic apparatus and from sperm tails show the tubule wall to be composed of linear chains or files of small particles. The reported spacing of the subunits along the chain has a rather wide range, but most reports seem to be close to either 40 or 80 Å. This sort of spacing has been seen in freeze-etched cells [17] as well, giving us confidence in the validity of the image seen after negative staining of isolated tubules. The relationship of the particles between adjacent files is not clear, but in most published micrographs there is a suggestion that the subunits form a spiral of low pitch about the central axis. The number of files which complete the circle is difficult to count from the negatively stained preparations, but appears to exceed 10. Particularly favorable cross sections of the tubules from plant cells indicates that 13 such files are present [13, 14], and it may be that this is a constant among the microtubules which have a 240 Å diameter.

It is clear that there are differences in the stability of microtubules found in different situations. For instance the tubules of flagella, cilia, and centrioles remain intact during fixation in osmium tetroxide, while those of the mitotic spindle and interphase cytoplasm tend to disperse unless previously stabilized by a dialdehyde. It may be that in all cases the fundamental skeleton of the tubules is very similar with more or less stability resulting from the kind and amount of accessory substances which may be associated with the tubule surface.

Microtubules have been shown to be subject to reversible dissolution under the influence of low temperature [30] and hydrostatic pressure of over 500 atmospheres [16, 31] and to the effects of colchicine and related drugs [6, 28]. These effects have been interpreted as resulting from the assembly and dissolution of the particles which constitute the tubules. It seems probable that reversible disassembly plays a dominant role in the changes in microtubule distribution during various stages of the mitotic cycle (Figs. 1–4, 6, 7), and that movement of individual microtubules from one place to another, as during chromosome movement at anaphase, is a less frequent event.

## References


1. Allen, R. D., and Bowen, C. C., *Caryologia* **19,** 299 (1966).
2. André, J., and Thiéry, J. P., *J. Microscopie* **2,** 71 (1963).
3. Bajer, A., *Exptl. Cell Res.* **13,** 493 (1957).
4. Bikle, D., Tilney, L. G., and Porter, K. R., *Protoplasma* **61,** 322 (1966).
5. Barnicot, N. A., *J. Cell Sci.* **1,** 217 (1966).
6. Brinkley, B. R., and Stubblefield, E., *Chromosoma* **19,** 28 (1966).
7. Byers, B., and Porter, K. R., *Proc. Natl. Acad. Sci. U.S.* **52,** 1091 (1964).
8. DeHarven, E., and Bernhard, W., *Z. Zellforsch. Mikroskop. Anat.* **45,** 378 (1956).
9. Flemming, W., *Arch. Mikroskop. Anat. Entwicklungsmech.* **18,** 141 (1880); transl. in *J. Cell Biol.* **25,** 1 (1965).
10. Harris, P., *J. Cell Biol.* **14,** 475 (1962).
11. Hepler, P. K., and Newcomb, E. H., *J. Cell Biol.* **20,** 529 (1964).
12. Inoué, S., *Exptl. Cell Res.* **2,** 305 (1952).
13. Ledbetter, M. C., *J. Agr. Food Chem.* **18,** 405 (1965).
14. Ledbetter, M. C., and Porter, K. R., *Science* **144,** 872 (1964).
15. Luyxs, P., *Exptl. Cell Res.* **39,** 643 (1965).
16. Marsland, D., *Intern. Rev. Cytol.* **4,** 199 (1956).
17. Moor, H., High Vacuum Report, No. 9. Balzers AG, Balzers, Principality of Liechtenstein, 1966.
18. Newcomb, E. H., and Bonnett, H. T., Jr., *J. Cell Biol.* **27,** 575 (1965).
19. Pease, D. C., *J. Cell Biol.* **18,** 313 (1963).
20. Pickett-Heaps, J. D., and Northcote, D. H., *J. Cell Sci.* **1,** 109, 121 (1966).
21. Porter, K. R., *in* "Ciba Foundation Symposium on Principles of Biomolecular Organizations" (G. E. W. Wolstenholme and M. O'Conner, eds.), pp. 308–345. Churchill, London, 1966.
22. Porter, K. R., and Machado, R. D., *J. Biophys. Biochem. Cytol,* **7,** 167 (1960).
23. Porter, K. R., Ledbetter, M. C., and Badenhausen, S., *Proc. 3rd European Reg. Conf. Electron Microscopy, Prague* Vol. B, 119 pp. CSAV, Prague, 1964.
24. Robbins, E., and Gonatas, N. K., *J. Cell Biol.* **21,** 429 (1964).
25. Roth, L. E., and Daniels, E. W., *J. Cell Biol.* **12,** 57 (1962).
26. Sabatini, D. D., Bensch, K. G., and Barnett, R. J., *J. Histochem. Cytochem.* **10,** 652 (1962).
27. Strasburger, E., "Zellbildung und Zelltheilung." Fischer, Jena, 1880.
28. Stubblefield, E., and Brinkley, B. R., *J. Cell Biol.* **30,** 645 (1966).
29. Tilney, L. G., and Porter, K. R., *Protoplasma* **60,** 317 (1965).
30. Tilney, L. G., and Porter, K. R., *J. Cell Biol.* **34,** 327 (1967).
31. Tilney, L. G., Hiramoto, Y., and Marsland, D., *J. Cell Biol.* **29,** 77 (1966).
32. Whaley, W. G., and Mollenhauer, M. H., *J. Cell Biol.* **23,** 327 (1964).

# ON THE ASSEMBLAGE OF MYOFIBRILS[1]

**DAVID SPIRO AND MARTIN HAGOPIAN**

*Department of Pathology, College of Physicians and Surgeons, New York, New York*

## INTRODUCTION

The object of this review is to assess some of the factors involved in the assemblage of myofibrils of striated muscle. Since certain features of differentiated myofibrils have a bearing on the problem of myofibrillogenesis, it seems pertinent to review briefly muscle fine structure

## STRUCTURE OF MYOFIBRILS IN VERTEBRATE MUSCLE

Myofibrils vary in dimensions from the large straplike contractile elements found in insect skeletal muscles to the myofibrils, approximately 1 μ in diameter, of vertebrate striated muscle. All striated muscle is characterized by the presence of regularly repeating units or sarcomeres (Fig. 1). A sarcomere is defined as that portion of a myofibril from one Z line to the next Z line. All sarcomeres are further subdivided into an A band and I bands. Other bands, such as the M–L complex, or pseudo-H zone, and H zone, may also be present and will be discussed subsequently. The major bands of the sarcomere, namely, the A and I bands, reflect the disposition of the filaments that comprise the sarcomeres. As first shown by the elegant studies of Hanson and Huxley [24] and Huxley [36, 37] on rabbit sartorius muscle, the contractile units consist of a partially overlapping array of thicker myosin and thin actin filaments (Fig. 2). Since Huxley and Hanson's original study, a similar filament disposition has been confirmed for other types of vertebrate skeletal and heart muscles [13, 15, 29, 55, 59, 72, 74]. The thick filaments, which measure about 120 Å in diameter and 1.5–1.6 μ in length, are hexagonally arrayed and confined to the A band. The thin filaments, which measure about 60 Å in diameter and 1 μ in length, on the other hand are connected to Z lines and course through the I band into the A band for a variable distance depending on muscle and sarcomere length. These thin filaments when interposed between the thick filaments of the A band oc-

[1] This work was supported in part by the General Research Support Grant and Grant HE-5906 of the National Institutes of Health of the United States Public Health Service.

FIG. 1. The band pattern of several myofibrils of cat heart papillary muscle is illustrated. The bands of the central sarcomere (Z to Z lines) are labeled. The M line shows lateral interconnections between filaments. The paired light lines adjacent to the M line are the L lines. M–L is the complex consisting of the central M and the paired L lines. The A band measures 1.5 $\mu$. $\times$ 32,000. *Insect*: Transverse section through the M line which discloses a hexagonal array of thick filaments and the presence of cross bridges between them. $\times$ 200,000.

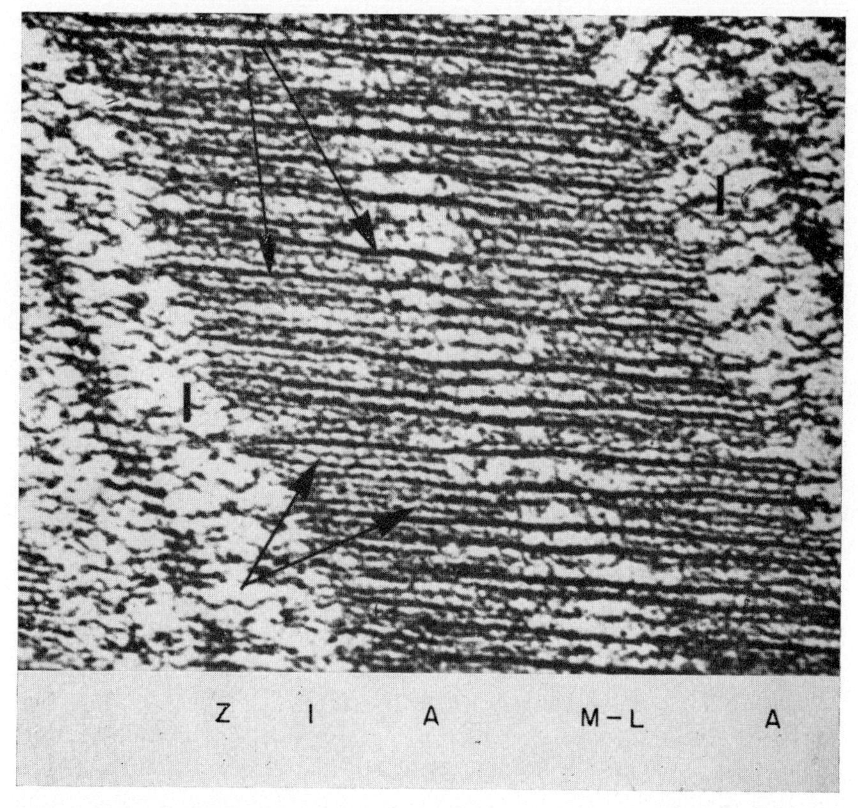


FIG. 2. Rat heart papillary muscle showing the presence of both thick and thin filaments within the A band. Only thin filaments are present in the I band, and only thick filaments are present in the M–L complex. Sarcomere bands are indicated at the bottom of the figure. × 125,000.

cupy the trigonal points in the primary hexagonal array of thick filaments. Therefore each thick filament is surrounded by 6 thin filaments and the resulting thin/thick filament ratio is 2 to 1. Huxley has demonstrated that lateral interconnections or cross bridges probably consisting of the heavy meromyosin moiety of myosin extend from the thick filaments to the adjacent thin filaments in the A band [36, 38]. The mid-portions of the thick filaments, however, lack these cross bridges for a distance of 0.15–0.2 $\mu$ [38]. In muscle which is fixed at sarcomere lengths of about 2.2 $\mu$, the thin filaments terminate at the edges of these central zones of the thick filaments that are devoid of the cross bridges [48, 72]. Thus the central areas of such sarcomeres exhibit a narrow zone (measuring 0.15–0.2 $\mu$) which is of lower electron density due to the absence of both thin filaments and cross bridges between the two sets of filaments (Fig. 1). The mid-portion of this zone, known as the M line, however, is somewhat more electron dense because of the presence of a different set of cross bridges, known as M line bridges, which interconnect the thick filaments in the center of the sarcomere (Fig. 1 and inset) [70]. This central portion of the sarcomere consisting of the M line with the two adjacent light or L lines has been referred to as the M–L complex or pseudo-H zone. It should be noted, however, that M lines are absent in certain types of vertebrate muscle, such as slow muscle fibers [47, 49]. In the I bands which are devoid of thick filaments the thin filaments show no particularly ordered arrangement. Where the thin filaments are attached to the Z line, however, they show a square array [14, 42, 54]. It is of interest that the set of thin filaments on one side of the Z line is out of register with respect to the set of thin filaments on the opposite side of the Z line in the adjacent sarcomere.

## STRUCTURE OF MYOFIBRILS IN INVERTEBRATE MUSCLE

The sarcomeres of invertebrate skeletal muscles exhibit the same basic structure as those of the vertebrate muscles inasmuch as they are also composed of a partially overlapping array of thick and thin filaments (Figs. 3–5). Invertebrate myofibrils, however, differ from those of vertebrates as regards filament lattices, ratios of thin to thick filaments, diameters of thick filaments, and the length of both the thick and thin filaments (Figs. 3–5). Table I tabulates some of these reported parameters in a variety of vertebrate and invertebrate muscles most of which were fixed in glutaraldehyde (see Table I for references). It would appear that the diameters of the thick myosin filaments and ratios of thin to thick filaments are related when striated muscles of various species are compared. In addition the thicker myosin filaments are generally longer. As noted in Table I these values are smallest for vertebrate muscles and largest for a number of invertebrate muscles. There is general agreement that in glutaraldehyde-fixed vertebrate muscles the length of the thick filament is about 1.5–1.6 $\mu$ and that the thin to thick filament ratio is 2. Recent measurements of thick filaments in similarly fixed human and frog sartorius skeletal and several species of heart muscle indicate that their diameters are about 120 Å [22, 58, 72]. In addition, Huxley [38] observed a thick filament diameter of 120 Å in rabbit psoas muscle fixed in formaldehyde and negatively stained. It should be noted, however, that thick filament diameters of greater than 120 Å have been reported in a fish muscle [15] and chick embryo skeletal muscle [13]. It is of interest that when verte-

FIG. 3. Longitudinal section of a cockroach femoral muscle. Sarcomeres measure 5.8 μ in length, thick filaments measure 4.5 μ, and the thin filaments measure 2.3 μ. × 18,000.

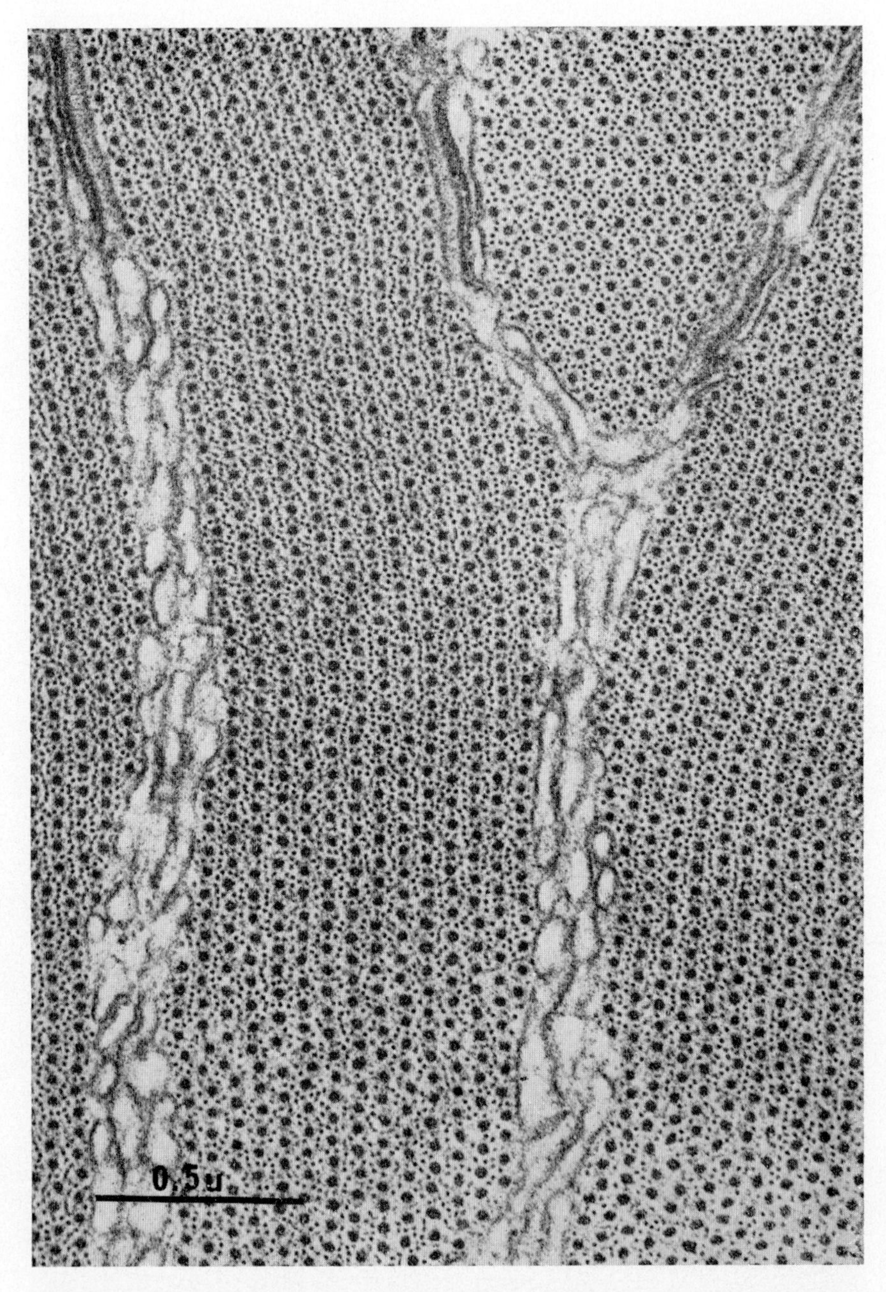


FIG. 4. Transverse section of cockroach femoral muscle which discloses a thin/thick filament ratio of 6. The diameter of the thick filaments is about 180 Å. × 56,000.

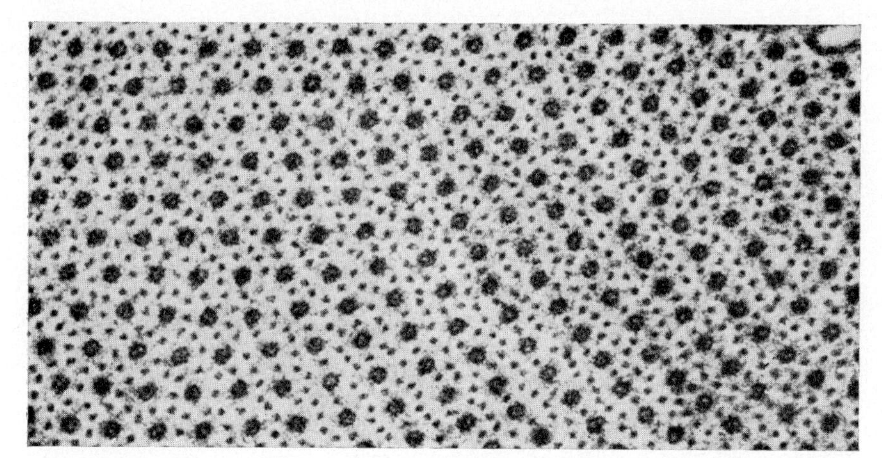

Fig. 5. Transverse section of cockroach flight muscle disclosing thin/thick filament ratio of 4 and thick filaments measuring 150–160 Å in diameter. × 158,000.

brate muscle thick filaments are reconstituted *in vitro* their maximal length (2.0 μ) approximates their *in vivo* lengths [38]. It would thus appear that the lengths of the thick filaments are dictated by the mode of packing and staggering of the myosin molecules.

In some insect flight muscles the thin filaments are located in the diadic position of the hexagonal array of thick filaments in contrast to vertebrates, where as stated above, the thin filaments are at the triadic position [3, 41, 62]. As a consequence of this filament packing, such flight muscle has a thin to thick filament ratio of 3 and the diameters of the thick filaments are somewhat greater than those of vertebrate muscles measuring about 140 Å [5]. Other types of flight muscle have thin to thick filament ratios of 4 : 1 [5, 22] and may have thick filaments measuring up to 150–160 Å in diameter (Fig. 5). The lengths of the thick filaments vary in different species of insect flight muscle, the range being 2.2–3.0 μ [23, 67], but they are invariably greater than those of vertebrate muscle thick filaments. In the last 5 muscles [20, 56, 68, 69, 76] listed in Table I the thin to thick filament ratios are 6, the diameters of the thick filaments are 160 Å or greater and their lengths are 3 μ or more (Figs. 3 and 4). It appears therefore that thicker filaments with larger numbers of myosin monomers, and thus more reactive sites per unit length, can interact with greater numbers of thin filaments. The relationship of greater filament diameter and length is not as clear. It is possible that either the larger numbers of myosin molecules or differences in their packing or stagger with respect to another copolymer results in a stable structure and a longer filament length [38]. The functional implications of these varying filament arrays

TABLE I. *Muscles Initially Fixed in Glutaraldehyde*

| Muscle | Length of thick filament ($\mu$) | Diameter of thick filament (Å) | Ratio of thin filament to thick filament | References |
|---|---|---|---|---|
| Vertebrate | | | | |
| Human flexor carpi radialis | 1.5 | 100–210 | 2 : 1 | [58] |
| Dog heart | 1.5 | — | 2 : 1 | [73] |
| Cat heart | 1.5 | 100–120 (Spiro, Unpublished) | 2 : 1 | [72] |
| Chicken heart | 1.5 | 110 | 2 : 1 | [22] |
| Chicken breast | 1.6 | — | 2 : 1 | [46] |
| Frog sartorius | 1.5–1.6 | — | 2 : 1 | [48, 72] |
| Frog semitendinosus | 1.6 | — | 2 : 1 | [48] |
| Rabbit psoas[a] | — | 100–120 | — | [38] |
| Fish muscle | — | 150 | 2 : 1 | [15] |
| Chicken leg muscle | — | 160–170 | 2 : 1 | [13] |
| Invertebrate | | | | |
| Dragonfly flight | 2.2[b] | — | 3 : 1 | [67] |
| Butterfly flight (*Phytometra*, *Minucia*, and *Abraxas* | — | 140 | 3 : 1 | [5] |
| Butterfly flight (*Vanessa*, *Pieris*) | — | 140 | 4 : 1 | [5] |
| Cockroach flight | 2.7 | 150–160 | 4 : 1 | [22] |
| Cockroach intersegmental | 3.6–4.1 | 160–180 | 6 : 1 | [68] |
| Cockroach femoral | 4.5 | 180–200 | 6 : 1 | [20] |
| Insect visceral | | 160–180 | 6 : 1 | [69] |
| Walking leg muscle of a crayfish | 3.0–6.0 | 200 | 6 : 1 | [76] |
| Somatic musculature of a nematode | 6.0 | 230 | 6 : 1 | [56] |

[a] Fixed in formaldehyde and negatively stained.
[b] Measured from published electron micrograph.

as regards active tension per unit area of contractile substance cannot be assessed at present. In accord with the studies of Hanson and Lowy [25, 26] it is noted that the diameters of the thin filaments are the same in various species although there is variation in thin filament length. Thin filaments measuring more than 1 μ in length (which is the length of these filaments in vertebrate muscles) are consistently found in invertebrates with long sarcomeres [20, 22]. The factors that dictate the length of thin filaments of various types of muscle cannot be assessed since, unlike reconstituted thick filaments, reconstituted thin filaments may be much longer than their *in vivo* counterparts [38]. We shall return to this point in muscle differentiation.

Other structural differences noted in invertebrate muscle are related to the Z lines and M lines. Auber and Couteaux [4] have demonstrated in insect muscle that the array of thin filaments in continuity with the Z line reflects their disposition in the A band rather than the square array observed in vertebrate muscle [14, 42, 54]. Z lines in obliquely striated muscle such as in *Ascaris* are relatively rudimentary [56]. In a bee flight muscle there are interconnections between the Z lines of adjacent myofibrils [16]. M lines are regularly found in asynchronous insect flight muscles [4, 41, 61, 65, 66] but are rarely present in other types of invertebrate muscle (Fig. 3).

## VARIATION IN SARCOMERE STRUCTURE AS A FUNCTION OF MUSCLE LENGTH

The sliding filament mechanism first proposed by H. E. Huxley and Hanson [40] and A. F. Huxley and Niedergerke [34] has been firmly established for several different types of striated muscle. This mechanism involves the sliding of thin filaments relative to the thick filaments as the sarcomere, and therefore the muscle, changes its length. The lengths of both the thick and thin filaments, however, remain constant except at very short sarcomere lengths. As a consequence of the changes in the disposition of the filaments with changes in sarcomere length, various band patterns are observed.

In vertebrate muscle a sarcomere length of 2.2 μ corresponds in general to the apex of the active length–tension curve, i.e., that muscle length where developed tension is maximal [18a, 38, 72] (Fig. 6). In such sarcomeres the A band measures 1.5–1.6 μ and each I band, 0.3–0.35 μ (Fig. 6). The A band remains constant in width at all sarcomere lengths greater than 1.5 μ while the width of the I band varies directly with the sarcomere length [34, 40] (Figs. 6–8). At sarcomere length

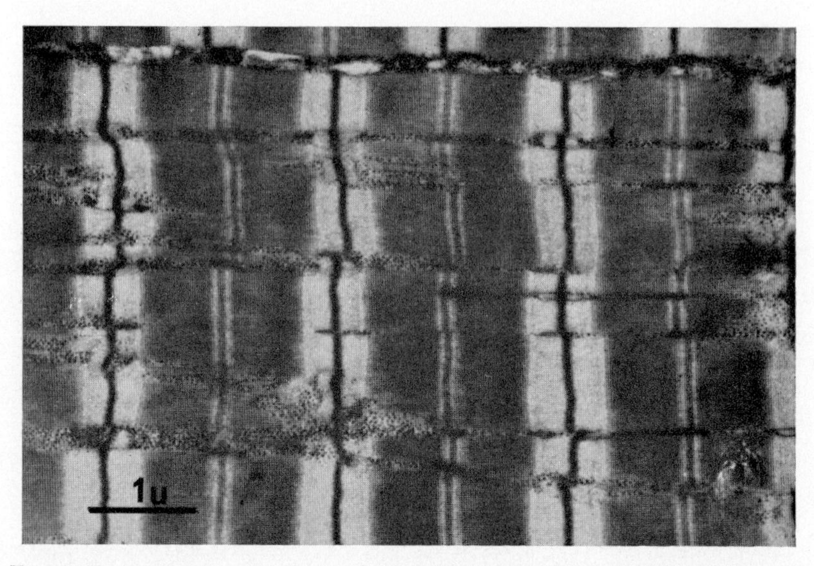


FIG. 6. Frog sartorius muscle fixed near the apex of the length–tension curve. A bands measure 1.5 μ in length and each I band measures 0.35 μ in width. × 14,000.

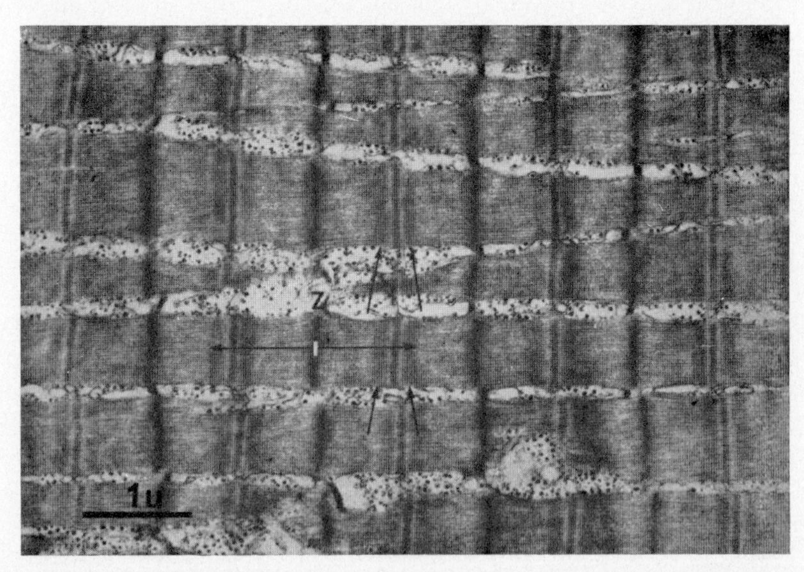


FIG. 7. Frog sartorius muscle fixed near the bottom of the ascending limb of the length–tension curve. No I band is seen. The sarcomere, which consists only of the A band and its central M line–L line complex, measures 1.5 μ in length. Note the A contraction bands (vertical arrows) flanking the L lines. The single longitudinal arrow extending from the limits of the A contraction bands of two adjacent sarcomeres measures 2.0 μ which is the combined length of two adjacent sets of thin filaments. × 14,000.

FIG. 8. Frog sartorius muscle fixed on the descending limb of the length–tension curve. The relatively long sarcomeres measuring about 3.0 $\mu$ in length have wide H zones (delimited by arrows) as well as wide I bands. A bands measure 1.5 $\mu$ in length. $\times$ 14,000.

below 2.0–2.2 $\mu$ there is a progressive fall in active tension with decreasing sarcomere length. This portion of the length–tension curve is known as the ascending limb. At sarcomere length greater than 2.2 $\mu$, active tension declines with increments in muscle and sarcomere length defining the descending limb of the length–tension curve. At sarcomere lengths of less than 1.7 $\mu$, two additional dark bands appear within the A band adjacent to the L lines (Fig. 7) [39, 47, 72]. These additional bands may be termed A contraction bands or double overlap bands, and their widths vary inversely with sarcomere length. The A contraction bands are formed by the penetration of thin filaments through the M–L complex into the opposite half of the sarcomere resulting in a double overlap of thin filaments. Accordingly, transverse sections through such A contraction bands disclose twice as many thin filaments there, or a thin-thick filament ratio of 4 rather than 2 [39, 71]. At sarcomere lengths less than 1.5 $\mu$ (which is less than the length of thick filaments) the Z lines broaden [24]. These broadened Z lines or Z contraction bands presumably reflect some folding of the thick filaments adjacent to the Z line at very short sarcomere lengths. At sarcomere lengths greater than 2.2 $\mu$ there is a progressive withdrawal of the thin filaments from the A band into the widening I bands. This results in a band of lower density in the A band known as the H zone

(Fig. 8) [24, 48]. The width of the H zone varies directly with sarcomere length. In skeletal muscle sarcomeres can be reversibly extended to about 3.7 μ with complete disengagement of thin filaments from the A band [35]. In mammalian heart muscle, however, the upper limit of sarcomere length which can be achieved is 2.7 μ [72].

Figure 9 schematically depicts the disposition of the filaments at various sarcomere lengths for vertebrate muscle. Similar band pattern changes as a function of muscle length have been observed in invertebrate muscles. In the femoral muscle of the cockroach a double overlap of thin filaments has been observed in sarcomeres shorter than 4.0 μ [20]. In these double overlap zones the thin to thick filament ratio is 12 : 1 rather than 6 : 1 (Fig. 10). Rosenbluth [57] has reported a

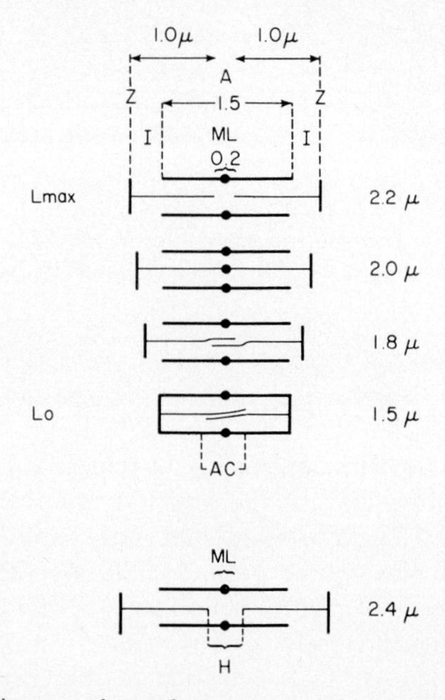


FIG. 9. Schematic diagram of vertebrate muscle which demonstrates the penetration of thin filaments into the opposite half of the sarcomere at short sarcomere lengths to produce the A contraction bands (*AC*). At 2.2 μ (*Lmax*) thin filaments terminate at the edge of the M–L complex (*ML*); at 2.0 μ the thin filaments meet head on in the center of the sarcomere. At sarcomere lengths of 1.8 μ the thin filaments have started to bypass one another largely within the confines of the M–L complex. At still shorter sarcomere lengths the zone containing twice as many thin filaments is proportionally longer, creating the A contraction bands. Note that in overextended muscle (2.4 μ) the withdrawal of thin filaments from the A band into the I band creates an expanding H zone (which includes the central M–L complex). At all muscle lengths depicted, the lengths of thick and thin filaments remain constant at 1.5 μ and 1.0 μ, respectively.

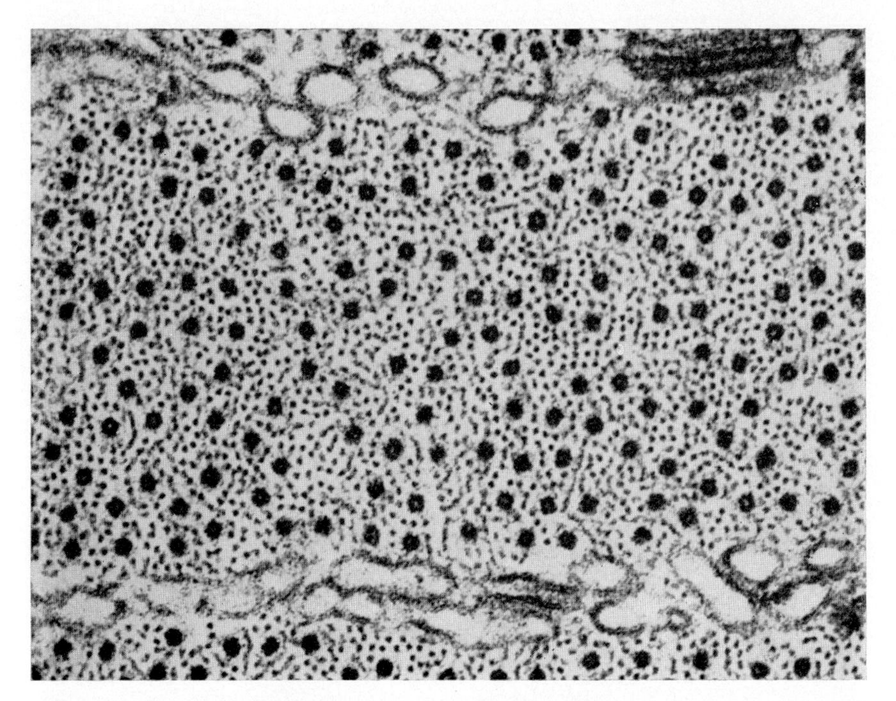

FIG. 10. Transverse section through A contraction band of shortened cockroach femoral muscle which discloses twice the number of thin filaments. Thin to thick filament ratio is 12 rather than 6. Compare to Fig. 4. × 100,000.

double overlap of thick as well as thin filaments in the obliquely striated muscle of *Ascaris*. Penetration of thick filaments through the Z lines in a shortened mollusc muscle has been observed [32]. H zones have been observed in a variety of stretched invertebrate muscles.

## MYOFIBRILLOGENESIS

There have been a number of electron microscopic studies dealing with myofibrillogenesis in vertebrate muscle [1, 2, 8, 9, 13, 27, 29, 31, 43, 52, 53, 78, 79]. However, despite these studies, numerous questions concerning the assemblage of myofibrils remain unanswered. The earlier electron microscopic papers dispelled classical concepts which postulated the origin of myofibrils from preexisting organelles such as mitochondria, Golgi apparatuses, centrioles [31, 53, 62, 79]. Indeed such postulates regarding the origin of myofibrils are naive in terms of current knowledge of the synthesis of cell-bound proteins. Several investigators called attention to the large numbers of nonmem-

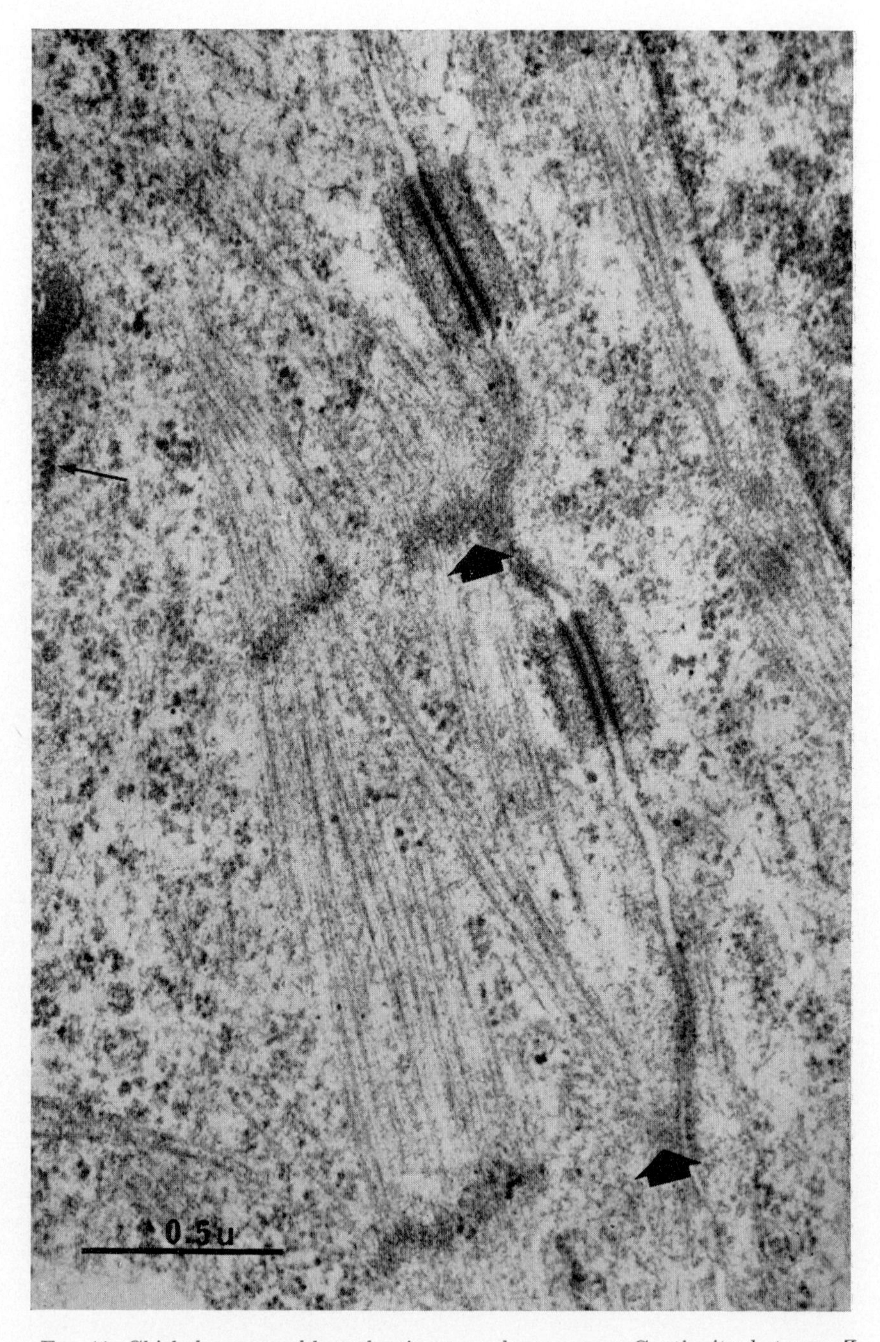


FIG. 11. Chick heart myoblast showing two desmosomes. Continuity between Z lines and zonulae adherentes are indicated by the broad arrows. Narrow arrow indicates polyribosome. × 54,000.

brane-associated ribosomes and, more recently, polyribosomes [2, 6, 8, 27, 28, 52, 53, 77] as the sites of myofilament synthesis (Figs. 11 and 16). Recently Heywood *et al.* (30) have characterized the polysome fractions associated with myosin synthesis in embryonic muscle. There is general agreement that the earliest recognizable myofilaments are randomly arrayed throughout the cell cytoplasm [2, 13, 27, 31] (Figs. 12 and 13). Some observers have indicated that in earlier stages there is a preponderance of randomly arrayed thin filaments to thick filaments [13, 52]. The significance of this observation as regards sequential synthesis of actin and myosin is unclear since it is conceivable that varying degrees of lateral aggregations of myosin molecules into filaments are observed in thin sections. Attention has also been called to microtubules in differentiating muscle cells [13, 29, 53], but again the significance of these structures to myofibrillogenesis is obscure. The length of the randomly arrayed filaments cannot be adequately assessed from thin sections, but thick filaments measuring up to 1.6 $\mu$ [2, 13, 29] and thin filaments measuring up to 1.1 $\mu$ [13] (which corresponds to their lengths in differentiated vertebrate sarcomeres) have been reported.

The next stage in myofibrillogenesis involves the aggregation of myofilaments into nonstriated bundles of myofilaments which occurs primarily but not invariably at the periphery of the cells [1, 13]. These myofibrils lack all the bands of striated muscle (Figs. 13–15), but it would appear that the normal compound filament lattice of the differentiated sarcomere is present at this stage [2, 6, 13, 29] (Fig. 12).

In the next stage of development the unbanded myofibrils are transected by Z lines at intervals of 1.5 $\mu$ [2, 13, 29]. Despite the fact that there are now repeating units or sarcomeres, no I bands or M–L complexes are observed [13, 31] (Figs. 16 and 17). The apparent absence of I bands may be due to the fact that the sarcomeres are uniformly contracted since the myofibrils cannot be loaded in an unshortened state in embryonic muscle. Transverse sections through these myofibrils at this stage definitely show the normal compound filament lattice.

The lengths of the thick and thin filaments comprising the primitive nonbanded myofibrils have not been adequately assessed. However, in recent studies utilizing chick embryo hearts, it has been noted that there are often abrupt changes in the directions of the myofilaments as well as discontinuities of the filament arrays at intervals of 1.5 $\mu$ [22] (Figs. 13, 14, and 18–20). This strongly suggests that the thick and possibly the thin filaments have achieved their ultimate lengths prior to the appearance of the Z lines. In other words, the elongation

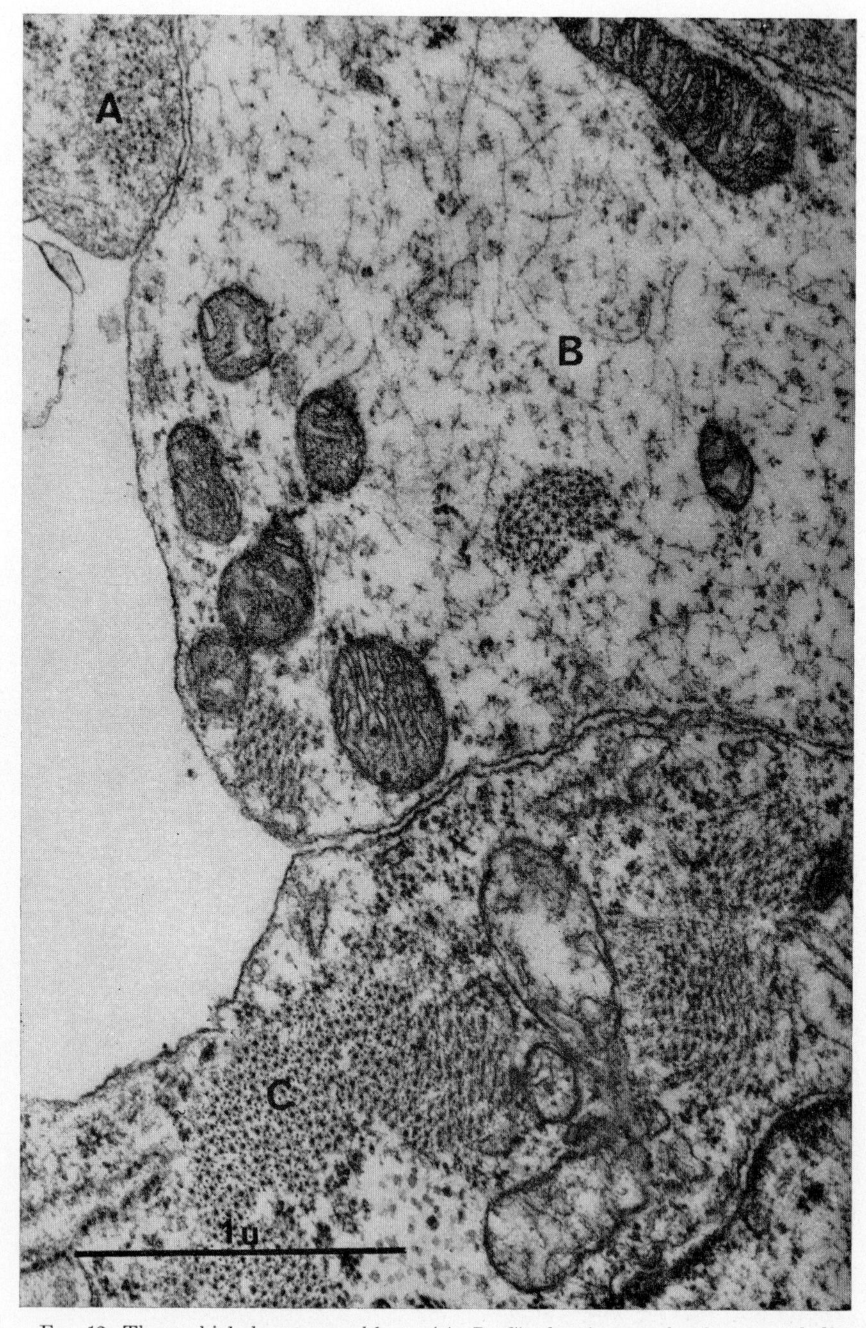


FIG. 12. Three chick hearts myoblasts (*A, B, C*) showing randomly arrayed filaments in cell *B*. The normal compound filament lattice of vertebrate muscle is best seen in cell *C*. × 40,000.

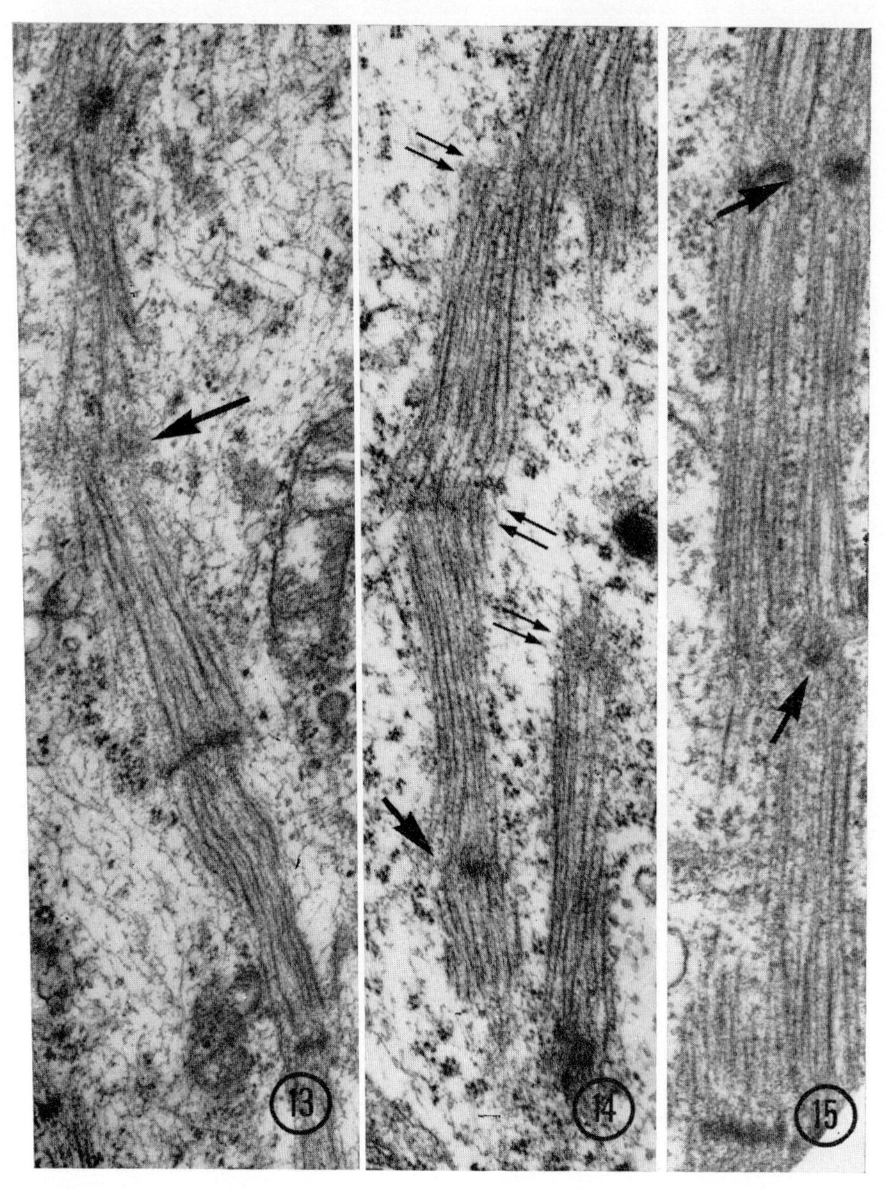


FIG. 13. Randomly arrayed filaments are also noted in this heart myoblast. The myofibril contains several Z lines but lacks other bands of striated muscle. At an interval of 1.5 $\mu$ from a Z line there appears to be discontinuity in the myofibril (arrow). $\times$ 31,000.

FIG. 14. Primitive myofibril similar to that seen in Fig. 13. A partial Z line is shown by the single arrow. At intervals of 1.5 $\mu$ along the myofibrils, discontinuities of the myofilaments are observed (double arrows). $\times$ 33,000.

FIG. 15. A complete as well as two incomplete Z lines (arrows) are observed. Each Z line is separated by an interval of 1.5 $\mu$. $\times$ 48,000.

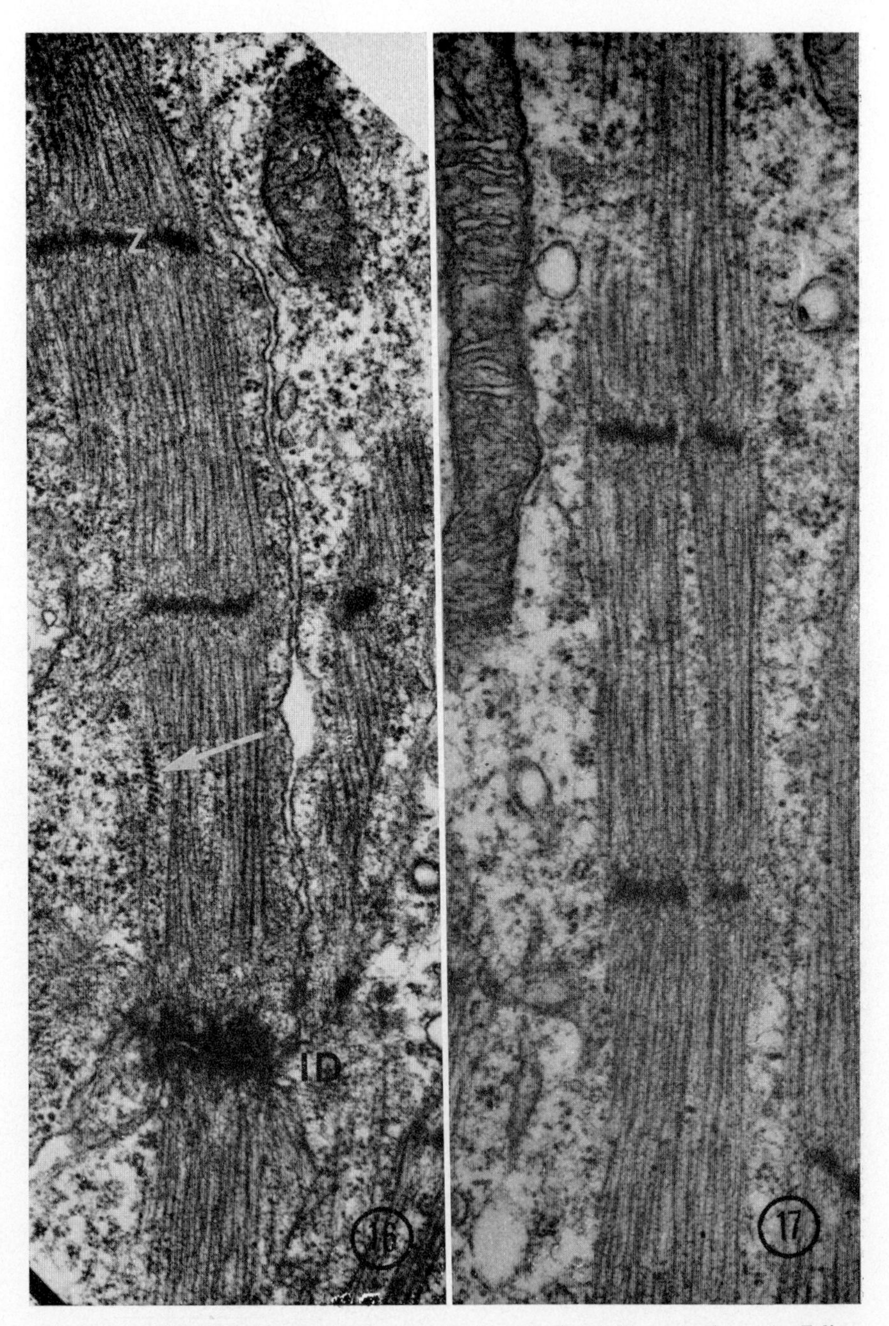


FIG. 16. A myofibril transected by an intercalated disc (*ID*) as well as two Z lines is noted. Spacing between intercalated disc and Z line and between two Z lines is 1.5 $\mu$. Other sarcomere bands are absent. Arrow points to a polyribosome. $\times$ 42,000.

FIG. 17. A myofibril similar to that seen in Fig. 16. $\times$ 50,000.

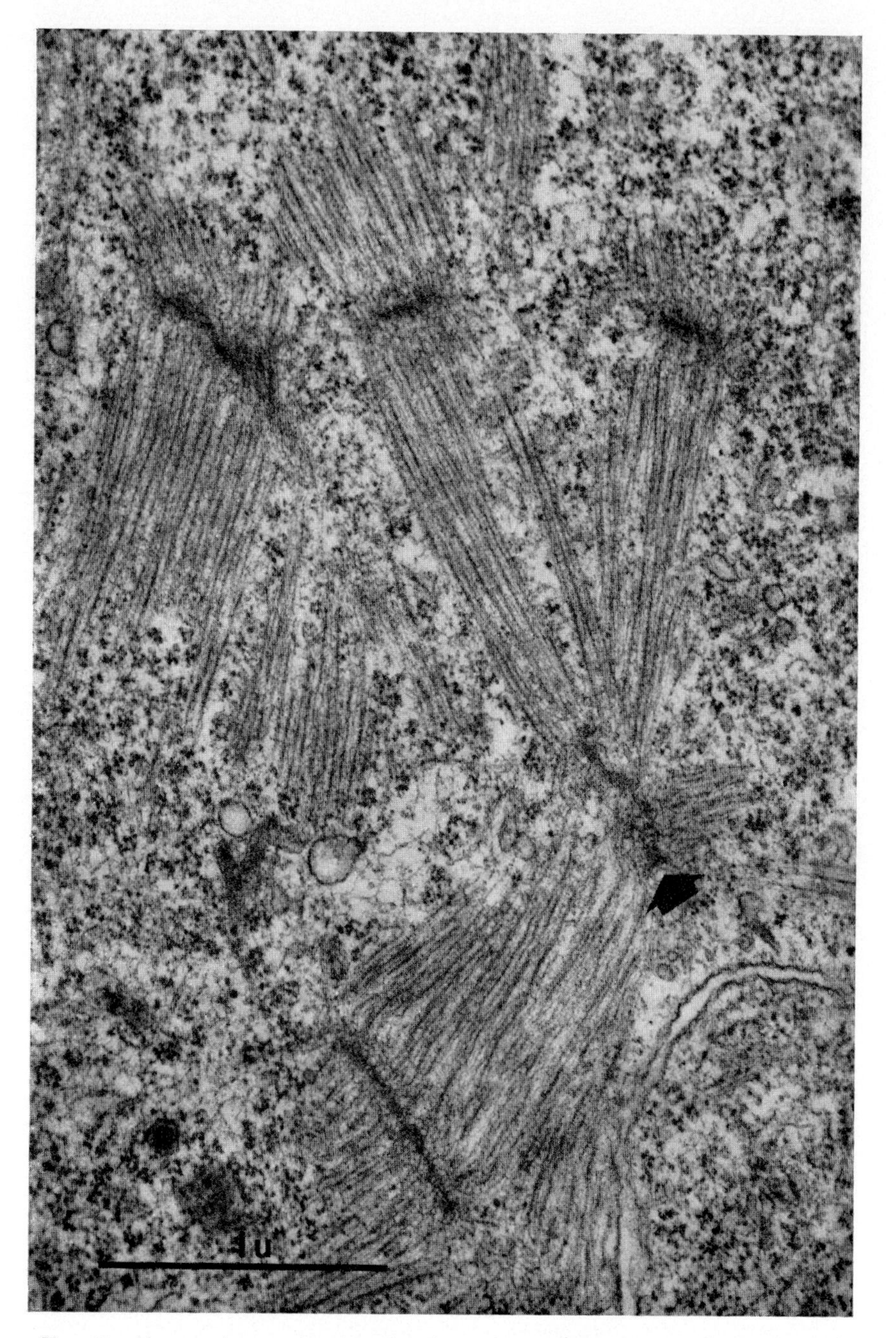


FIG. 18. Abrupt changes in the direction of myofilaments are observed at the level of a Z line (arrow). × 39,000.

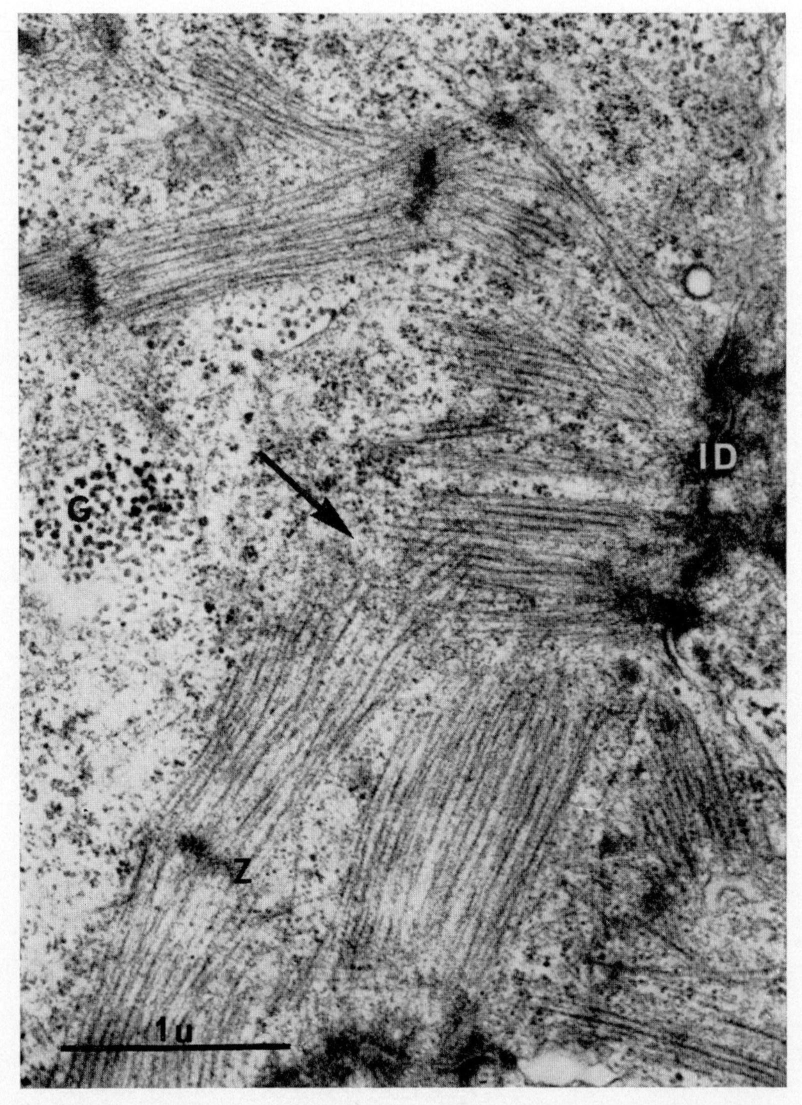


FIG. 19. A change in the orientation of myofilaments is noted at a distance of 1.5 $\mu$ (arrow) from both the adjacent intercalated disc (*ID*) and the adjacent Z line (*Z*). *G*, glycogen granules. × 33,000.

of myofibrils could involve the sequential addition of sets of filaments, the myosin filaments of which measure 1.5 $\mu$ in length.

The origin of the Z line has not been definitely established. However, our recent studies of embryonic chick heart indicate that Z lines arise

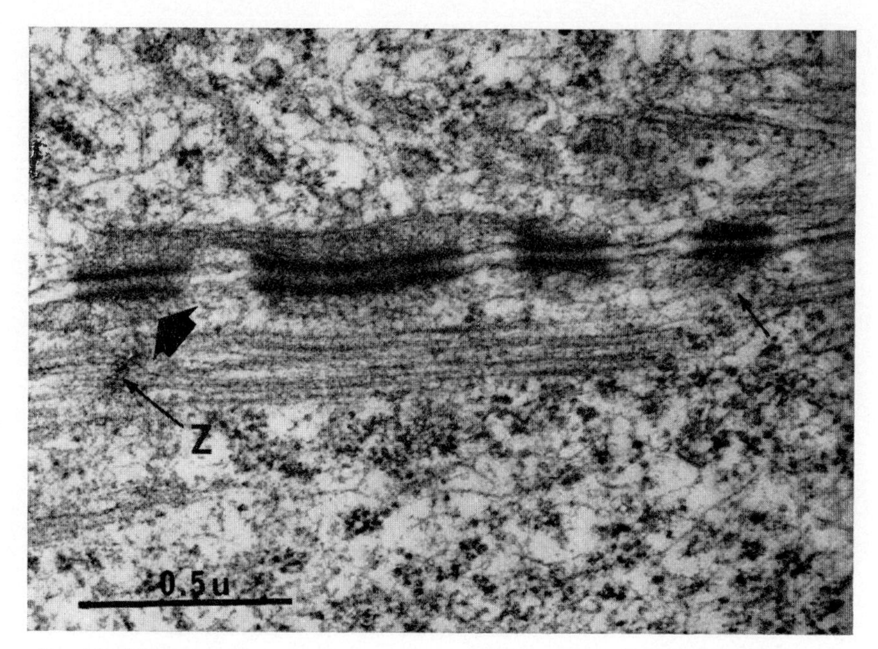


FIG. 20. Numerous desmosomes are seen parallel to a primitive myofibril. A Z line is noted in close proximity to one of the desmosomes (large arrow). The small arrow points to a region of discontinuity of myofilaments which is 1.5 $\mu$ in distance from above-mentioned Z line and which will be the probable site of another Z line. $\times$ 57,000.

from either typical desmosomes (maculae adherentes) or the closely related zonulae adherentes [22]. In embryonic heart muscle numerous desmosomes [78] or zonulae adherentes are observed on the cell surfaces parallel to as well as at right angles to the long axis of the myofibrils (Fig. 20). These two types of surface structures are often continuous with underlying Z lines which either completely or partially transect the myofibril (Figs. 11, 20, and 21). The electron density of the Z lines appears similar to that of the cytoplasmic meshwork of the desmosomes and zonulae adherentes. Similar zonulae adherentes characterize the major portion of the intercalated disc, where the thin filaments terminate at a level which corresponds to the next Z line [12, 45, 51, 64, 74] (Figs. 16 and 19). Therefore the zonula or fascia adherens [11] of the intercalated disc may be considered a modified Z line. Grimley and Edwards [19] and Heuson-Stiennon [29] called attention to the possible relationship between desmosomes and Z lines. In addition to the desmosomes and zonulae adherentes at the cell surfaces, these structures are also observed along invaginations of

FIG. 21. Arrows point to a Z line which is continuous with a zonula adherens structure at the cell surface. × 80,000.

the cell surface membrane where they are also in continuity with Z lines (Figs. 22–24). Thus it would appear that zonulae adherentes and desmosomes are carried inward or form along invaginations of the cell surface membranes which may represent the developing T system,

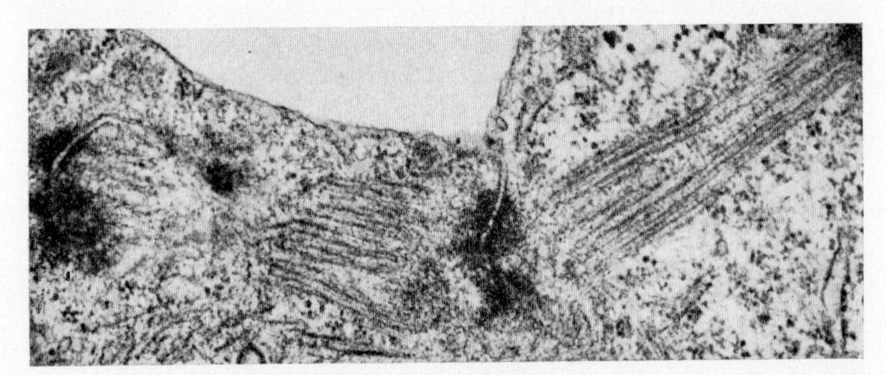

FIG. 22. Two cytoplasmic tubules, one of which appears to be continuous with the cell surface. Dense material resembling that of zonula adherens is seen adjacent to these tubules where they transect the myofibril where Z lines are apparently formed. × 33,000.

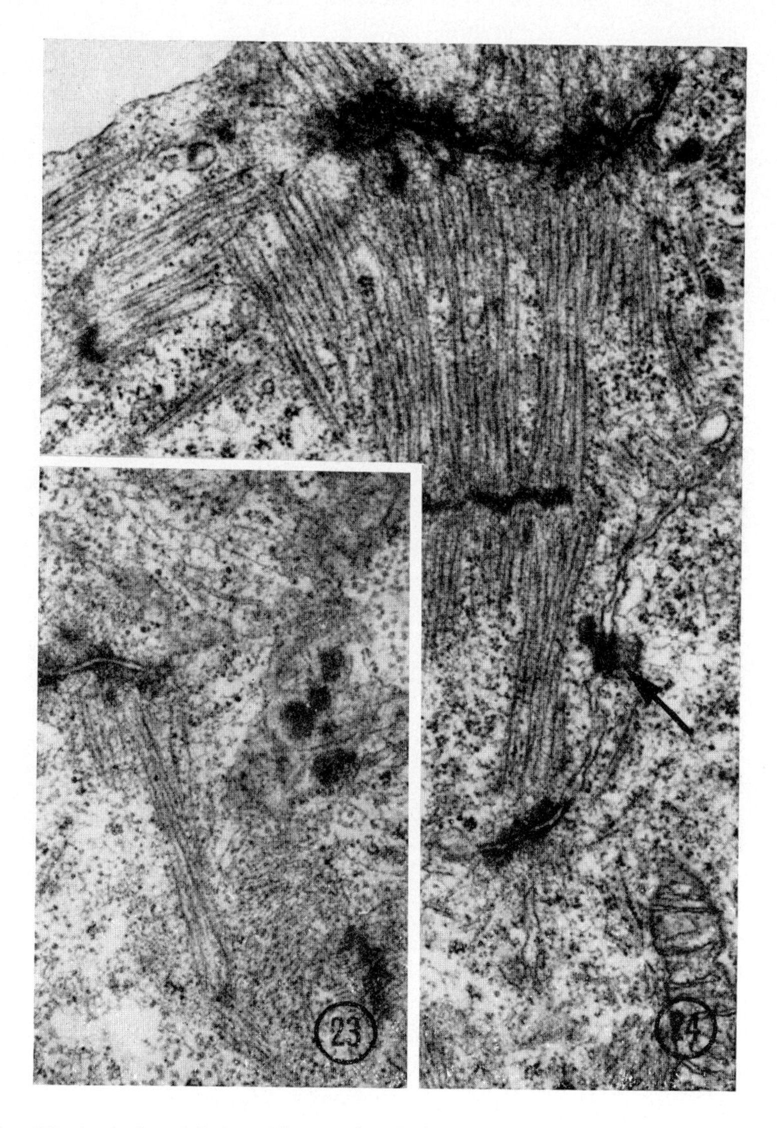

FIG. 23. A similar tubule with associated dense material is observed transecting a myofibril. × 31,000.

FIG. 24. Two tubules with associated dense material appear to be forming Z lines 1.5 μ in distance from a well-formed Z line. Note the desmosome in relation to the lower tubule (arrow). × 33,000.

where they give rise to Z lines deeper within the differentiating cell. It is of interest that Allen and Pepe [2] observed what appeared to be developing T system in relation to myofibrils prior to the development of Z lines. Continuity between Z lines and both T tubules and cell surface membranes have been seen in the adult femoral muscle of a cockroach [21]. It is intriguing to note that the cytoplasmic filaments converging on typical desmosomes such as those occurring in epidermal cells have the approximate diameters of actin filaments [10, 11].

The thick and thin filaments are probably self-assembled from myosin and actin molecules, respectively, inasmuch as similar filaments have been reconstituted *in vitro* from these monomeric units [38] in a manner similar to other fibrous proteins such as collagen or fibrin. The aggregation of thick and thin filaments to form the lattice of the primitive myofibril probably also represents a process of self-assemblage. This most likely involves the steric interaction of the two sets of filaments at predetermined sites on the actin molecules of the thin filaments with the heavy meromyosin cross bridges of the thick filaments. In this regard an ordered association of actin filaments with heavy meromyosin molecules has been demonstrated *in vitro* [38]. As Fischman [13] points out, neither the M line cross bridges nor the Z lines play a role in early myofibrillar assemblage since the latter occurs prior to the appearance of either of these two sarcomere lines. In addition, muscles which have discontinuous Z lines or which lack M lines, such as the obliquely striated muscle of *Ascaris* [56] or the cockroach femoral muscle [20] respectively, have highly ordered filament arrays. Since there are apparent discontinuities in primitive myofibrils at 1.5 $\mu$ intervals, the repeating sarcomere unit is probably established at this stage prior to the development of the Z line. While the myofilaments, myofibrils, and primitive sarcomeres appear to be self-assembled, this does not seem to be the case for the differentiated sarcomere. Undifferentiated sarcomeres acquire further substructure by the superimposition of Z lines at the regularly repeating discontinuities in the primitive myofibrils. These Z lines, presumably composed of the protein tropomyosin [38], originate in relation to desmosomes or zonulae adherentes of the cell surface membrane or invaginations thereof.

We discussed previously that the lengths and diameters of the thick filaments are characteristic for the type of packing of the staggered array of myosin molecules. Therefore the length of the A band is determined by the type of steric interaction of the myosin molecules which constitute the thick filaments. The factors that govern the lengths of the thin filaments are more obscure because, as previously pointed out, actin filaments reconstituted *in vitro* may be many times longer than they are within the sarcomere [38]. The lengths of the thin filaments might be determined by their interaction with the neighboring thick filaments. Huxley has shown that the thin and thick filaments in opposite halves of the sarcomere are polarized in opposite directions precluding interaction of the two sets of filaments when the thin filaments have passed into the opposite side of the sarcomere [38]. Therefore the distance from one end of the thick filaments to the far edge of the M–L complex in the opposite half of the sarcomeres could dictate the length of the thin filaments.

The most accurate measurements of the lengths of thick and thin filaments in vertebrate muscle have been made by Page and Huxley [48]. Their measurements indicate a length of 1.6 $\mu$ for thick filaments and about 2.0 $\mu$ for adjacent sets of thin filaments, which also includes the width of the Z line. The length of one set of thin filaments would be somewhat under 1 $\mu$ (excluding the width of the Z line). If one accepts 0.2 $\mu$ as the width of the M–L complex, the length which includes one-half of the thick filaments plus one-half of the M–L complex is 0.9 $\mu$. This figure certainly approximates the actual lengths of the thin filaments of vertebrate muscle. In the cockroach femoral muscle, the length of the thick filaments is 4.5 $\mu$ and the length of 1 set of thin filaments is 2.3 $\mu$ [20] (Fig. 3). It is obvious that one-half of a thick filament measures 2.25 $\mu$, or almost the length of the thin filaments. Similarly in the cockroach flight muscle one-half of a thick filament measures 1.35 $\mu$ and the thin filaments are about 1.5 $\mu$ in length [22]. In these cockroach muscles the widths of the zone in the center of the thick filaments that are devoid of the heavy meromyosin cross bridges and which correspond to the M–L complex of vertebrate muscle are unknown. Accordingly, one-half the width of this zone cannot be added to the half lengths of the thick filaments, possibly accounting for the discrepancy between the above thick filament measurements and thin filament lengths in cockroach muscle. Thus it is possible that the length of the thin filaments is determined by the length of that portion of the thick filaments where interaction between actin and myosin is not precluded. This portion would include the half length of the thick filament as well as one-half the width of the M–L complex.

The lateral ends of the thin filaments are attached to the Z line where filaments from adjacent sarcomeres terminate. Whether the spacing of the Z line is determined by the cell surface or some parameter of the myofibril such as the length of the thick or thin filaments is a matter of conjecture. As regards the latter possibility it is noted that the Z lines arise at intervals corresponding to the length of thick filaments in embryonic muscle. As previously stated, this may be attributed to the fact that the sarcomeres in embryonic muscle are contracted. The presence of myofibrils radiating in several directions from areas where Z lines are destined to form or are forming has been noted (Figs. 18 and 19). This suggests that the Z line plays a role in the subsequent parallel orientation of the myofibrils [27, 52, 78]. In view of the possible role of microtubules in cytoplasmic streaming, it has been postulated that these structures are also related to the orientation of myofilaments [13].

## PATHOLOGICAL ALTERATION OF MYOFIBRILS

In certain diseased states changes related to both the assemblage and fate of myofibrils are observed. Therefore some of the fine structural changes that are observed in myofibrils in pathological conditions will be briefly mentioned. It has been apparent for many years from both clinical and experimental observations that overloading the heart results in hypertrophy of myocardial fibers. Experimentally when the heart is subjected to an excessive pressure load there is a prompt increase in RNA and protein synthesis [17, 44]. Electron microscopic studies of hypertrophied heart confirm that there is an increased mass of contractile elements. Myofibrils from such hearts are larger but exhibit a normal filament array [55]. The size of the levator ani muscle is influenced by circulating levels of androgen and will undergo hypertrophy or atrophy depending on hormonal conditions. This system might provide a convenient model for studying the origin and fate of myofibrils. In muscle degeneration due to the deprivation of the vascular supply [75] and in neurogenic atrophy [50] as well as in other diseased states [18], it has been observed that the Z lines and thin filaments in the I bands are the first to disappear while the A band remains intact initially. It would appear that the thick filaments in the A band stabilize the interdigitated thin filaments in the earlier stages of myofibrillar degeneration. In some muscle diseases changes in the orientation of myofibrils are noted [60]. These abnormally oriented myofibrils are often circumferentially disposed with their long axes at right angles to the long axis of the muscle fiber. In other instances myofibrils may be randomly oriented throughout the muscle cell. In McArdle's disease [59] there is a deposition of large amounts of glycogen in muscle fibers. These glycogen deposits greatly distort the I band and may play a role in the muscle weakness which occurs in this disease. Finally, in a muscle disease known as nemaline myopathy [63] the major alteration consists of widening of the Z lines.

REFERENCES

1. Allbrook, D., *J. Anat.* **96,** 137 (1962).
2. Allen, E. R., and Pepe, F. A., *Am. J. Anat.* **116,** 115 (1965).
3. Auber, J., *Compt. Rend.* **261,** 4845 (1965).
4. Auber, J., and Couteaux, R., *J. Microscopie* **2,** 309 (1963).
5. Auber, J., *Compt. Rend.* **264,** 621 (1967).
6. Bergman, R. A., *Bull. Johns Hopkins Hosp.* **110,** 187 (1962).
7. Brandt, P. W., Reuben, J. P., Girardier, L., and Grundfest, H., *J. Cell. Biol.* **25,** 233, (1965).
8. Cedergren, B., and Harary, I., *J. Ultrastruct. Res.* **11,** 428 (1964).
9. Dessouky, D. A., and Hibbs, R. G., *Am. J. Anat.* **116,** 503 (1963).
10. Fawcett, D. W., *Exptl. Cell Res.* **8,** 174 (1961).
11. Fawcett, D. W., *in* "The Cell, Its Organelles and Inclusions" (D. W. Fawcett, ed.), p. 376. Philadelphia, Pennsylvania, 1966.
12. Fawcett, D. W., and Selby, C. C., *J. Biophys. Biochem. Cytol.* **3,** 261 (1957).
13. Fischman, D. A., *J. Cell Biol.* **32,** 557 (1967).
14. Franzini-Armstrong, C., and Porter, K. R., *Z. Zellforsch. Mikroskop. Anat.* **61,** 661 (1964)
15. Franzini-Armstrong, C., and Porter, K. R., *J. Cell Biol.* **22,** 675 (1964).
16. Garamvolgyi, N., *J. Microscopie* **2,** 107 (1963).
17. Gluck, L., *Science* **144,** 1244 (1964).
18. Gonatas, N. K., Perez, M. C., Shy, G. M., and Evangelista, I., *Am. J. Pathol.* **47,** 503 (1965).
18a. Gordon, A. M., Huxley, A. F., and Julian, F. J., *J. Physiol.* (*London*) **167,** 42 (1963).
19. Grimley, P. M., and Edwards, G. A., *J. Biophys. Biochem. Cytol.* **8,** 305 (1960).
20. Hagopian, M., *J. Cell Biol.* **28,** 545 (1966).
21. Hagopian, M., and Spiro, D., *J. Cell Biol.* **32,** 535 (1967).
22. Hagopian, M., and Spiro, D. (1967). In preparation.
23. Hanson, J., *J. Biophys. Biochem. Cytol.* **2,** 691 (1956).
24. Hanson, J., and Huxley, H. E., *Symp. Soc. Exptl. Biol.* **9,** 228 (1955).
25. Hanson, J., and Lowy, J., *J. Mol. Biol.* **6,** 46 (1963).
26. Hanson, J., and Lowy, J., *Proc. Roy. Soc.* (*London*) **B160,** 449 (1964).
27. Hay, E. D., *Z. Zellforsch. Mikroskop. Anat.* **59,** 6 (1963).
28. Heuson-Stiennon, J. A., *J. Microscopie* **3,** 229 (1964).
29. Heuson-Stiennon, J. A., *J. Microscopie* **4,** 657 (1965).
30. Heywood, S. M., Dowben, R. M., and Rich, A., *Proc. Natl. Acad. Sci. U.S.* **57,** 1002 (1967).
31. Hibbs, R. G., *Am. J. Anat.* **99,** 17 (1956).
32. Hoyle, G., McLear, J. H., and Selverston, A., *J. Cell Biol.* **26,** 621 (1965).
33. Huxley, A. F., *Proc. Roy. Soc.* (*London*) **B160,** 486 (1964).
34. Huxley, A. F., and Niedergerke, R., *Nature* **173,** 971 (1954).
35. Huxley, A. F., and Peachey, L. D., *J. Physiol.* (*London*) **156,** 150 (1961).
36. Huxley, H. E., *J. Biophys. Biochem. Cytol.* **3,** 631 (1957).
37. Huxley, H. E., *in* "The Cell" (J. Brachet and A. E. Mirsky, eds.), Chapt. 7. Academic Press, New York, 1960.
38. Huxley, H. E., *J. Mol. Biol.* **7,** 281 (1963).
39. Huxley, H. E., *in* "Muscle," Proc. Symp., Fac. of Med., Univ. of Alberta (W. M. Paul, E. E. Daniel, C. M. Kay, and G. Monckton, eds.), pp. 3–28. Macmillan (Pergamon), New York, 1965.

40. Huxley, H. E., and Hanson, J., *Nature* **173,** 973 (1954).
41. Huxley, H. E., and Hanson, J., *Electron Microscopy, Proc. Stockholm Conf., 1956* (F. S. Sjöstrand and J. Rhodin, eds.), Academic Press, New York, 1957.
42. Knappeis, G. G., and Carlsen, F., *J. Cell Biol.* **13,** 323 (1962).
43. Lindner, E., *Anat. Record* **136,** 234 (1960).
44. Meerson, F. Z., *Circulation Res.* **10,** 250 (1962).
45. Moore, D. H., and Ruska, H., *J. Biophys. Biochem. Cytol.* **3,** 261 (1957).
46. Page, S. G., *Proc. Roy. Soc.* (*London*) **B160,** 460 (1964).
47. Page, S. G., *J. Cell Biol.* **26,** 477 (1965).
48. Page, S. G., and Huxley, H. E., *J. Cell Biol.* **19,** 369 (1963).
49. Peachey, L. D., and Huxley, A. F., *J. Cell Biol.* **13,** 177 (1962).
50. Pellegrino, C., and Franzini, C., *J. Cell Biol.* **17,** 462 (1963).
51. Porter, K. R., and Palade, G. E., *J. Biophys. Biochem. Cytol.* **3,** 269 (1957).
52. Price, H. M., Howes, E. L., and Blumberg, J. M., *Lab. Invest.* **13,** 1279 (1964).
53. Przybylski, R. J., and Blumberg, J. M., *Lab. Invest.* **15,** 836 (1966).
54. Reedy, M. K., *Proc. Roy. Soc.* (*London*) **B160,** 458 (1964).
55. Richter, G. W., and Kellner, A., *J. Cell Biol.* **18,** 195 (1963).
56. Rosenbluth, J., *J. Cell Biol.* **25,** 495 (1965).
57. Rosenbluth, J., *J. Cell Biol.* **34,** 15 (1967).
58. Schotland, D. L. (1967). In preparation.
59. Schotland, D. L., Spiro, D., Rowland, L. P., and Carmel, P., *J. Neuropathol. Exptl. Neurol.* **24,** 629 (1965).
60. Schotland, D. L., Spiro, D., and Carmel, P., *J. Neuropathol. Exptl. Neurol.* **25,** 431 (1966).
61. Shafiq, S. A., *J. Cell Biol.* **17,** 351 (1963).
62. Shafiq, S. A., *J. Cell Biol.* **17,** 363 (1963).
63. Shy, G. M., Engel, W. K., Somers, J. E., and Wanko, T., *Brain* **86,** 793 (1963).
64. Sjöstrand, F. S., Andersson-Cedergren, E., and Dewey, M. M., *J. Ultrastruct. Res.* **1,** 271 (1958).
65. Smith, D. S., *J. Biophys. Biochem. Cytol.* **11,** 119 (1961) .
66. Smith, D. S., *J. Cell Biol.* **19,** 115 (1963).
67. Smith, D. S., *J. Cell Biol.* **28,** 109 (1966).
68. Smith, D. S., *J. Cell Biol.* **29,** 449 (1966).
69. Smith, D. S., Gupta, B. L., and Smith, U., *J. Cell Sci.* **1,** 49 (1966).
70. Spiro, D., *Trans. N.Y. Acad. Sci.* **24,** Sect. II, 879 (1962).
71. Spiro, D., *in* "The Myocardial Cell: Structure, Function, and Modification by Cardiac Drugs (S. A. Briller and H. L. Conn, eds.), pp. 13–61. Philadelphia, Pennsylvania, 1966.
72. Spiro, D., and Sonnenblick, E. H., *Circulation Res.* **14,** 14 (1964).
73. Spotnitz, H. M., Sonnenblick, E. H., and Spiro, D., *Circulation Res.* **18,** 49 (1966).
74. Stenger, R. J., and Spiro, D., *J. Biophys. Biochem. Cytol.* **9,** 325 (1961).
75. Stenger, R. J., Spiro, D., Scully, R. E., and Shannon, J., *Am. J. Pathol.* **40,** 1 (1962).
76. Swan, R. C., *J. Cell Biol.* **19,** 68A (1963).
77. Waddington, C. H., and Perry, M. M., *Exptl. Cell Res.* **30,** 599 (1963).
78. Wainroch, S., and Sotelo, J. R,. *Z. Zellforsch. Mikroskop. Anat.* **55,** 622 (1961).
79. Van Breeman, V. L., *Anat. Record* **113,** 179 (1952).

# THE STRUCTURE AND COMPOSITION OF CILIA

**I. R. GIBBONS**[1]

*The Biological Laboratories, Harvard University, Cambridge, Massachusetts*

The aim of this paper is to review the properties of two principal ciliary proteins, to consider their relationship to the structures seen in electron micrographs of whole cilia, and to explore the implications of this evidence to the question of self- or patterned-assembly in the formation of ciliary structure. Much of this work has been undertaken in collaboration with Dr. Fernando Renaud and Dr. Arthur Rowe.

Cilia have been isolated from *Tetrahymena pyriformis* by an ethanol-calcium method [3, 13]. The principle of this method is to suspend the *Tetrahymena* in a solution containing 9% ethanol, 2.5 m*M* EDTA, and Tris buffer pH 8.2. The cells remain alive and motile in this medium. Addition of excess $CaCl_2$ (12 m*M*) causes immediate detachment of the cilia. The point of breakage lies between cilium and basal body, so that the basal body remains with the cell body. Cilia and cell bodies are then easily separated by differential centrifugation. Beginning with a 40-liter culture of *Tetrahymena,* one obtains a yield of about 400 mg of cilia, dry weight.

The differing solubilities of the various structural components of cilia make it possible to fractionate them and study them separately [2, 3]. For example, by first dialyzing the cilia against EDTA and then extracting with 0.6 *M* KCl, one obtains a preparation of pure ciliary membranes (Fig. 1). An alternative procedure is to extract the intact cilia with digitonin; under suitable conditions this selectively removes the membrane, leaving a preparation of pure axonemes (Fig. 2). This preparation of axonemes can be further fractionated by dialyzing them against EDTA at low ionic strength. Approximately one-third of the axonemal protein (fraction 1) passes into solution during this dialysis. Examination of the insoluble residue (fraction 2) in the electron microscope shows that it consists of a fairly pure preparation of outer fibers (Fig. 3); the other axonemal components, including the arms on the outer fibers, have been almost completely removed.

Assay of these various fractions for protein and for ATPase activity (Table I), shows that most of the ATPase activity of the cilia

[1] Present address: Pacific Biomedical Research Center, University of Hawaii, Honolulu, Hawaii.

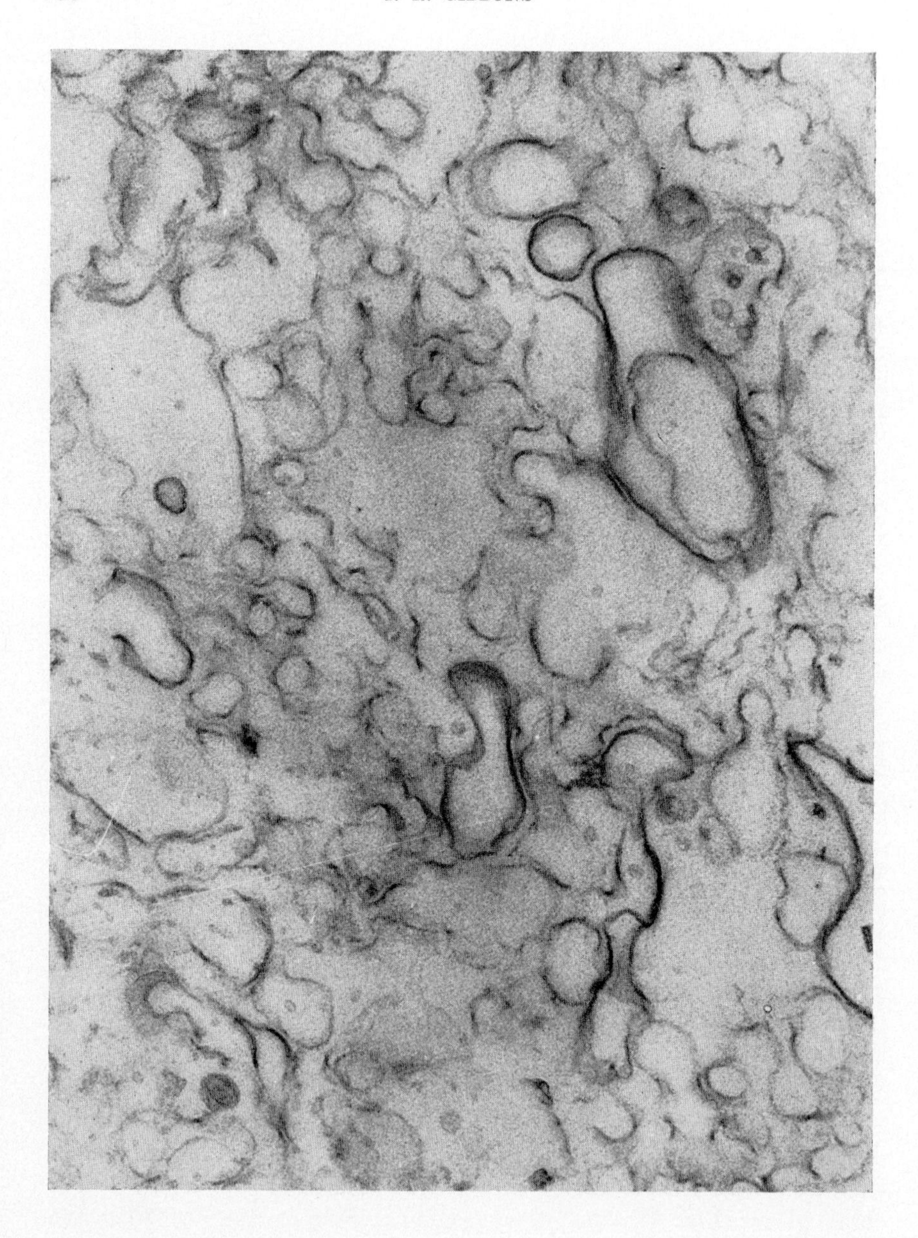

FIG. 1. The membrane fraction obtained by dialyzing cilia against Tris-EDTA solution, and then extracting with 0.6 *M* KCl. × 55,000. From Gibbons [3].

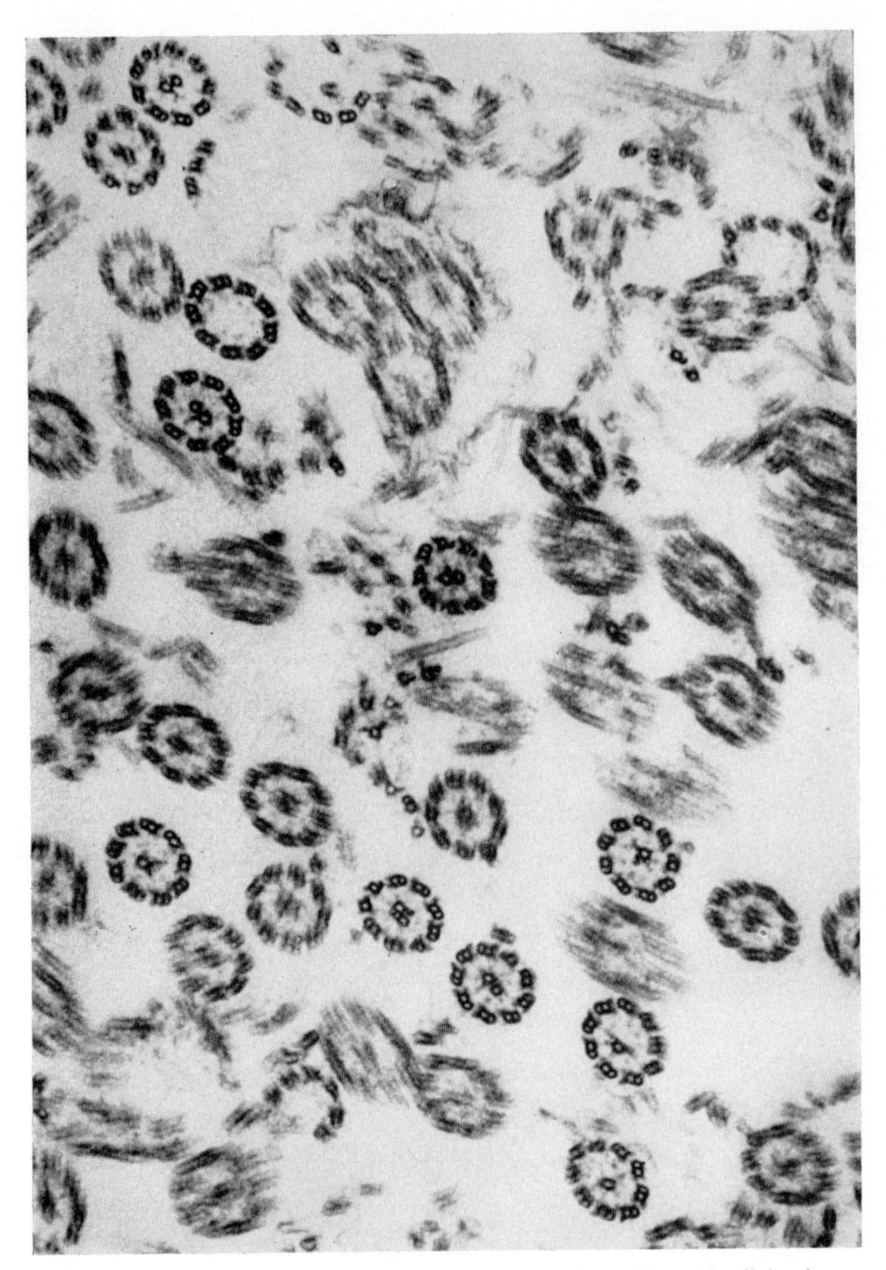

FIG. 2. The axoneme fraction obtained by extracting cilia with digitonin. × 50,000. From Gibbons [3].

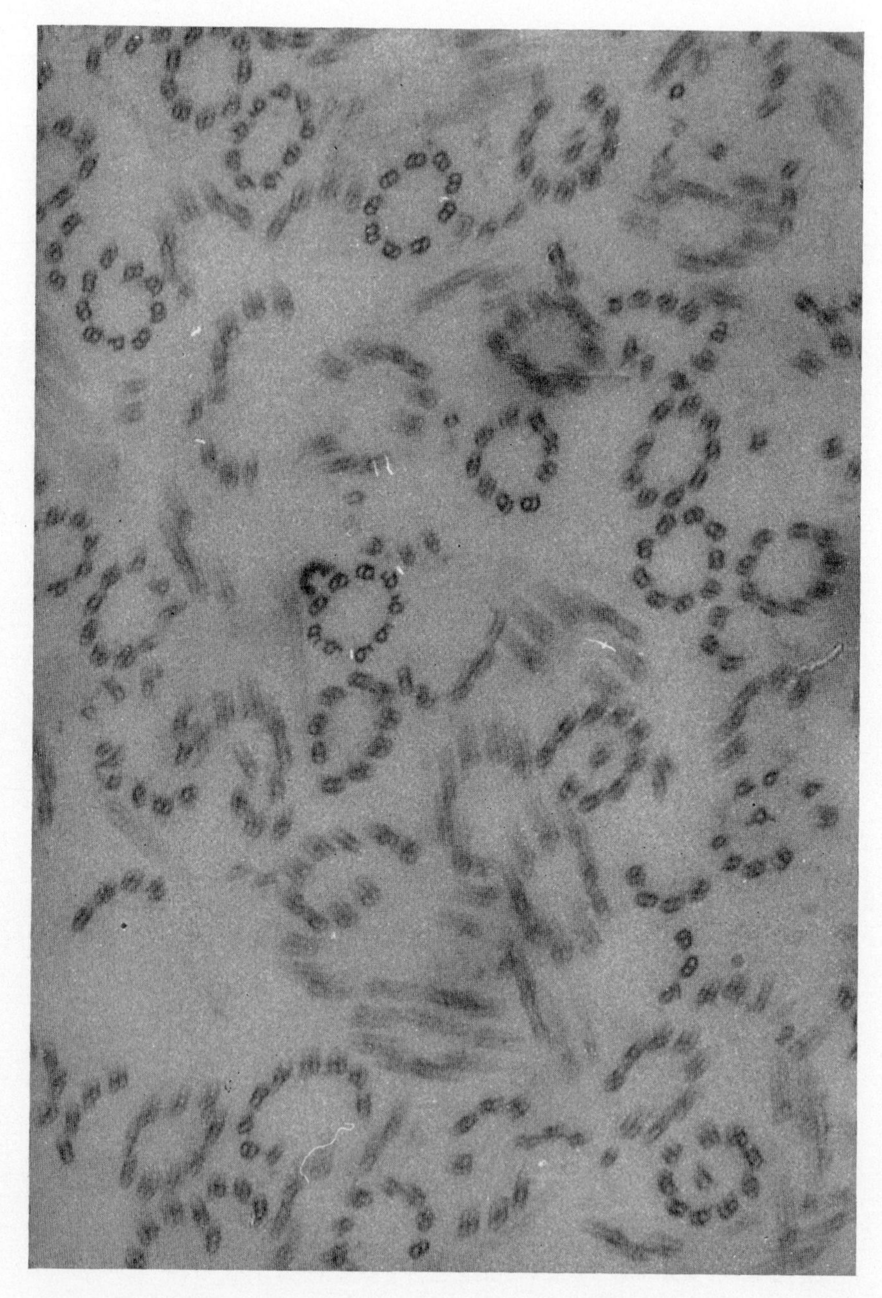

FIG. 3. Preparation of outer fibers obtained by dialyzing the axoneme fraction against Tris-EDTA solution. × 65,000. Gibbons [3].

TABLE I. *Distribution of Ciliary Protein and ATPase Activity*

| Protein | | Protein | ATPase activity[a] |
|---|---|---|---|
| Membrane-bound protein | | 22 | 26 |
| Soluble matrix protein | | 28 | |
| Axonemal protein, Fraction 1 | | 18 | 70 |
| 30 S dynein | 7 | | |
| 14 S dynein | 3 | | |
| 4 S protein[b] | 8 | | |
| Axonemal protein, Fraction 2 (outer fibers) | | 32 | 4 |
| | | 100 | 100 |

[a] Values were obtained under standard assay conditions with $Mg^{++}$ activation [see reference 3].
[b] Non-ATPase.

is in the axonemal fraction, with just a small amount associated with the membranes. Essentially all the axonemal ATPase activity passes into fraction 1 of the axonemal protein.

Examination of fraction 1 in the analytical centrifuge shows three principal components (Fig. 4). These three components have been separated by zonal centrifugation through a sucrose density gradient. The two fast components represent two forms of the axonemal ATPase protein, and they have been named 14 S dynein and 30 S dynein on the basis of their sedimentation constants ($S^{\circ}_{20w}$) [5]. The 4 S component has no ATPase activity. Preliminary characterization of the 14 S and 30 S forms of dynein has suggested that they are related as monomer and polymer, and that the 14 S dynein arises from partial breakdown of the 30 S form during preparation [4].

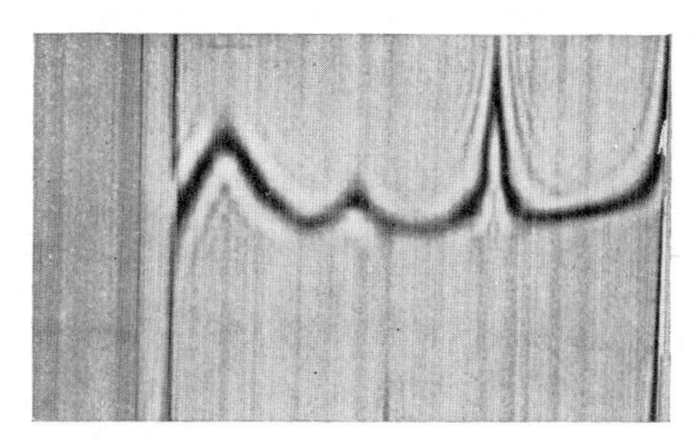

FIG. 4. Analytical ultracentrifugation of the supernatant fraction. Cilia were dialyzed for 48 hours against Tris-EDTA–5m$M$ KCl solution, and centrifuged. Supernatant consists of fraction 1 of the axonemal protein plus some matrix protein. From Gibbons [3].

We have two pieces of evidence regarding the localization of dynein in the axoneme. Since the dynein passes into fraction 1 it must be located in one of the structural components that are solubilized during dialysis, and not in the outer fibers (fraction 2). Comparison of the rate at which dynein appears in solution with the rate of disappearance of the various structural components has indicated that the dynein correlates best with the arms on the outer fibers [3]. This suggests that the arms are the principal site of the dynein. A more positive localization is provided by reconstitution experiments in which purified dynein is made to recombine with the outer fibers. This recombination is performed simply by adding 14 S or 30 S dynein to a suspension of outer fibers in the presence of $Mg^{++}$. The results of a typical experiment (Table II) show that most of the 30 S dynein recombines with the outer fibers but that only a small fraction of the 14 S dynein does so. Electron micrographs of this experiment (Fig. 5) show that recombination with 30 S dynein restores arms to many of the outer fibers. The arms appear to have returned with a remarkable degree of precision to the same position they had in intact cilia, with one or a pair of arms on subfiber A of most outer fibers. A count of the average number of arms visible after recombination with 30 S dynein indicated that

TABLE II. *Recombination of 14 S and 30 S Dynein with Fraction 2*[a]

| Sample | Concentration of fraction 2 (mg/ml) | Concentration of dynein added (mg/ml) | Percent of added dynein that became bound to fraction 2 |
|---|---|---|---|
| 30 S dynein | 1.2 | 0.26 | 66 |
| 14 S dynein | 2.1 | 0.15 | 15 |
| Control | 4.6 | None | — |

[a] 14 S and 30 S dynein were prepared by dialysis of whole cilia against Tris-EDTA-5 KCl solution (1 m*M* Tris buffer, 0.1 m*M* EDTA, 5 m*M* KCl, pH 8.3 at 0°C), followed by density gradient fractionation. Fraction 2 was prepared from a second batch of cilia by extracting with digitonin, and dialyzing against Tris-EDTA solution containing 3 m*M* KCl. Recombination was carried out for 90 minutes at 0° in a medium containing 2.5 m*M* $MgSO_4$, 10 m*M* KCl, 15 m*M* Tris-HCl buffer, pH 8.3. Pellets from the centrifugation were fixed for electron microscopy (see Fig. 5). All concentrations are the final ones in the recombination mixture. Dynein preparation was 5 days old (from isolation of cilia). Fraction 2 preparation was 1 day old (from isolation of cilia).

FIG. 5. "Reconstituted" cilia obtained by mixing purified 14 S and 30 S dynein with fraction 2 in the presence of $Mg^{++}$. See Table II for further details. Modified from Gibbons [3].

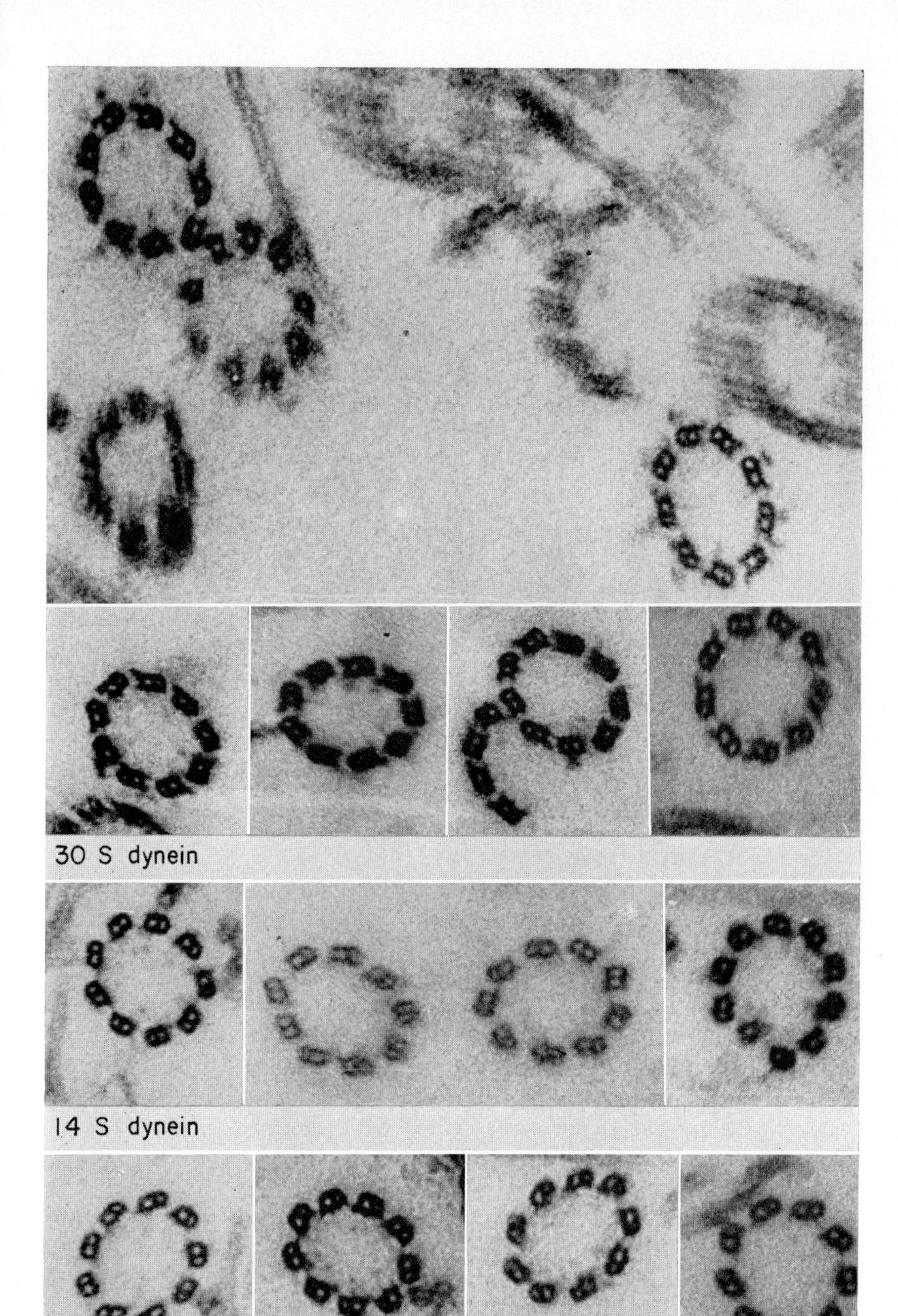
30 S dynein
14 S dynein
Control

around 60% of the original number were present. The preparation of outer fibers treated with 14 S dynein did not appear significantly different from the control [3].

The results of this reconstitution experiment show that adding back purified 30 S dynein restores the arms on the outer fibers. It seems very probable, therefore, that the arms consist of 30 S dynein. The fact that the arms can be put back by simple manipulation of the ionic environment indicates that the arms and outer fibers together possess the necessary structural specificity for their correct assembly, and is of obvious significance for the normal embryological formation of cilia.

Electron micrographs of 30 S dynein, shadowcast with platinum by the method of Hall [7], show that it consists of rod-like particles of variable length (Figs. 6 and 7). The particle height is 70–90 Å. The lengths of the particles in a typical preparation varied between 400 and 5000 Å, with a weight-average length of 1700 Å. When the particles lie obliquely with respect to the direction of shadowing, a repeating globular structure (period about 140 Å) can often be seen along their length. Composite pictures, obtained by linear transposition, reveal this periodicity even more clearly.

The same structural features can be seen when 30 S dynein is examined by the negative-contrast method (Fig. 8). However, this method of preparation seems to result in a somewhat smaller value for the longitudinal periodicity (about 100 Å), and many of the dynein particles appear severely disrupted.

Electron micrographs obtained by shadow-casting 14 S dynein show globular particles (Fig. 9). The heights are in the range 70–100 Å, and the widths 90–140 Å. These dimensions suggest that the 14 S dynein molecule can be represented as an ellipsoid with approximate axes 85 Å, 90 Å, and 140 Å [5].

Molecular weights of several preparations of dynein have been determined by centrifugation according to the methods of Klainer and Kegeles [9] and Van Holde and Baldwin [12]. The accuracy of these determinations has so far been limited by heterogeneity, but approximate values of 600,000 and 5,400,000 have been obtained for 14 S and 30 S dynein, respectively.

The two forms of dynein have similar enzymatic properties [4]. They both require the presence of a divalent cation for ATPase activity: $Mg^{++}$, $Ca^{++}$, $Mn^{++}$, $Fe^{++}$, $Co^{++}$, and $Ni^{++}$ can all function as activator, but $Be^{++}$, $Cd^{++}$, $Sr^{++}$, and $Zn^{++}$ cannot. Specificity for ATP or deoxy ATP is moderately high; other nucleoside triphosphates are hydrolyzed at only about 10% of the rate of ATP; while AMP,

FIG. 6. Electron micrograph of 30 S dynein shadowcast with platinum by the method of Hall. $\times$ 18,000. From Gibbons and Rowe [5].

pyrophosphate, and *p*-nitrophenyl phosphate are not hydrolyzed. The two forms of dynein both have a pH optimum at pH 8.5–9. When assayed under standard conditions with $Mg^{++}$ activation, 14 S and 30 S dynein had maximal velocities of 3.5 and 1.3 $\mu$mole $P_i$/(minutes $\times$ milligrams of protein), respectively, and Michaelis constants of $3.5 \times 10^{-5}$ and $1.1 \times 10^{-5}$ *M*. The general similarity in enzymatic properties is consistent with a monomer-polymer relationship between 14 S and 30 S dynein, the differences that occur being the result of a configurational change attendant upon the state of polymerization. Modification of 30 S dynein, by treatment with high concentrations of

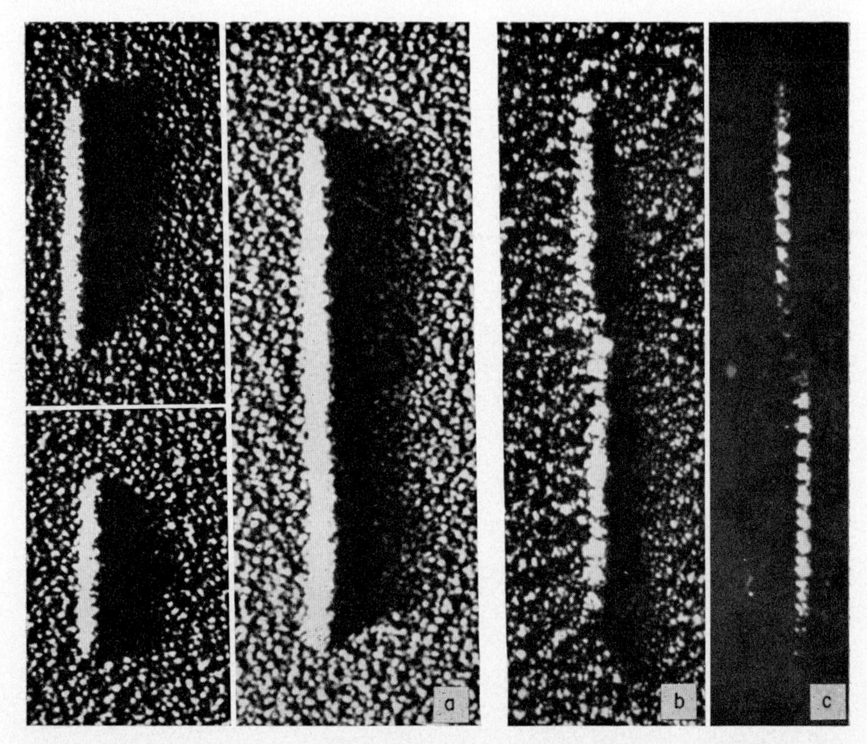


FIG. 7. Selected particles of 30 S dynein at higher magnification. (A) Three particles shadowed approximately normal to their long axes. (B) Two particles shadowed obliquely. (C) Same particles as in (B) with six images superimposed, translating through a distance equivalent to 140 Å between exposures. Modified from Gibbons and Rowe [5].

neutral salt, or by brief digestion with trypsin, causes the specific activity to rise toward values typical for 14 S dynein [4].

Solutions of fiber protein have been obtained either by dissolving the isolated outer fibers in 0.6 *M* KCl pH 8.3, or by aqueous extraction of an acetone powder of whole cilia [10, 11]. Examination of these solutions in the analytical centrifuge shows a component sedimenting at about 6 S together with some heterogeneous aggregate sedimenting in the range 8–30 S. The presence of this aggregated material, which probably consists of partly denatured protein, makes direct study of these solutions difficult. Pending the refinement of techniques to avoid the denaturation we have resorted to use of urea or guanidine hydrochloride to disperse the aggregates. Analytical centrifugation of the protein dissolved in 5 *M* guanidine hydrochloride yields a single symmetrical peak with a sedimentation constant of about 2.1 S (Fig. 10).

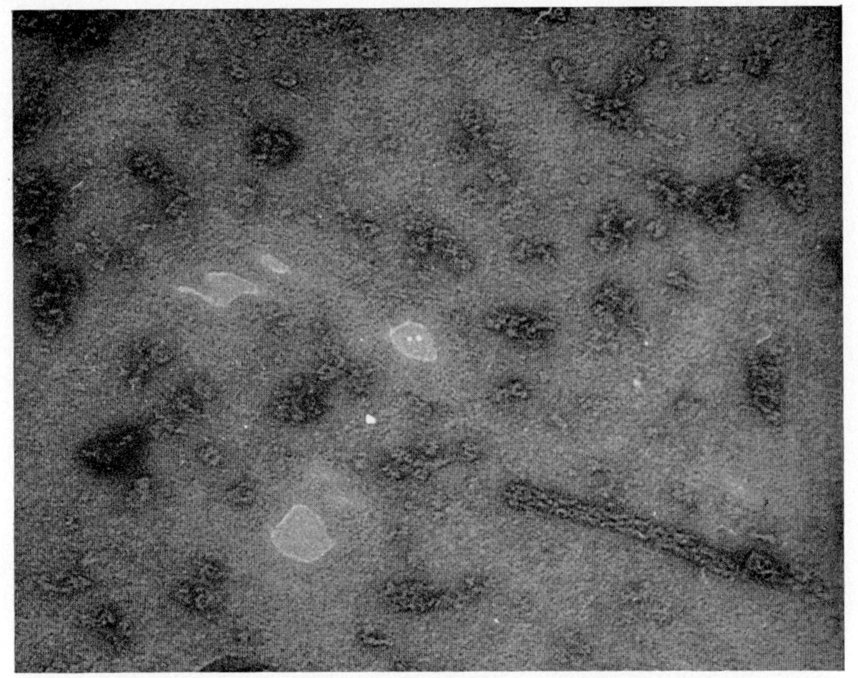

FIG. 8. 30 S dynein negatively contrasted with uranyl acetate.

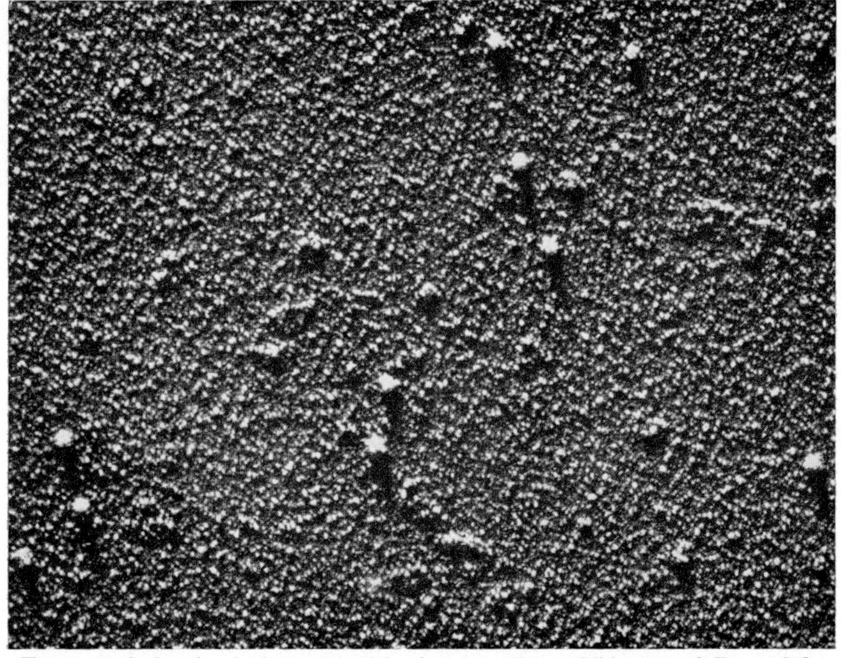

FIG. 9. 14 S dynein shadow-cast with platinum. From Gibbons and Rowe [5].

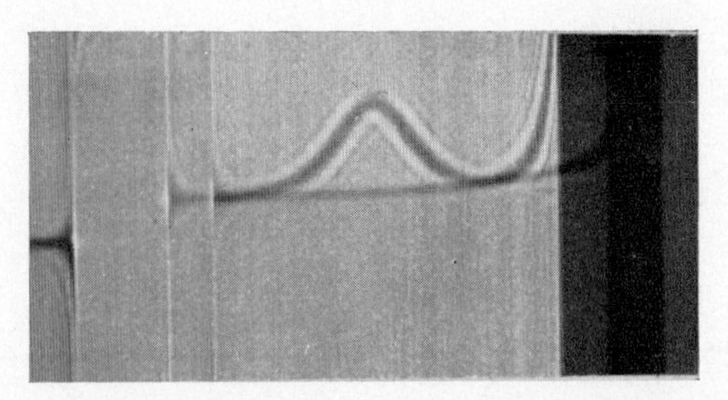

FIG. 10. Analytical ultracentrifugation of outer-fiber protein dissolved in 5 *M* guanidine hydrochloride–0.12 *M* mercaptoethanol. From Renaud *et al.* [11].

The homogeneity of the protein can be tested more critically by gel electrophoresis [10, 11]. After reduction and alkylation in 8 *M* urea, most of the protein migrates as a single band on electrophoresis in polyacrylamide gel at pH 8.9 (Fig 11). In addition to this densely staining major band there are a number of minor bands which stain much less intensely. About five of these minor bands are obtained from

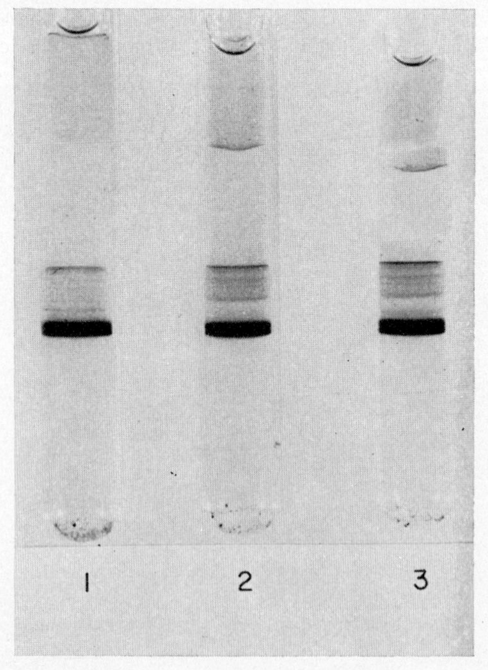


FIG. 11. Disc electrophoresis in polyacrylamide gels made with 8 *M* urea. Gels 1, 2, and 3 represent samples of outer-fiber protein prepared by three different procedures. All samples have been reduced and alkylated. See Renaud *et al.* [11] for further details.

solutions of outer fibers; extracts from acetone powders show only one minor band. Visual comparison of the relative intensity of the major and minor bands suggests that about 90% of the total protein is in the major band. In some experiments the major band has appeared to be resolved into two closely spaced subbands of equal intensity, but we have not yet excluded the possibility of this being an artifact.

The molecular weight of the fiber protein, dissolved in 5 *M* guanidine hydrochloride, has been determined as 55,000 ± 5000 by the Archibald and sedimentation-diffusion methods [11]. Essentially the same value was obtained whether the protein was prepared by dissolving isolated outer fibers or by extracting acetone powder. However, molecular weights determined in guanidine-hydrochloride solution are subject to possible systematic error from binding to the protein.

The amino acid composition of fiber protein has been determined in samples hydrolyzed in 6 *N* HCl at 110°C [11]. The results (Table III) show a relatively high content of the polar amino acids, especially glutamic and aspartic acids. Assay with dithiobis(nitrobenzoic acid)

TABLE III. *Amino Acid Composition of the Outer Fibers of Cilia*

| Amino acid | Residues per $10^5$ gm of protein[a] |
|---|---|
| Lysine | 51 |
| Histidine | 22 |
| Arginine | 41 |
| Aspartic acid | 94 |
| Threonine[b] | 46 |
| Serine[b] | 54 |
| Glutamic acid | 117 |
| Proline | 39 |
| Glycine | 80 |
| Alanine | 56 |
| Cysteine[c] | 13 |
| Valine | 53 |
| Methionine | 26 |
| Isoleucine | 49 |
| Leucine | 66 |
| Tyrosine | 29 |
| Phenylalanine | 39 |
| Tryptophan[d] | 7 |

[a] Most values represent the average of 3 hydrolysis times on each of two different samples.

[b] Extrapolated to zero hydrolysis time.

[c] Determined by reaction with DTNB.

[d] Determined from the spectrum in 0.1 *N* NaOH.

showed the presence of about 7 moles of cysteine per 55,000 molecular weight. The overall amino acid composition of fiber protein shows some resemblance to that of muscle actin, such differences as occur being of the same order as those between samples of actin from different species of animal [1].

The evidence of homogeneity from centrifugation and from electrophoresis indicates that the outer fibers consist largely of a single protein component. It is not known whether the minor components observed on electrophoresis are a functional part of the outer fiber, or whether they are impurities absorbed from the matrix.

The resemblance in preparation procedure, in physical chemical properties, and in amino acid composition suggest that the outer fiber protein is fairly closely related to muscle actin. However, the proteins are clearly not identical for the arrangements of molecular subunits in an outer fiber [6] is quite different from that in an F-actin fila-

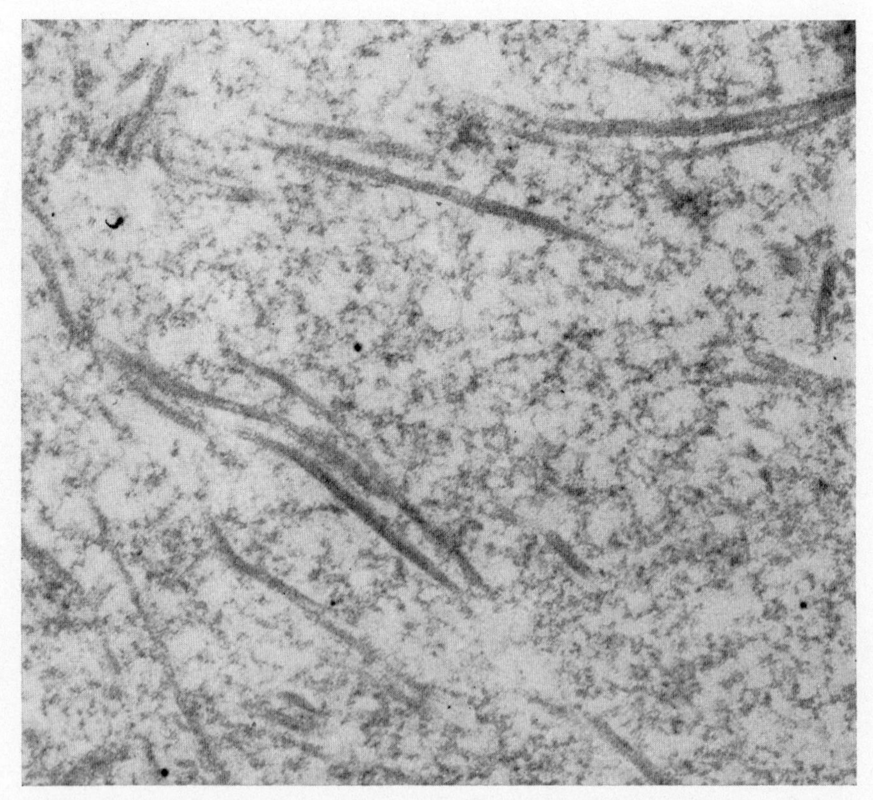

FIG. 12. The mixed fibrous and amorphous precipitate obtained by dialyzing a solution of outer-fiber protein against 0.03 *M* phosphate buffer, pH 6.5.

ment [8]. To examine the relationship of the two proteins in more detail, it will be necessary to make both preparations from a single species of animal.

Preliminary attempts to reconstitute the ciliary fibers from solutions of fiber protein have met with only partial success [10]. Dialysis of an aqueous solution of fiber protein against 0.03 *M* phosphate buffer pH 6.5 yields a mixed fibrous and amorphous precipitate (Fig. 12). Examination of the precipitated fibers shows that they are about the same size as native outer fibers and are composed at least in part of parallel protofilaments 40 Å in diameter. However, the reprecipitated fibers are much less regular in form, and they appear solid rather than tubular. We cannot exclude the possibility that our isolation technique partially denatures the protein, but our results so far suggest that a preexisting "seed" may be necessary to initiate the tubular form.

## Acknowledgments

We thank Dr. Guido Guidotti, and Dr. Ray Stephens for assistance with the amino acid analyses and Dr. J. T. Edsall and Dr. K. R. Porter for the use of equipment in their laboratories. This work was supported by United States Public Health Service grants GM 12124, 1-K3-GM 21,937, and 1-F2-GM 19204.

## References

1. Carsten, M. E., and Katz, A. M., *Biochim. Biophys. Acta* **90,** 534 (1964).
2. Gibbons, I. R., *Proc. Natl. Acad. Sci. U.S.* **50,** 1002 (1963).
3. Gibbons, I. R., *Arch. Biol. (Liège)* **76,** 317 (1965).
4. Gibbons, I. R., *J. Biol. Chem.* **241,** 5590 (1966).
5. Gibbons, I. R., and Rowe, A. J., *Science* **129,** 424 (1965).
6. Grimstone, A. V., and Klug, A., *J. Cell Sci.* **1,** 351 (1966).
7. Hall, C. E., *J. Biophys. Biochem. Cytol.* **7,** 613 (1960).
8. Hanson, J., and Lowy, J., *J. Mol. Biol.* **6,** 46 (1963).
9. Klainer, S. M., and Kegeles, G., *J. Phys. Chem.* **59,** 952 (1955).
10. Renaud, F. L., Rowe, A. J., and Gibbons, I. R., *J. Cell Biol.* **31,** 92A (1966).
11. Renaud, F. L., Rowe, A. J., and Gibbons, I. R. *J. Cell Biol.* **36,** 79 (1968).
12. Van Holde, K. E., and Baldwin, R. E., *J. Phys. Chem.* **62,** 734 (1958).
13. Watson, M. R., and Hopkins, J. M., *Exptl. Cell Res.* **28,** 280 (1962).

# PROBLEMS OF MORPHOPOIESIS AND MACROMOLECULAR STRUCTURE IN CILIA

D. E. HOOKES, SIR JOHN RANDALL, AND J. M. HOPKINS

*Department of Biophysics, University of London King's College, and Medical Research Council Biophysics Research Unit, London, England*

## THE PURPOSE OF THIS PAPER

This laboratory has been interested for some time in the formation or morphopoiesis of the ciliary organelle and more recently in the genetic control of this process [39, 40, 52]. In addition, a number of biochemical studies of ciliary proteins have been made [53, 54, 55, 56].

The purpose of this paper is essentially fourfold:

(1) To review in rather broad terms some of the main problems of organelle formation.

(2) To summarize the work of this laboratory on the formation of the ciliary organelle and on biochemical studies of its proteins.

(3) To present electron micrographs of fragmented axonemes of the cilia of *Tetrahymena pyriformis* (strain S).

(4) To examine possible interpretations of these observations in terms of the known physical theory of contrast in the electron microscope and a mathematical (computer) analysis of the types of image to be expected from certain model structures and negative stain distributions and profiles.

These various investigations form part of a coherent attack on the problems of the formation of the ciliary organelle. The paper is essentially a report in progress, and no comprehensive picture can possibly emerge at this stage.

## PROBLEMS OF ORGANELLE MORPHOPOIESIS

The study of differentiation, morphopoiesis, and development in higher organisms has been a feature of classical biology for generations. It is clear that the phenomena involved are complex in the extreme. Success in the understanding of developmental processes in molecular terms in higher organisms must surely depend in part on whether it is possible to break down the problems into simpler component parts. That certain very interesting possibilities exist in this direction is already clear from the work of Beermann [1] and of

Gurdon [24] to give only two examples. In general, however, the chances of obtaining a good overall picture of development in a higher organism at a molecular level still seem remote.

Several factors combine to make the investigation of organelle formation particularly prescient at this time.

1. Great advances, familiar to all, have taken place during the last decade or so in the understanding of the molecular basis of the hereditary process. These investigations have emphasized the importance of genetic factors in problems of morphogenesis and have led to other studies with important consequences for the understanding of biological control processes, especially in microorganisms [28].

2. A cell organelle is but part of the whole. It is therefore reasonable to assume that a single organelle will present fewer (but not necessarily simpler) problems of morphopoiesis than the whole cell or organism, since only a fraction of the genome is involved. On a rough estimate perhaps 100 genes would be required to carry the information necessary to build an organelle. For any higher organism the corresponding figure would be at least tens of thousands.

3. It is an accepted principle of the experimental method that the investigation of change in any system, physical, chemical, or biological, provides a means of obtaining information about that system. Thus the absorption of energy by an assembly of atoms or molecules leads to knowledge of spectra and in turn of atomic and molecular structure. Likewise, the change on the genome of an organism by means of mutation may lead to a modification of both structures and function of part of an organism, e.g., of an organelle. In the Protista there are organisms with suitable organelles (cilia, flagella, mitochondria, chloroplasts) and life cycles for such investigations.

4. Techniques for the isolation of certain organelles (e.g., mitochondria and cilia) are already available or show promise of development. This opportunity for the biochemical characterization of the molecular species of which a particular organelle is composed can therefore now be explored. However, the mere enumeration of the molecular characteristics of an organelle is of comparatively small account compared with the possibility of distinguishing between possible processes of self-assembly and those of more complex origin, involving perhaps sequential events in the genesis and formation of the organelle. In general terms one would expect both types of process to be present.

5. The preparation of large numbers of organelles (or parts thereof) in a pure state may provide a unique opportunity for the investigation of molecular organization by physical means, such as X-ray diffraction.

On the other hand only small amounts of material are required for electron microscope or electron diffraction studies.

6. The disposition of numbers of cilia and flagella over precise regions of cortex in the Protista suggests that the investigation of organized patterns of organelles may be of wider significance. The phenotypic appearance of the ciliates may in part be determined by the genetically controlled molecular pattern of the cell membrane. In the Metazoa the precise nature of this pattern could influence and control organ and tissue genesis and function. In ciliates this pattern could determine where particular organelles are generated and thus lay out in molecular terms the taxonomic blueprint for the species and thereby its recognition. At the same time, studies of the transposition of cortical entities in Protozoa (e.g., the work of Tartar [49] on *Stentor* and Beisson and Sonneborn [2] on *Paramecium*) pose new problems which have not yet been explained, nor will they be until many more facts are available.

7. The question whether cytoplasmic inheritance is involved in organelle morphopoiesis is obviously cogent. On the other hand, in *Chlamydomonas* the flagellum *structure* on present evidence is controlled by chromosomal genes [40, 52]. The finding of DNA at basal body sites in *Tetrahymena* [38] and more recently in *Paramecium* [47] may also be of significance in the genesis of single or multiple organelles, as has been discussed elsewhere, but proof of its mode of action (assuming it to be generally present in such systems) is lacking at present.

## THE CILIARY ORGANELLE

Klug has discussed the structure of "simple" viruses in terms of self-assembly processes. The morphopoiesis of more complex viruses such as the T-even bacteriophages are being actively studied by Epstein, Kellenberger, and Favre and their colleagues [15, 16, 29, 30]. This approach is as important, if not more so, than that of the investigation of orthodox organelles. The recent investigation of Edgar and Wood [14] on the morphopoiesis of bacteriophage T4 in extracts of mutant-infected cells is a valuable pointer to future progress.

Cilia and flagella are well-defined organelles capable of characterization in structural and biochemical terms. They may be found on organisms with a life cycle appropriate for genetic studies. Ignoring for the present that techniques immediately available would allow us to examine reasonably precisely the biochemistry of the external flagellum rather than that of the whole organelle, it is appropriate to consider what kind of information has already been obtained.

## *Electron Microscope and Other Results*

### *The Mature Organelle*

From a hydrodynamic standpoint cilia (and flagella) may be regarded as the microscopic external motile appendages of a great variety of small organisms, particularly the Protista. (This description also applies to bacterial flagella, which are not considered in this paper.) Biologically speaking, however, the organelle of which the cilium is the external part, consists not only of the cilium, but also the cortical structure known variously as the *basal body, kinetosome,* and *blepharoplast,* together with other complex structures that may be attached to it. In *Tetrahymena,* as we shall see, there is an appendage to the basal body known as a root which appears in the electron microscope as a striated tapering fiber, the function of which is unknown (see Fig. 12). In *Stentor,* there are also roots, but in this organism these structures are of much greater complexity; the basal bodies of the kineties have additional structures attached to them known as M-bodies which serve to contract the organism and alter its shape [37]. Differentiation thus exists between the ciliary organelles of the main cortex and those of the membranelles. The organelle with which this paper deals is thus a complex, but usually well-defined entity. The external moiety, the cilium, may be detached and its nature and properties examined independently of the rest of the organism. Recently, Randall and Disbrey [38] using an extension by Watson and Hopkins [53] of the Mazia and Child [36] technique, have isolated the cortex of *Tetrahymena* and shown cytochemically and by tracer techniques that DNA is associated with the sites of basal bodies in this organism. More recently Smith-Sonneborn and Plaut [47] have extended these findings to *Paramecium* using similar methods.

The fine structure of the ciliary organelle, particularly the cilium itself, has been investigated rather thoroughly from its appearance in thin sections in the electron microscope, particularly by Gibbons and Grimstone [22]. Much of the earlier work has been summarized by Fawcett [17]. In consequence, knowledge of the now well-known $(9 + 2)$, more accurately $[(9 \times 2) + 2]$ axoneme structure of the cilium as exemplified by this technique will be assumed. Less certain features of the structure are the secondary fibrils, possibly both radial and longitudinal, lying in the matrix between the outer nine pairs of tubules and the central pair. While the sidearms or chelae seem to be present in most preparations from a variety of organisms, the above secondary structures are more elusive. It is difficult to say whether the uncertainty about the existence of these apparently minor features arises from the use of faulty techniques, or whether they are genuinely absent from the cilia of some organisms or even from some strains of the same organism. Only extensive reinvestigation using improved techniques will resolve this problem.

Meanwhile we shall concentrate on the main features already referred to. It should be noted that electron microscope studies of thin sections of cilia and flagella indicate the following features. (1) The inner or center pair of tubules are identical in size, separate from each other, and lie off the axis of the cylindrical structure in a diametral plane. (2) The twin tubules forming each of the outer nine pairs are not in general strictly identical in size. The tubule A, to which the sidearms are attached, is in some circumstances slightly larger. (3) The tubules of each outer pair are either adjacent or possibly share material (see Fig. 12), as will be discussed subsequently.

The dimensions of the tubules are of course relevant to the discussion of the macromolecular organization of cilia (pages 139–164). Their accurate determination is difficult and will be discussed below (page 131).

In a cognate field, Tilney and Porter [50] find that the cytoplasmic tubules observed in Heliozoa (*Actinospherium nucleofilium*) are 210–240 Å in diameter. In plant material Ledbetter and Porter [33] have observed tubules 230–270 Å in diameter.

The whole axoneme system of tubules in cilia is coherent and not readily fragmented or broken, even when the external membrane has been removed. Since the distances between outer fibrils and the center pair are substantial, it seems likely that the axoneme as a whole is held together by structures and material not yet adequately demonstrated, such as the radial fibrils already referred to, possibly embedded in an amorphous matrix.

### *Basal Body Structure*

The first point to notice is that, in all instances of which we are aware, the outer boundary of the basal body is continuous with the outer nine pairs of ciliary tubules and is usually about 0.5 μm in length and 0.1–0.2 μm in diameter. Toward the lower end of this cylindrical squirrel-cage structure the twin tubules become triplets, but it does not seem to be known precisely at what point this occurs for any particular organism. Reference to Fig. 1 makes clear that chiefly in other respects readily observable differences in the fine structure of

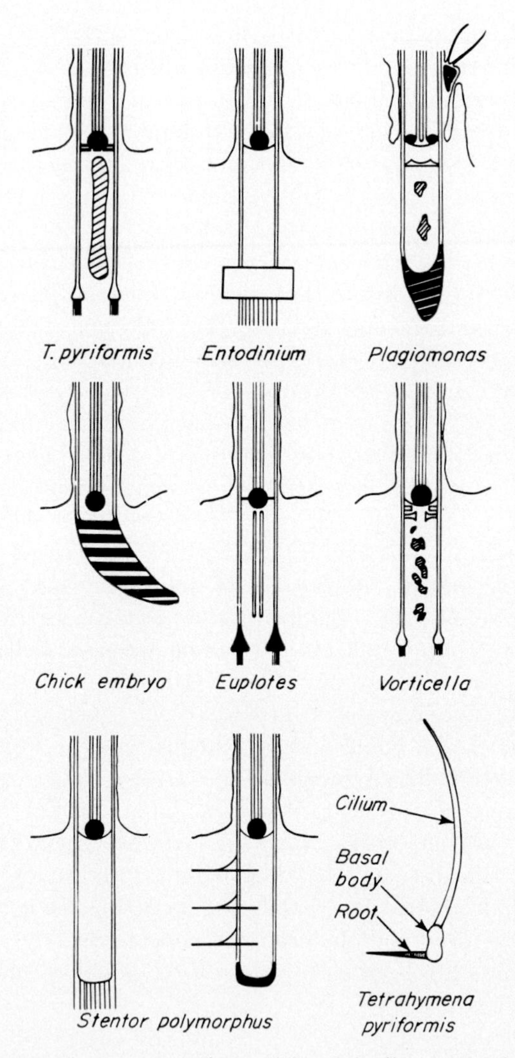


FIG. 1. Diagrammatic representations of basal bodies taken from a variety of plant and animal sources.

basal bodies exist between various organisms. Apart from the so-called root structures and their particular point of attachment, there are notable distinctions in the complexity of the contents of the squirrel-cage framework (see Figs. 1–3). The central pair of tubules (*CP*) of the cilium or flagellum is also variously attached. In all instances the central pair of tubules appears to terminate at the surface of the organism, usually in a quasiglobular structure some hundreds of

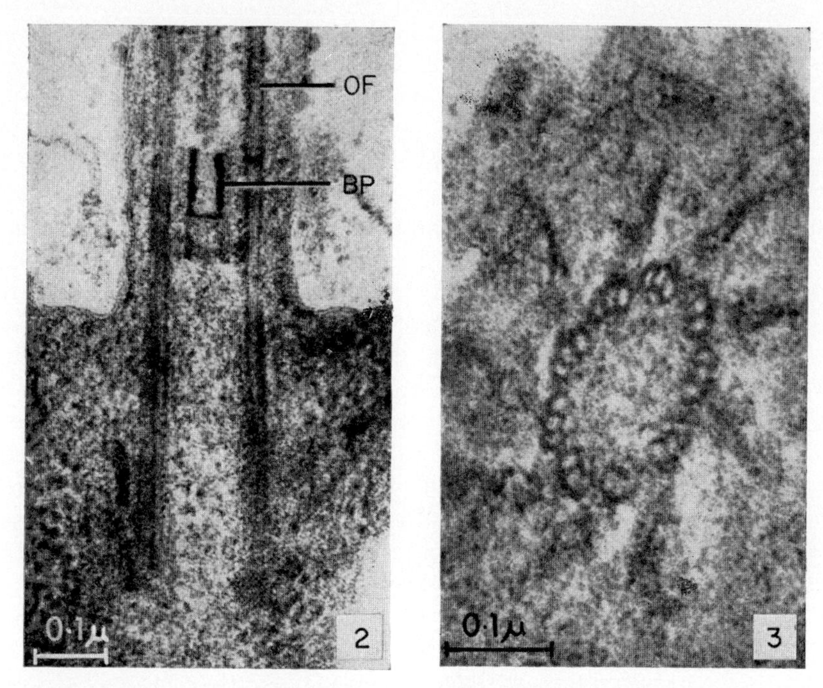


FIG. 2. Longitudinal section through a basal body of *Chlamydomonas reinhardii;* the central pair fibers terminate at the basal plate (*BP*). Electron micrograph. × 92,200.

FIG. 3. The nine outer pairs of fibers (*OF* of Fig. 2) of the cilium are continuous with those of the basal body, and these eventually radiate out to form the typical triplet formation of fibers. Electron micrograph. × 150,500.

Ångstroms in diameter which itself is very close to a plate or ring structure defining the "upper" end of the squirrel cage.

*In Development*

The examination of the formation of the ciliary organelle, and particularly of the basal body is difficult. Our own experience derives chiefly from a study of the regenerating membranelles of *Stentor coeruleus* in the electron microscope [37]. Really good fixation in this earlier work eluded us. Moreover, the first stages of development are ill-defined morphologically. It seems that light microscope observations can be misleading. The extensive work by Chatton and Lwoff [9] led Lwoff [35] to his well-known conclusion that the basal body (or kinetosome as he called it) is endowed with genetic continuity and that a new basal body is always formed by the division of a pre-existing one. Neither this work on *Stentor* nor that of Grassé on *Trypanosoma equiperdum* confirms the division hypothesis [23]. Nor does it seem that basal bodies in ciliates and flagellates always arise *de novo*. There are however certainly instances in which no preformed basal body seems likely to be involved, such as that of *Naeglaria* examined by Schuster [44] and that of fibroblasts reported by Stubblefield [48]. These instances are, however, ones in which a physiological change induces the ciliated or flagellated condition. In regenerating *Stentor* membranelles it is not clear whence come the first one or few basal bodies. Are they products of existing body organelles—perhaps a functionally differentiated group—from the longitudinal kineties, or do they arise *de novo?* Tartar [49] has shown clearly how the membranelles in *Stentor* begin to form at a rather precise region of the cortex, the so-called stomatogenic field. Similar considerations apply to mouth formation in *Tetrahymena*. What is clear from our own studies of *Stentor* is that once one or a few membranelle basal bodies have been formed the remainder derive from those already there, but not by division. The pre-existing or parent organelle provides something essential for the genesis of a new one. Thus there appears to be generative production of new organelles which can only take place with the assistance of "parent" organelles. Without unduly stretching our incomplete observations, there is substantial electron microscope evidence in *Stentor* that the basal body and the external cilium can develop simultaneously. It does not seem necessary for the basal body to be complete before that of the cilium can begin. Since both basal body and cilium contain a geometrically identical cylindrical arrangement of nine pairs of tubules, genesis could possibly proceed both externally (cilium) and internally (basal body). The striking fact is that both appear to develop (in *Stentor* at least) in the "correct" position of the cortex. Migration of one or other component does not seem to be usual, although the whole membranelle later moves from the position in which it first begins to form. New basal bodies and cilia arise adjacent to pre-existing organelles. In *Stentor* the generation of the ciliary organelle, as far as can be structurally observed, takes place in the cortex. On the other hand Dirksen has observed the presence of presumptive basal bodies throughout the cytoplasm in developing ciliated epithelium [13].

Other features of ciliary or related organelle genesis are likely to be revealed as work proceeds. Already Renaud and Swift have shown that in *Allomyces* a new centriole begins as a small one by the side of the old [42]. Thus as it matures each element of the structure (presumably a tubule) not only increases in length, but also changes its position with respect to its neighbors.

### *Genetics*

In a series of extensive investigations Kellenberger and his colleagues have explored the genetic control of phage genesis and structure [14, 29, 30]. In this laboratory cognate studies of the flagella of *Chlamydomonas reinhardii* are being carried out utilizing mutant strains with structural abnormalities of this organelle. This approach provides a possible means of enumerating and mapping the genes controlling the formation of this organelle. To succeed, it is essential to find suitable mutations, and for phage the discovery of amber and temperature-sensitive mutants has been valuable. In *Chlamydomonas* mutations affecting the central pair of tubules, the overall appearance of the flagellum, its length and its motility have been discovered. In addition, a partial suppressor of some of the mutations has been isolated and shown to act with differing degrees of efficiency on different mutations [52]. In all, some fifty mutants are now known. It has been demonstrated that mutation at any of four genes may lead to lack of organization in the central pair. This suggests that normally at least four genes interact to form the central pair of tubules. No mutations, however, have yet been discovered which appear to affect directly either the basal body structure or the outer nine pairs of tubules of the external flagellum. It is reasonable to assume that future research will reveal many new (nonlethal) mutations and thus help to clarify the genetic aspects of organelle control and morphopoiesis.

### *Chemical Studies*

Unlike the flagella of bacteria which have a much simpler structure, protistan cilia and flagella have not been the subject of very extensive biochemical investigation. In the beginning it was found that the methods then available for the isolation of cilia from large cultures of the organism were at an early stage of development; the yields were low and the cilia frayed and structurally incomplete. By 1961–1962, methods for obtaining tens of milligrams of cilia had been published. The cilia were shown by electron microscopy to contain all the main elements of structure normally revealed by this instrument. Beginning with cilia isolated by the method described, Gibbons was able to isolate intact preparations of ciliary fibers after the removal of the membrane with digitonin [21]. These isolated fibers are soluble in certain salt solutions at neutral pH. Moreover, it seems that structurally distinct parts of the fiber complex can be separated by differential extraction. When these protein extracts from the outer nine pairs of ciliary tubules are examined in the ultracentrifuge, the material sediments with a very broad boundary (4 S to 50 S), presumably indicating various stages of aggregation of the proteins. Using methods that have been successfully applied to other protein structures we have first treated the fibrillar extracts with mercaptoethanol to break disulfide bonds, then blocked the liberated sulfhydryl groups with iodoacetamide, dissolved the products in 8 *M* urea, and subsequently fractionated them on urea-starch gels. This treatment would be expected to liberate the individual polypeptide chains of which the tubules are ultimately composed. The electrophoretic results were tentatively interpreted as indicating that the longitudinal tubules are composed mainly of a set of rather similar polypeptide chains. An unexpected feature of the electrophoresis experiments was a close similarity between the patterns from different protein extracts. It seems that the separation by differential extraction is not so clear-cut as would appear from electron micrographs. Perhaps the different parts of the $(9 + 2)$ complex—central pair and tubules of the outer nine pairs—are composed of similar proteins.

### *Possible Morphopoietic Processes*

Investigations of the kind just described are necessary contributions to an overall picture of organelle morphopoiesis. In addition, however, the contribution each makes to the organization and function of the whole must be in due course elucidated. Detailed analysis of possible morphopoietic processes is not practicable at present, but some considerations, which it is useful to bear in mind, can be made. The arguments are tacitly put forward in terms of proteins, but could also apply to other macromolecular systems.

Some useful aspects are apparent from the results of experiments with T4 bacteriophage already referred to. These investigations show that only a limited number of the genes associated with morphopoiesis control actual synthesis of protein components. The head of the bacteriophage contains about 300 identical protein units built into a precise structure. If there are other different proteins present, it must be in very small amounts. But the number of genes so far implicated in the production of phage heads is 8. The inference is that seven of these perform functions of assembly rather than synthesis. The nature of these functions is quite unknown. It is not of course impossible that the function of some of these genes is to produce small numbers of proteins with special attributes. These molecules might serve to hold the structure together. Although many of the morphopoietic genes of T4 phage are grouped in clusters there is yet no evidence that a cluster behaves as an operon.

These remarks about phage genesis indicate that a step-by-step examination of mutants and their structure accompanied by appropriate ancillary investigations of function may lead in time to a reasonable understanding of some morphopoietic processes. Similar hopes may be expressed for the ciliary organelles although it is not yet clear whether this system offers an equally fruitful field of study.

The primary structure of each protein is genetically determined and the secondary and tertiary structures are physical consequences of that dependence, modified to some extent by the environment in which they exist. The overall structures and other properties of any assembly of the individual protein molecules therefore stem in direct line from the genetic code. According to the nature and disposition of surface groups (e.g., positive and negative charges, hydrophobic or hydrophilic groups), different types of structure, with different stabilities with respect to their environment, will be formed, such as fibers, tubules, sheets, or other geometrically recognizable figures when the concentration and other conditions necessary for aggregation are met. The term *self-assembly* is often used to describe such a process. It should be noted that processes of this kind are not limited to a single species of molecule, as the well-known case of tobacco mosaic virus illustrates, where large numbers of protein molecules are assembled round a single RNA helix. Such assemblies can be dispersed and reassembled. The point made here, which is often ignored, is that even self-assembly processes involve genetic control through its influence on the final configuration of the macromolecule. By self-assembly is meant the spontaneous aggregation of biological macromolecules into a specific structure without the intervention of some other—generally biochemical—process to bring it about. For example the heads and tails of T4 bacteriophage do not spontaneously join to form the complete virus particle; apparently the action of about 4 genes is required to complete the structure. Time is involved in processes of self-assembly only insofar as all physical (and chemical) processes take a finite period to complete—this might be called the interaction interval.

In morphopoiesis, time is involved in this elementary sense, but also conceivably in a series of sequential operations, the intervals $t_1$, $t_2$, . . . between which may well be small and difficult to observe. One important aspect can be illustrated by a further reference to T4 phage: clearly the processes necessary for head-to-tail attachment cannot begin until the tail structure has reached the appropriate stage of completion at one end. More formally, imagine a set of molecular species $\alpha$, $\beta$, $\gamma$, . . . required to make parts A, B, C, . . . FG, . . . L of an organelle O in which these parts are related to each other in some particular way (Fig. 4). Various fairly obvious possibilities then arise. Individual parts of O could be built from one (or more) of the molecular species $\alpha$, $\beta$, $\gamma$ . . . either (1) by self-assembly or (2) by the intervention of specific gene products leading to stronger bonds than in (1). If A is a symmetric structure, it is possible that part B, if made simultaneously, could become attached to part A before the latter has been completed. In general, however, one should allow for orderly temporal processes, but they

| (a) | (b) | (c) |
|---|---|---|
| A B C A | A B C A | A B C A |
| D | D | D |
| E F | D | E′ F′ |
| G H | D | G H |
| I J K I | E F | I J K I |
| L L | G H | L L |
| M N P | I J K I | M N P |
| | L L | |
| | M N P | |

FIG. 4.

may not always be strictly necessary, in which case some at least of the values $t_1$, $t_2$, etc. will tend to be zero. If part B were to envelop part A the latter would have to be formed in full before B could be completed. Mutations could give rise to changes in the number and nature of individual components (as in Fig. 4 b and c) with consequences for the overall structure organization and function of the organelle.

The coming together of two parts of an organelle or other biological structure may not be an addition process analogous to the positioning of one brick with respect to its neighbor. Reorganization may take place:

$$A + B \rightarrow AB \rightarrow A'B'$$

It is also conceivable that some components of an organelle (not necessarily "structures" in the above more formal sense) may be but temporarily associated with it for specific purposes. Enzyme X could act on its substrate, part Y, in order to effect the union of Y with part Z. Once this is achieved the purpose of Z has been served. There are many theoretically possible variations on this theme, including the shedding of more formally organized components than a thin layer of enzyme molecules.

In the genesis of the ciliary organelle the plasma membrane has been observed to protrude, and the assembly of the axoneme begin, before the basal body structure is complete [39]. While in general and at present it seems desirable to postulate sequential processes, it should be recognized that construction of various parts may go on simultaneously; the truly sequential processes, if present, may be therefore chiefly involved in the final assembly.

An organelle is only a part of a cell, yet many organelles take up characteristic positions within the cell. A cell is no more a bag of organelles than it is a bag of enzymes. It was shown in the early part of this paper that ciliary organelles form species-specific patterns, which in this instance involve parts of another organelle: the membrane system. Without elaboration, it seems reasonable to postulate organelle interaction as a factor in the positioning of single or multiple organelles within the cell.

In summary, present knowledge suggests that the following may be postulated as important features of organelle morphopoiesis:

1. A well-defined fraction of the genome is associated with the genesis of each organelle. In this fraction two main groups of genes may be distinguished: those associated with the synthesis and control of the constituent molecules of the organelle and those associated with the actual morphopoiesis of these constituents.

2. Protein synthesis has been shown to require an initiator (*n*-formylmethionine) which could possibly be unique [6]. One consequence of a unique initiator is its subsequent removal by enzymatic means.

Once the initiator has been removed, some constituents of an organelle may form into necessary structures by virtue of their physical properties and may thus be denoted as processes of self-assembly.

3. While simultaneous production of essential parts is not precluded, the final construction of the organelle will probably depend on detailed timing and control of particular processes.

4. The *construction* of an organelle probably requires small numbers of specialized macromolecules in particular positions, e.g., at the sites of attachment of chelae and radial subfibrils in cilia. A natural consequence of this is that idealized models of organelle structure (necessary as a basis for discussion and calculation) are unlikely to be strictly correct. The principle of quasi-equivalence of molecules enunciated by Caspar and Klug for virus structures is at least as relevant for organelle formation and structure [8]. Quasi-equivalence in ciliary tubules could

result in distortion of the idealized structure into a randomly puckered circular or elliptic cylinder.

5. The position of single and multiple organelles within the cell will involve interactions with other organelles and gene products.

6. The *functioning* of an organelle may involve the attachment of other important but structurally insignificant molecular species to the observable parts of the organelle.

## ELECTRON MICROSCOPE OBSERVATIONS ON CILIARY TUBULES

### *Introduction*

This part of the paper records electron microscope (EM) studies of isolated cilia of the protozoan *Tetrahymena pyriformis* (strain S) from most of which the plasma membrane has been removed. Direct observations of the EM image of whole tubules negatively stained with uranyl acetate reveals a longitudinal array of several rows of closely spaced intensity maxima. This immediately suggests that these components of the structure are composed of discrete units, possibly some 40 Å in size which we shall refer to for the time being as globules. However, as will be shown below, the interpretation of a three-dimensional image of either globular or fibrillar components interspersed with staining material is a complex matter which requires very careful analysis. This section of the paper presents the observations as simply as possible. Analysis, interpretation, and discussion are deferred to the final section.

### *Cultures, Materials, and Methods*

The procedures used in most experiments were twofold: (a) the removal and isolation of cilia from organisms and the subsequent purification of the sample, followed by (b) the dissolution and disposal of the outer plasma membrane.

Most of the work to be described has been carried out on the cilia of *Tetrahymena pyriformis:* a few experiments have also been done on the flagella of *Chlamydomonas reinhardii*; see also Ringo [43]. Preliminary indications are that the results on *Chlamydomonas* are substantially the same as those on *Tetrahymena,* and these experiments will not be discussed further.

#### *Isolation of Cilia*

*Tetrahymena pyriformis,* strain S was cultured at 27°C in proteose-peptone medium (1% Difco proteose-peptone; 0.2% Difco yeast extract; 0.1% Gurr's bacteriological glucose) for 3–4 days until the stationary phase of growth was reached. The method of removal and isolation of the cilia was essentially the same as that reported by Watson and Hopkins [53]. There were two slight modifications on the earlier techniques: smaller quantities of animals were used (1 liter instead of 8 liters) and the isolation of the cilia was carried out in one step instead of two. One-liter cultures of *T. pyriformis* strain S in stationary phase were centrifuged at 100 *g* for 5–6 minutes at 4°C. It was found essential to carry out all subsequent steps on the preparation at this temperature. The animals were washed once in cold 0.025 *M* sodium acetate solution (pH 7.0), resuspended in 50 ml of 0.025 *M* sodium acetate and concentrated at 100 *g* for 5–6 minutes. The ciliary isolation medium contained 12% v/v ethanol, 3 m*M* EDTA, about 25 m*M* $Ca^{++}$, 25 m*M* $Cl^-$, 25 m*M* $Na^+$, and 25 m*M* $Ac^-$ at pH 7.3. Of this cold solution 15 ml was added to the concentrated ciliates and the detachment of the cilia began immediately. The resuspension was gently stirred and allowed to stand in the cold room for about 3–5 minutes. The cell bodies were removed by centrifugation at 100 *g* for 10 minutes at 4°C, the detached cilia remaining in the supernatant. The cilia were concentrated at 10,000 *g* for 20 minutes in the cold and washed once in a Tris-Mg buffer, pH 8.2 (3.6 gm of Tris-HCl, 615 mg of $MgSO_4$ per liter adjusted to pH 8.2 with 1 *N* HCl).

### *Removal of Ciliary Membrane*

The cilia were reconcentrated and then suspended in 2 or 3 ml of a solution containing 0.005% to 0.5% digitonin in Tris-Mg buffer, pH 8.2. In some instances this digitonin extraction was repeated to ensure the complete removal of the ciliary membrane, then the cilia were washed in Tris-Mg buffer, pH 8.2. On occasions the central pair of tubules was removed according to the method of Gibbons [21].

### *Preparation of Specimens for Examination in the Electron Microscope*

In general the shaft of a cilium prepared in the manner described is a coherent structure, even when the membrane has been removed, and dispersal or fragmentation of the tubules is necessary if macromolecular detail is to be seen at the higher magnifications. No outstandingly successful method has been found for this purpose, but the best results have been obtained by the application of the following procedure.

Isolated cilia were suspended in a cold (4°C) solution containing $2 \times 10^{-4}$ *M* ATP, 0.14 *M* KCl, *M*/150 phosphate buffer at pH 7.0 and then gently crushed by means of a hand homogenizer. This solution in which the cilia are suspended is referred to as a fragmenting buffer.

A drop of the ciliary suspension was transferred to a carbon-coated copper grid and the fragments were allowed to settle for 5 seconds. The liquid was then removed by touching with filter paper and replaced by a drop of 1% uranyl acetate in distilled water (pH 4.0). This also was withdrawn after 5 seconds and the specimen then allowed to dry for 1–2 hours before examination in the electron microscope.

All specimens were examined in an RCA EMU 3G electron microscope (accelerating voltage, 100 kV) fitted with a double condenser, 30 $\mu$m objective aperture and high-resolution specimen holder. Initial magnifications were about 58,000, and all electron micrograph prints used in this paper were prepared from reversed negatives. In consequence the appearance of biological material surrounded by negative stain is identical in reproduction with that of the original plates: viz. black on a white background.

## *Results*

In this section will be described the electron microscope observations on various types of negatively stained preparations of ciliary tubules and other components. At the outset a question of nomenclature should be mentioned. One of the striking features of the micrographs is the apparent globular nature of the subunits from which the tubules appear to be constructed. The complexities of interpretation suggest that a more neutral term should be used to describe the image; one such term is an array of *intensity maxima*. We have found it difficult to use repeatedly this somewhat clumsy nomenclature and frequently revert to the more convenient terms *apparent globules* or *subunits*. In fact this merely anticipates our final conclusions. The point to be made here is that globular subunits cannot be inferred directly from mere visual inspection of the micrographs.

### *Specimens with Membrane Attached*

In a few instances portions of cilia have been observed with pieces of membrane still attached. An example of this is shown in Fig. 5 and appears to represent the tip of a cilium. The apparent particulate nature of the tubules and of the membrane—which may be partly disintegrated—is evident. Although many micrographs of this type have been obtained, it is not possible from observations of negatively stained specimens alone to be certain whether the tubules are enclosed in a membrane sheath or rest beneath, or above, a single membrane fragment.

### *Unfragmented Cilia*

In spite of the fragmentation treatment many of our preparations are of the kind shown in Fig. 6. The membrane component has been completely removed, but the whole complement of $[(9 \times 2) + 2]$ tubules remains intact and coheres as if it were a single fiber. The overall width of each specimen in such preparations averages about 2000 Å, which is rather larger than one would expect from the dimensions of undistorted sectioned material, but is consistent with a flattened assembly of tubules.

### *Fragmented Cilia: Data on Tubules*

The most important of the specimens examined have been frayed at the ends or fragmented by the treatment already described. Consequently it has been possible to observe single pairs of the outer nine system and also single tubules of the center pair. Examples of these are shown in Figs. 7 and 8, respectively. Figures 9 and 10 show a more comprehensive selection of tubules, indicating the variety of images that has been obtained. Measurements with a micrometer eyepiece of tubules in thin transverse sections of cilia indicate that each tubule has an overall diameter of about 240 Å. (Although in *Tetrahymena* there are small differences in overall diameter between tubules A and B of the outer pairs, i.e. $A > B$, and between these and the center pair, $CP > A$ or B, these differences are ignored in this presentation owing to the real difficulties of precise measurement.) The overall dimensions of the negatively stained tubules illustrated in Figs. 7–11 show a variation from 240 Å to 340 Å sometimes along the length of a single tubule (see Fig. 11). A mean value of about 300 Å can be chosen, and is also most frequently found. This indicates that most of the tubules are somewhat flattened or compressed as a result of the staining or other preparative procedures.

The three most dominant features of the images of the tubules in Figs. 7–11 are: (a) the discrete intensity maxima (or apparent globules) already referred to; (b) The arrangement of these maxima into longitudinal arrays (or columns) parallel to the axis of the tubule; and (c) the large amount of disorder or confusion that is found; small regions of an approximately regular arrangement of the intensity maxima are interspersed with regions of disorder. The regions of order very rarely extend across the whole width of the tubule, or for more than 3 or 4 maxima in an axial direction, (see Fig. 10a, region *1*). The remainder of this description is in three parts which describe more precise parameters of the images.

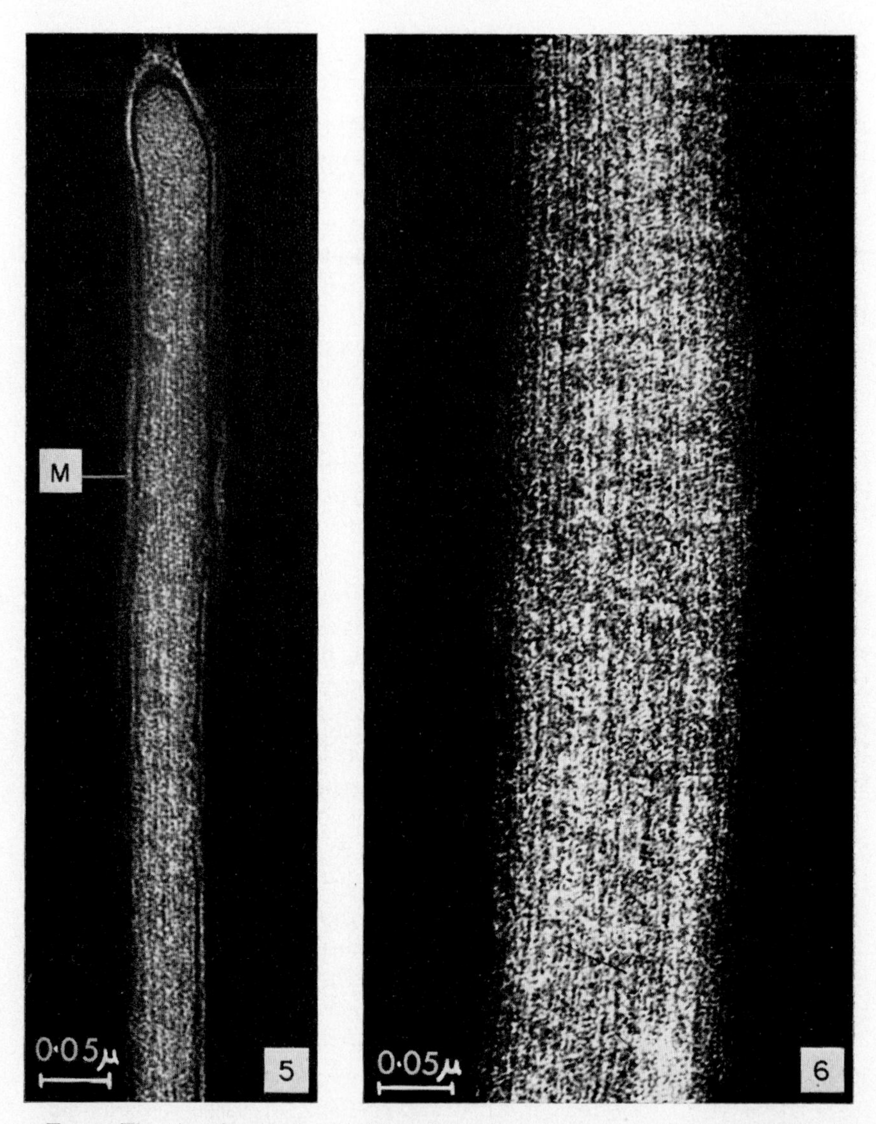


FIG. 5. The tip of an isolated cilium from *Tetrahymena pyriformis* stained with uranyl acetate. The outer membrane (*M*) forms a continuous sheath around the tubules. Electron micrograph. × 184,400.

FIG. 6. Portion of an unfrayed isolated cilium negatively stained with uranyl acetate. The apparent globular nature of the tubules can be clearly seen. Electron micrograph. × 184,400.

---

FIG. 7. An example of a typical central pair of tubules (*CP*) from a fragmented isolated cilium of *Tetrahymena pyriformis*. In this electron micrograph the tubules show no interconnecting components. Stained with uranyl acetate. × 350,000.

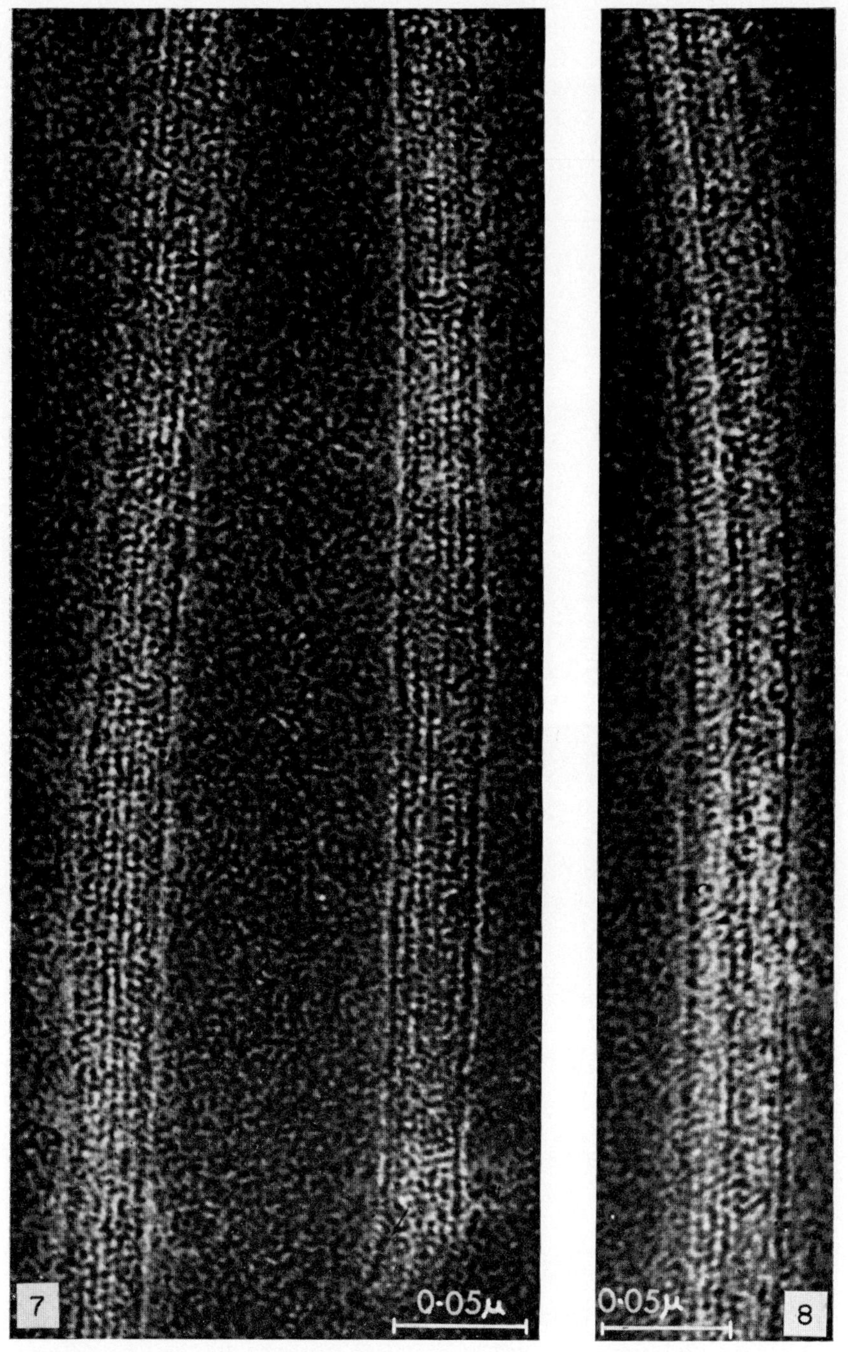


FIG. 8. An outer pair of tubules from a fragmented isolated cilium of *Tetrahymena pyriformis,* uranyl acetate stained. The tubules are connected and apparently share a common wall. (See also Fig. 12). Electron micrograph. × 350,000.

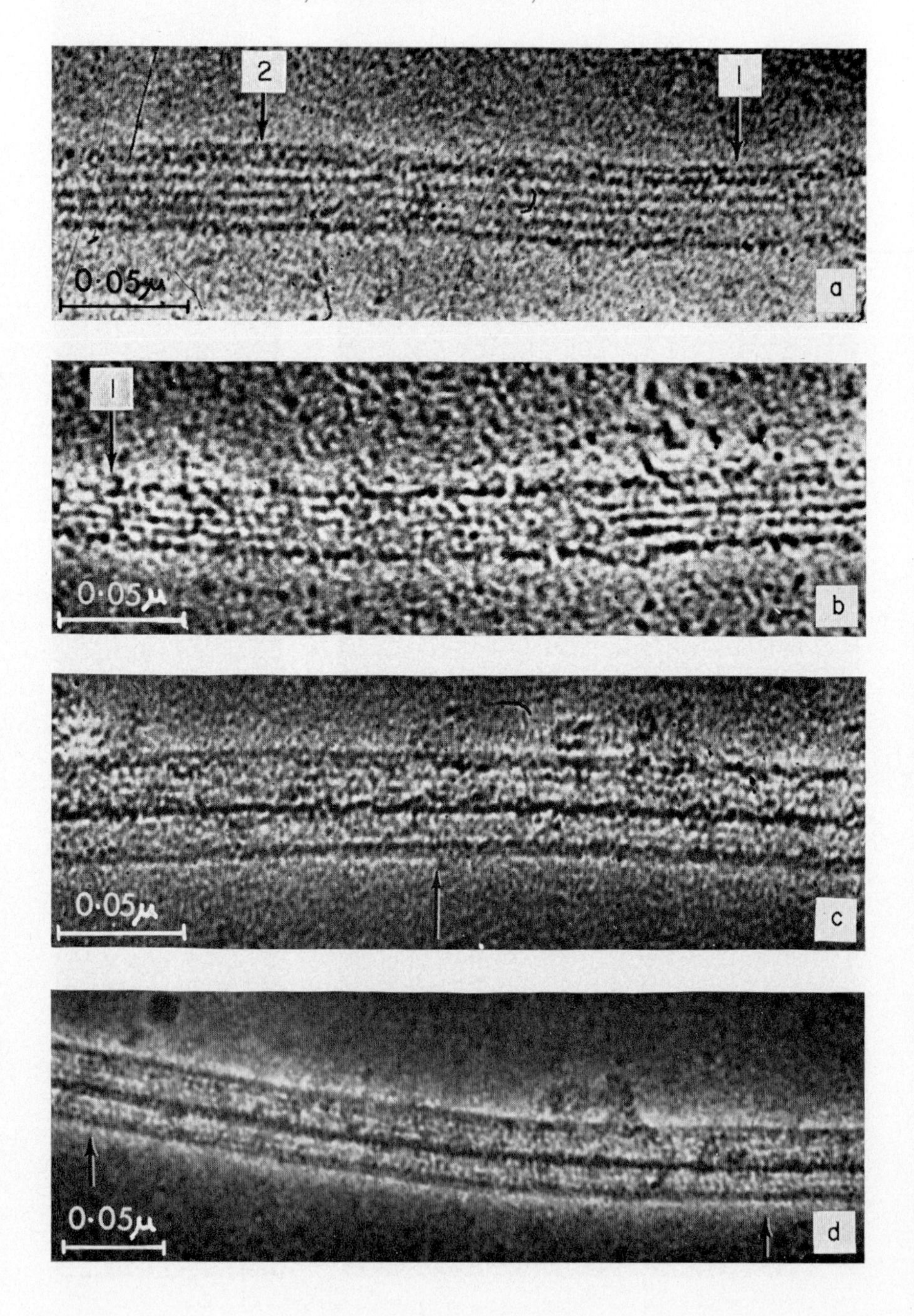
2
1
0·05μ
a
1
0·05μ
b
0·05μ
c
0·05μ
d

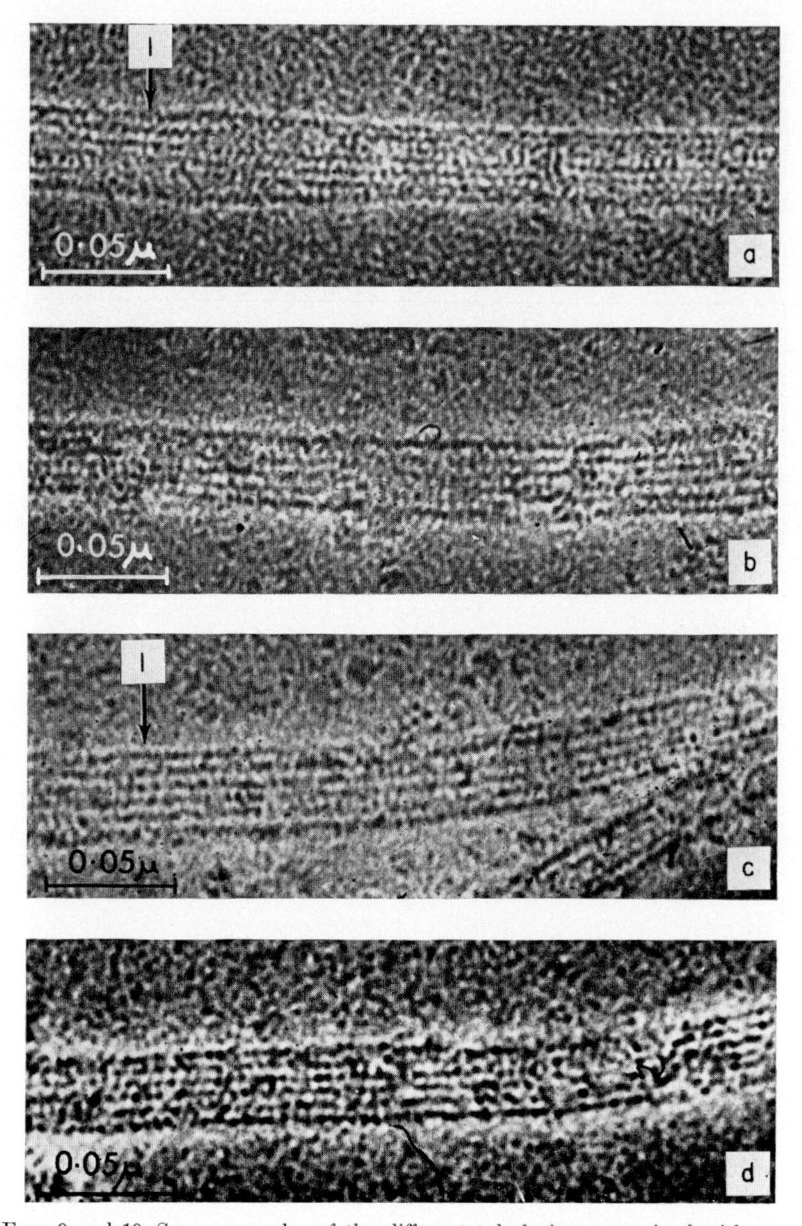


FIGS. 9 and 10. Some examples of the different tubule images stained with uranyl acetate. In particular they show a variation in the number of longitudinal columns of apparent globular subunits. Fig. 9, a–c, × 350,000. Fig. 9 d, × 277,000. Fig. 10, a–c × 350,000. Fig. 10d, × 369,000.

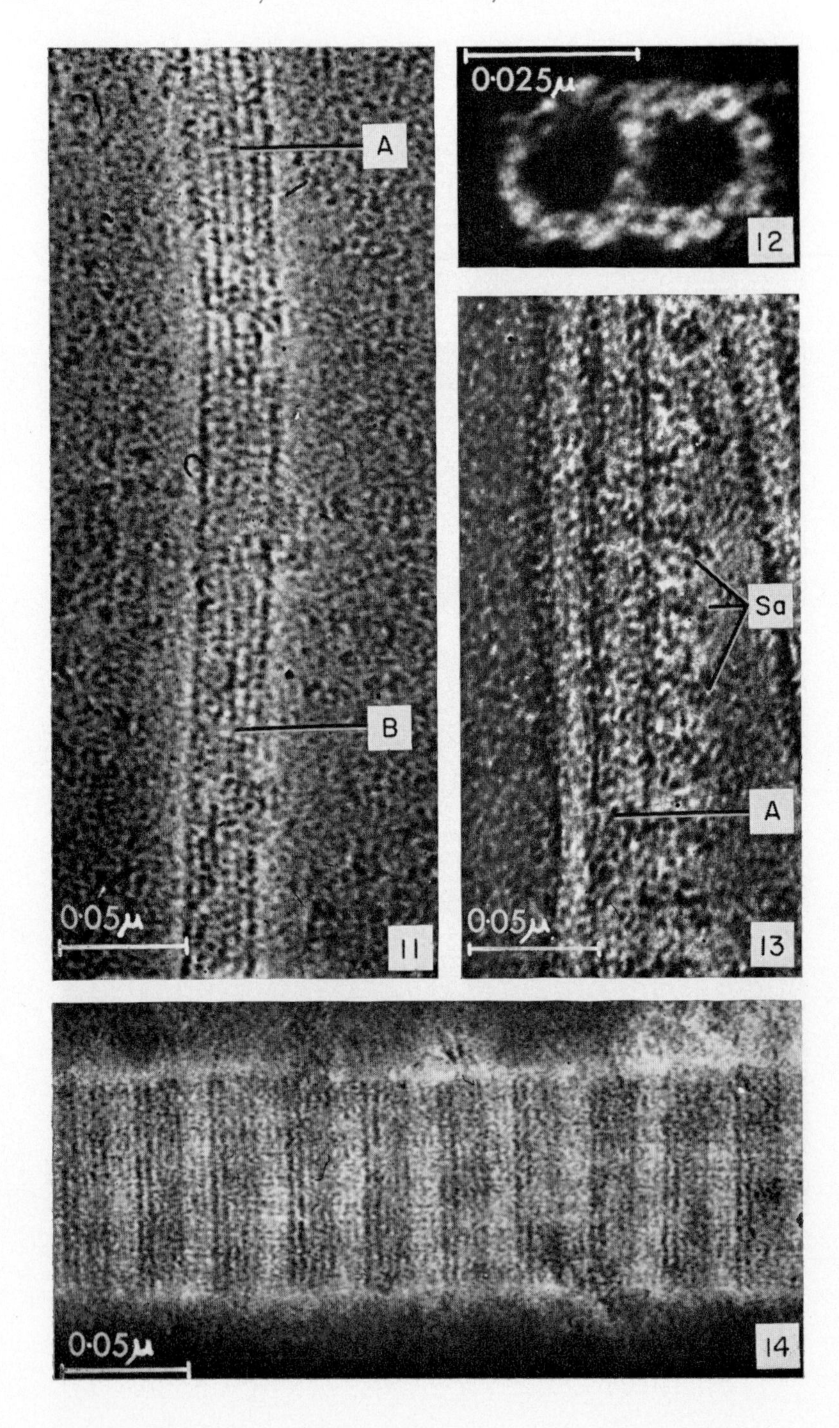
0·025μ
12
A
B
0·05μ
11
Sa
A
0·05μ
13
0·05μ
14

*The Number of Longitudinal Columns Visible across the Width of the Image of a Single Tubule.* This varies from one tubule to another, and even along the length of a single tubule. By far the most frequent value is 6 (see Fig. 9a, Fig. 10, a–d). Values of 4 and 5 are also found, and, very occasionally, 7 (see Fig. 9a, region *2*). Four-column images are found only in one or both of the component tubules of an outer pair (see Fig. 9 c and d). These are more frequent than 5 columns, which are relatively rare (see Fig. 9b, region *1*).

*Spacing of the Apparent Globules along each Column.* This also shows considerable variation (measurements made with a micrometer eyepiece). The average values measured over suitable length of tubule range from 22 Å to 60 Å with most values laying in the range 28–50 Å. A value of approximately 40 Å is both a mean and also the most frequent value. The range of values observed for a given tubule tends to vary from one tubule to another; for example, some tubules contain values in the range 28–42 Å and some in the range 38–60 Å.

*Type of Order.* Wherever order occurs in the image the apparent globules or intensity maxima not only lie in rows parallel to the tubule axis, but also in rows inclined at an angle to the tubule axis. These two sets of lines define the order within a region containing regularly arranged maxima. The order can be described by an angle, $\psi$, which the oblique (or nonaxial) rows make with a line perpendicular to the axis. The value of $\psi$ tends to vary from one region of order to the next. The mean and most frequent values lie in the range 10°–20°, but values of 0° or 30° may also be found (see region *1*, Fig.

---

FIG. 11. One tubule from a central pair of tubules (*Tetrahymena pyriformis*): uranyl acetate stained. Longitudinal rows of apparent globular subunits are clearly seen. There is a variation in tubule width ranging from about 243 Å (*A*) to 315 Å (*B*) indicating variable flattening along the length. Electron micrograph. × 350,000.

FIG. 12. Transverse section through a pair of outer tubules of a flagellum of *Chlamydomonas reinhardii.* The lack of a smooth outline in this section suggests that the tubules are composed of smaller units, which on this evidence alone could be either globular or fibrillar. This electron micrograph and many others also indicate that the junction between the two tubules is a single common wall. × 933,500.

FIG. 13. A uranyl acetate-stained preparation of a pair of outer tubules from *Tetrahymena pyriformis.* Note the appearance of regularly spaced sidearms (*Sa*) along the length of one of the tubules (*A*). Each arm is composed of about 5 discrete globular subunits. Electron micrograph. × 350,000.

FIG. 14. Portion of a ciliary root isolated from *Tetrahymena pyriformis.* The root is composed of globular subunits similar to those of the ciliary tubules but differentiated into alternating "light" and "dark" bands. Uranyl acetate stained. Electron micrograph. × 346,800.

10c). The spacing between columns does vary, but not to the extent of spacing within columns. Values in the range 45–60 Å have been recorded.

### *Attachments to Ciliary Tubules: Sidearms (Chelae) or Radial Subfibrils*

In transverse sections of embedded cilia subunit A of each of the outer nine pairs of tubules usually has a pair of arms attached. (When viewed from its base, subunit A of each pair appears on the right-hand side and the arms point in a clockwise direction). It is also probable that each subunit A of the outer pairs has attached to it a radial fibril which terminates in the matrix. Longitudinal sections of embedded cilia show fairly uniformly spaced transverse striations which could represent either or both of the above types of attachment. In section the average distance apart of the striations is about 150 Å.

In the present work we have observed in negatively stained preparations twin fibrillar attachments to the outer nine pairs of *Tetrahymena* in the form of strings of about 5 discrete globular subunits each some 40–45 Å in size, and a consequent overall length of ~200 Å. Existing preparations do not allow of any accurate measurements, and the values just given are indicative, not definitive. These sets of twin fibrils are also ~150 Å apart. It is possible that both radial fibrils and sidearms occur in our micrographs (Fig. 13), but the overall appearance suggests the latter. The attachments were not present in the majority of our preparations and also appear to be fragile and difficult to preserve. In view of the later discussion (see section on computer approach below), it is interesting to note the apparent globular nature of the attachments in a situation where no superposition effects are likely to occur.

### *Ciliary Roots*

In some preparations in which whole organisms were used, ciliary roots have been exposed. These are about 2.6 μm in length and taper with a gradient of approximately $3 \times 10^{-2}$, i.e., 300 Å in 1 μm.

The following description (Fig. 14) refers to the structure apparent in the image and cannot be a true description of the three-dimensional object, since as we have frequently indicated, the image is the result of the passage of the electron beam through a three-dimensional object in and around which heavily absorbing stain is dispersed.

The ciliary roots display an assembly of apparent subunits arranged in approximately orthogonal array. Proceeding from the proximal portion to the tip the roots is differentiated into alternating "light" and "dark" bands. Each light band is composed of three rows of subunits and each dark one of five rows, the distal two of which can be distinguished from the remainder as rather larger subunits. The longitudinal periodicity of this overall pattern is about 325 Å. In a longitudinal direction the subunits are on average about 40 Å apart; transversely the corresponding figure is about 35 Å. If the observed image were known to be due to one layer only of the root structure, it would be reasonable to infer a distinct biochemical identity for the three electron-optically distinguishable units. This knowledge is not available at present, and the structural interpretation of electron micrographs of ciliary roots will not be considered further in this paper.

## A COMPUTER APPROACH TO THE INTERPRETATION OF THE ELECTRON MICROGRAPHS

### *Preliminary Discussion of the Image of a Negatively Stained Object Composed of Regularly Arranged Subunits*

The immediate object of electron microscope studies of the tubules of the ciliary axoneme is to determine as accurately as possible the distribution of matter within them. From the longer-term standpoint of morphopoietic processes it is necessary to understand how such structures are arrived at. There is also the further point of the possible connection between structure and function—in this instance, motility. For the inferences of such studies to be valid the structures determined from biological objects in the electron microscope should closely resemble those in the living state. The possibility of artifacts of preparation in electron microscope specimens of all kinds is a continual hazard that cannot be wholly circumvented. At the present time there are no certain cross-checks that can be applied. X-ray diffraction examination of undried material is perhaps the most hopeful. It is, however, by no means certain that the type of organization within each cilium and the degree of organization as a whole would be sufficient to give the required information. The intensity distribution in the image of an object in the electron microscope under conditions of negative staining depends not only on the distribution of matter in the object, that is, the shape, size, and distribution of the basic subunits, but even more sensitively on the distribution of stain within and about the object. The image of the tubule consists in broad terms of a distribution of closely spaced intensity maxima along lines parallel to the longitudinal axis of the tubule. The most obvious clues to the structure will come from the regularities or periodicities in the intensity distribution of the image. The variation in intensity from one set of maxima to another will reflect the shape of the subunits as well as the geometry of the structure and the distribution of the stain. Many electron micrographs of the supposedly periodic structures often appear confused and show little structure in the intensity distribution. Klug and Berger [31] have devised an elegant technique for analyzing such electron microscope images in terms of the periodicities present. The method uses the electron microscope image as a diffraction object under conditions of Fraunhofer diffraction. A plane beam of monochromatic light is passed through the electron microscope plate, and the resultant diffraction pattern is focused and photographed. Thus, any underlying periodicities in the intensity distribution that cannot be detected by eye give rise to a series of diffraction spots. Even if the periodicities are not perfect, but contain random small deviations from perfect order, the diffraction pattern will give the best average periodicity.

There are several reasons why the electron microscope image of a three-dimensional (3-D) object may be difficult to resolve. First, the image is a transmission image and consequently is the result of the superposition of the upper and lower parts of the structure. The extent to which any particular part of the object is "seen" in the image depends sensitively on the distribution of stain about it. For instance, if the stain does not completely surround the object, but fills only the grooves in the upper surface while immersing completely the lower surface, then the lower surface will be contrasted more strongly than the top. Thus, the fully contrasted lower part of the object would be superimposed on the only partially contrasted upper part. As a result, the image may present a very confused intensity distribution in which discrete maxima are difficult to observe. Even if the object were completely immersed, irregularities in stain thickness and distribution would also cause confusion. Similar effects could also be produced by distortion of the object during preparation or as a result of bombardment in the electron beam. (In this description the adjective "lower" is used to denote the position of part of the object lying on a carbon-coated grid on the bench. When in the electron microscope the object is usually positioned on the side of the grid remote from the source of electrons. In normal parlance the terms "upper" and "lower" would then be reversed).

Klug and Finch have discussed these problems in terms of the location of negative stain in spherical virus particles and in some cylindrical structures. [18, 19, 32]. They conclude from stereoscopic studies and from interpretation of the optical diffractograms that differential contrast of the object often takes place. If the upper part is only weakly contrasted the structure of the lower part may be seen quite clearly. On the other hand, if a spherical virus particle is completely immersed in stain the electron microscope image is often confused and shows little structure. Under these conditions the structure in the image was found to be a function of the orientation of the object with respect to the electron beam. If we imagine a virus particle to be rotated about a given axis there will be (according to the geometry of the particle) a restricted number of azimuths for which the superposition effects will produce clear images with distinct intensity maxima. It is these images, rather than the less distinct ones produced from intermediate values of azimuth, that can be most clearly interpreted. Finch and Klug used a shadowgraph technique to help interpret such images [19]. Caspar devised a more realistic analog model technique for the same purpose [7]. A polythene model of the supposed virus structure was constructed and embedded in a matrix of plaster of Paris and X-radiographs of the analog in various orientations with respect to the X-ray beam were then obtained. Comparison of the radiographs with micrographs of the actual virus particles showed distinct similarities; as expected, the radiograph image was found to be very sensitive to the orientation of the object. By the use of this technique it proved possible to confirm the general organization of the subunits in the virus structure. However, the technique has three definite disadvantages. First, the construction of the model is very laborious; and second, it is not easy to adjust the orientation of the model to that of the X-ray beam with the necessary precision. Consequently the appreciation of this technique to the solution of an unknown structure would involve the construction of a range of different models, and this would in general be impracticable. A similar objection would arise if one wished to apply the technique to the investigation of the effect of subunit shape on the intensity distribution in the image. Third, there is the difficulty of matching the relative X-ray absorbing powers of the polythene subunits and the mineral matrix to those of the object and of the negative stain of the actual electron microscope specimen.

From this discussion it emerges that the problems of interpreting the electron microscope images of three-dimensional objects composed of regular subunits are very considerable. It can be concluded that:

1. Visual inspection of the electron microscope image alone is in general unsatisfactory. Only in those instances where one side of the object is completely immersed in stain and the remainder virtually unstained may one expect to obtain reliable clues by use of eye alone.

2. A confused or comparatively structureless image cannot in general be taken to indicate an absence of structure in the object.

3. Within the electron microscope itself the employment of a tilting stage can provide useful information about the structure of the object.

4. The use of X-radiographs of plastic models of the object embedded in plaster of Paris is also of value.

5. The electron micrograph may be used as a diffracting object and the spectra analyzed for evidence of periodicity and structure in general.

6. It should be noted that distortion of the object, e.g., by the artifacts of preparation, can but increase the task of detailed interpretation of the image, whatever method be used.

In addition to the various approaches mentioned above, two others not so far employed for the interpretation of electron microscope images of macromolecular objects, should be considered:

7. Low-angle electron diffraction of the object could also provide evidence analogous to that in (5) above. However, the technical difficulties of providing high-beam intensity over a selected and very limited small area are considerable.

Finally, the computer approach should be included here.

8. From a physical standpoint one possible approach is the direct evaluation of the intensity distribution in the image plane of various models of the object under investigation. Such studies are ideally suited to the use of a computer. Furthermore, it should prove possible to incorporate into the subsumed models a number of different stain distributions.

Since, however, as far as we are aware, the computer approach indicated in (8) has not previously been used for this purpose, it will be set out in some detail in the appropriate sections below.

It is hoped eventually to combine the use of the computer-calculated image with optical diffraction studies of the observed image, possibly analyzed in terms of optical density.

The limitations and scope of the determination of image parameters for two particular models by computer will also be discussed and the preliminary results presented. Finally, these results will be compared with the observed images, and, as far as possible, conclusions will be drawn.

## *The Models*

Two models were considered in some detail using the computer approach. They represent two distinct possibilities for the form of the subunits which comprise the tubule structure. They were chosen first as simple, but radically different, models on which to test and develop suitable programs for the study of the effect of various parameters on contrast in the image; second, to provide a starting point for the determination of the tubule structure.

*Model 1* (M1) consists of a cylindrical array of 12 columns of *spheres* hexagonally close-packed and arranged parallel to the cylinder axis (Fig. 15a).

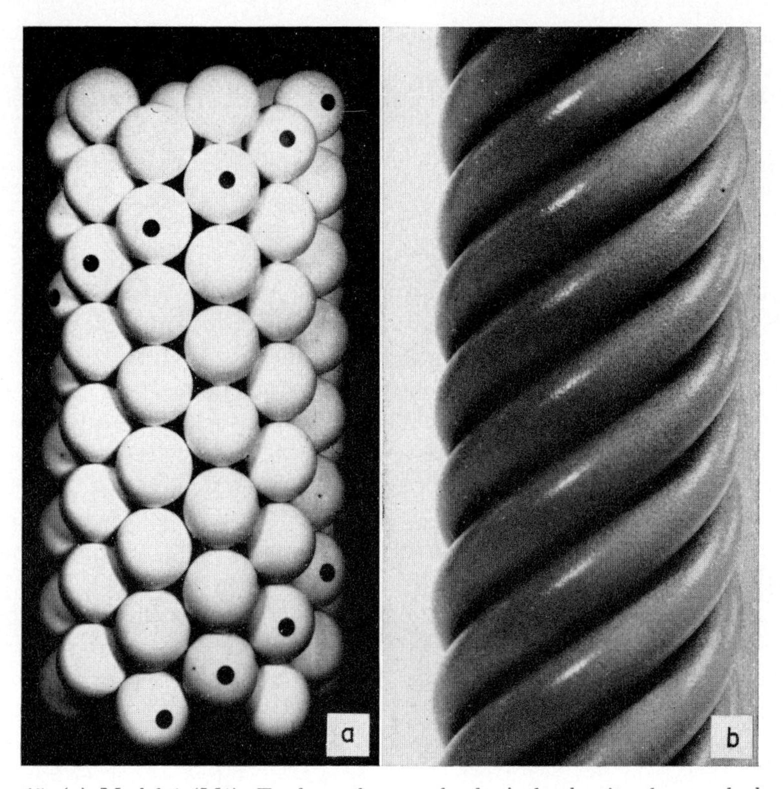


FIG. 15. (a) Model 1 (M1). Twelve columns of spherical subunits close-packed on the surface of a cylinder. (b) Model 2 (M2). Six helically wound filaments; the diameter of the helix and the cross section of the filaments (see text) are identical with the diameters of the cylinder and of the spheres, respectively, in M1.

*Model 2* (M2) (Fig. 15b) is in contrast composed of helically wound thick *filaments*. There are 6 of these continuous filaments per pitch of the helix, equally spaced along the $Z$-axis of the helix. The radius of each filament has been chosen to be equal to the radius of the spherical globules of M1.

In the section below on calculation of distribution we shall examine the variation of the normalized intensity distribution in the image of an M1 tubule as a function of the azimuthal angle of orientation with respect to the electron beam. The effects of compression on both models will also be studied. Six possible distributions of negative stain within and about the model tubules will be examined to determine the effect of each on the calculated intensity distributions. It must be emphasized that we shall be concerned only with changes in the distribution of intensity in the image and shall not attempt in this paper a direct quantitative comparison with measured values of the normalized intensity in the electron micrographs. (Normalized intensity $I_N$ is defined as the ratio of transmitted intensity, $I$, to that of the incident beam, $I_0$.

It will be shown later by the comparison of the computer results with the electron microscope image that a model composed of globular subunits (as in M1) is very much more probable than one composed of helically wound thick filaments (as in M2). It is proposed to describe here a more general set of models with globular subunits, of which M1 is a member and a special case. This set will provide a broader basis for subsequent discussion of the electron micrographs. A brief description of a preliminary study of superposition effects in a restricted subset of this set, using a graphical method, is given below (pages 163–164). Even this more general set of models cannot contain one that is correct in detail. The reasons for this will be given in the discussion. However, there is reason to suppose that one member of this set may be a good approximation to the actual structure, perhaps requiring only small periodic perturbations in its geometry. As we shall see, an approximate model correct, perhaps, in some general structural features, is all that can be expected at this stage, since neither the quality of the electron micrographs nor the method of measurement are sufficiently good.

It has been pointed out above that one of the most dominant features of tubule images is the high degree of longitudinal order; that is, the intensity maxima tend to lie in columns parallel to the tubule axis. This is a dominant feature of most electron micrograph images; often the columns of maxima (or apparent globules) are in register on either side of a region of confusion in the image. It is reasonable to assume, therefore, that the subunits are arranged, at least approximately, in columns parallel to the tubule axis. There may be some degree of artifact in this longitudinal linearity of the intensity pattern which could possibly arise as follows. Suppose that the more transverse bonds made by each subunit were selectively ruptured possibly by the strain or by distortion (e.g., a compression distortion) this could easily result in an enhancement of longitudinal arrangement of subunits and a tendency for them to "line up" parallel to the tubule axis. Since this linear order is consistently observed in electron micrographs it will be assumed that it represents a real feature of the underlying structure of the tubules.

The set of simplified or approximate models will now be described. They are based on the following assumptions: (1) all subunits are structurally identical; (2) each model is based on columns of lattice points; (3) each column is parallel to the axis of a cylinder and lies on the surface of the cylinder; (4) there is one subunit associated with each lattice point. Each model can now be described by four parameters: (a) $n$, the number of columns on the surface of the cylinder; (b) $R$, the radius of the cylinder; (c) $c$, the separation of each lattice point; within a column; (d) an integer $m$ which determines the relative translation ($\Delta Z$) of each column parallel to the axis of the cylinder as follows:

$$\Delta Z = mc/n$$

This formula follows from the fact that the $(n+1)$th column must correspond to the first column, i.e., $n$ successive translations must result in a total translation which is an integral multiple ($m$) of the spacing within each column ($c$). Therefore, each relative translation of adjacent columns must be $1/n$th of the total translation ($mc$). Clearly $m$ may have values $0, 1, \ldots n-1$. The significant values will be $0, 1 \ldots n/2$ (or $n/2 - 1$ if $n$ is odd) if the direction of the Z-axis is not defined. The choice of a particular set of parameters ($R, m, n, c$) will depend on structural properties of the subunits. If it is supposed that the subunits of the actual structure are approximately spherical, a special subset of these models may be even more useful. This is obtained by placing a sphere at each lattice point and assuming that it makes contact with nearest spheres in adjacent columns. The radius of the cylinder (or any of the other parameters) may then be calculated from the remaining three. M1 is a member of this subset with $n = 12$ and $m = 6$. Since the apparent globules are mostly spherical in appearance, one of the members of this subset may provide a good approximation to the structure.

### *Contrast in the Electron Microscope*

If there is an aperture-limiting diaphragm in front of the objective lens some electrons scattered by the object will be removed from the beam. The denser parts of the object will scatter electrons more strongly and will have more scattered electrons removed by the aperture diaphragm than do the less dense regions. Thus the image of the more strongly scattering regions will appear darker on the fluorescent screen than that of the more weakly scattering regions. The intensity of scattering from a given region will depend on the density of the atoms and on their individual scattering power. It is assumed that this is the main source of contrast in the electron microscope image. In other words, contributions to the contrast from spherical and chromatic aberration, and out-of-focus phase effects can be neglected if suitable operating conditions are chosen for the electron microscope [Hall, 25; Burge, 3; Silvester, 45]. These conditions will vary depending on the range of mass-thicknesses (see later) that are used. It is probably advisable to leave $\alpha$, the semi-angle subtended by the aperture diaphragm at the object, less than $6 \times 10^{-3}$ radians.

Considering this source of contrast only, it is now generally accepted that for a thin amorphous film, the effective transmitted intensity, $I$, of the electron beam is given by

$$I = I_0 \exp(-S\rho x) \tag{1}$$

Where $I_0$ is the incident intensity, $\rho$ is the density of the film, and $x$ the thickness. $S$ is called the mass-scattering coefficient and can be shown to be given by

$$S = N\sigma/A \tag{2}$$

when the film is thin enough for each electron to be scattered, on the average, once. $N$ is Avogadro's number, and $A$ is the atomic weight of atoms in the film. $\sigma$ is the total scattering cross section for scattering outside the angle $\alpha$. $\sigma$ can be written $\sigma = \sigma_i + \sigma_e$, i.e., the sum of the inelastic and elastic atomic scattering cross sections (similarly, $S = S_i + S_e$). Equation (1) has been verified experimentally by several workers [Hall, 25; Coupland, 12]: the product $\rho x$ is called the mass thickness and will be referred to as such in this paper.

There is strong theoretical evidence to suppose that $S$ is approximately independent of $Z$, the atomic number of the constituent atoms of the film [Lenz, 34; Burge and Smith, 5]. This has also been verified experimentally by many workers [Hall, 26; Burge and Silvester, 4; Reimer, 41]. There is also good experimental and theoretical evidence that Eq. (1) and the independence of $S$ from $Z$ (i.e., the linearity of log $(I/I_0)$ with the mass thickness) will hold for film thickness greater than the thickness in which electrons will be scattered on the average once [Hall and Inoué, 27; Cosslett, 10; Smith and Burge, 46]. It can be concluded that for the mass thickness anticipated (perhaps $\sim 10^{-5}$ gm

$cm^{-2}$), Eq. (1) and the approximate independence of $S$ from $Z$ will hold good [Cosslett, 11; Valentine and Horne, 51].

### *The Calculation of Normalized Intensity Distributions*

In general an object together with its stain will be compounded of a number of materials of varying density and thickness. The assessment of normalized intensity at a single point of the image plane therefore requires a detailed knowledge of $(\rho x)$ along the whole length of the path the beam takes through the object.

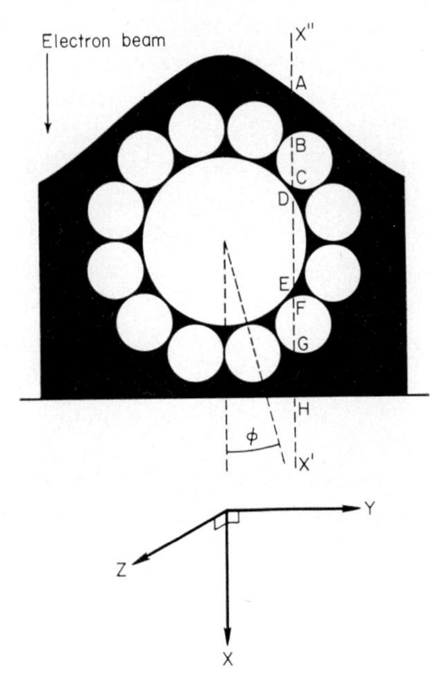


FIG. 16. A transverse cross section of a model structure (M1) embedded in negative stain. The axes $(X,Y,Z)$ are also those used in the calculation of the normalized intensity. $X'X''$ is a typical electron path through the object.

In Fig. 16 is illustrated in section a cylindrical (M1) tubule embedded in negative stain (the black region). There are several assumptions implied in this diagram.

First, it assumes a sharp boundary between the stain and the tubule and thus the absence of any positive staining. In practice this may not be so. We shall call this the assumption of *ideal negative staining.* Second, Fig. 16 implies that the structural components of the tubule can be represented by geometrically simple shapes, e.g., spheres. In fact, of course, the units are likely to be irregularly shaped protein complexes that only approximate to regular geometrical forms. However, it would clearly be meaningless to calculate transmitted intensities for points whose separation is less than the limit of resolution of the microscope, or less than the distance over which spherical aberration may be expected to cause fluctuations, since the irregularities in the dimensions and shape of the units would be expected to occur over distances of this order (i.e., ~5 Å). Hence they would not be detected due to the smoothing out effect of these aberrations and diffraction effects. In addition, we are more interested in changes in distribution of intensity to be expected from large differences in shape and arrangement of the units. The shape of the stain has also been drawn as geometrically smooth and symmetrical about a plane through the center, and is excluded from the central core. In practice one may expect variations in depth of the penetration and, perhaps, symmetry of the stain. Even more important is the case in which the stain only partially immerses the upper part of the object. This has not yet been investigated because of the difficulty of calculation. With these assumptions in mind we may now proceed to the detailed analysis of normalized intensity distribution in the image. This means evaluating Eq. (1) at a large number of points in the image plane so that intensity contours may be plotted for comparison with the electron micrographs. In practice the intensities have been determined at the points of a square lattice of constant ~6 Å.

The detailed nature of the calculation can be seen from Fig. 16. The effective values of mass thickness ($\rho x$) have to be determined along a large number of paths such as $X'X''$, and the resultant intensity ratio of Eq. (1) has to be determined. The calculations were carried out on an Elliott 803 computer.

We shall now set out in more detail the necessary equations.

In Fig. 16 consider electrons passing through the object along the path $X'X''$. From Eq. (1) the intensity transmitted at H in the plane $Y'Y''$ is given by:

$$I_H = I_0 \exp[-S\{\rho_o(BC + DE + FG) + \rho_s(AB + CD + EF + GH)\}] \quad (3)$$

where $\rho_o$ is the density of the unstained part of the structure, i.e., the protein of the tubule, and $\rho_s$ is the density of the negative stain.

It has been assumed for simplicity that the unstained core of the tubule has the same density as the material of the tubule. This assumption can be justified on the grounds that any actual difference in density will be small compared with the difference in density between the tubular structure and the stain. Equation (3) can be rewritten:

$$I_H = I_0 \exp[-S\{\rho_o(x_c + x_o) + \rho_s(x_s - x_c - x_o)\}] \quad (4)$$

where $x_c = DE =$ thickness of core; $x_o = (BC + FG) =$ thickness of the structure; $x_s = AH =$ the height of the stain envelope above $Y'Y''$. It has been assumed that both stained and unstained regions have the same value of $S$.

Rearrangement of Eq. (4) gives:

$$I_H/I_0 = \exp[-S\{\rho_s x_s - (\rho_s - \rho_o)x_c - (\rho_s - \rho_o)x_o\}] \quad (5)$$

The intensity transmitted by the supporting film will be given by the expression $I_T = I_H \exp(-Sw)$ where $w$ is the mass-thickness of the film. Since our interest is in relative values of intensity it is sufficient to calculate the values of $I_H/I_0$ already referred to as the normalized intensity and which will subsequently be denoted as $I_N$. Any fluctuation in the value of $w$ over the area of the object will influence the distribution of transmitted intensity; this possibility has not been taken into account in the present calculations.

For the purposes of this analysis it is assumed that $x_s$, the height of the stain envelope, is a function of $Y$, but not of $Z$. Similarly it is assumed that $x_o$ is also a function only of $Y$. For both models, $x_o$ will be a function of both $Y$ and $Z$. Thus Eq. (5) can be written:

$$I_N = \exp[-S\{\rho_s x_s(Y) - \rho' x_c(Y) - \rho' x_o(Y,Z)\}] \quad (6)$$

where

$$\rho' = \rho_s - \rho_o$$

From Eq. (6) it can be seen that the two-dimensional intensity distribution will be controlled by $x_o(Y,Z)$. Both $x_s(Y)$ and $x_o(Y)$ will affect the relative intensities in a transverse direction only. For model M1 (Fig. 16) $x_o(Y,Z)$ will also be a function of the azimuthal angle $\phi$; in fact it will repeat exactly every 60 degrees. In model M2, rotation about the axis of the tubule will produce only a translation of the distribution parallel to this axis. For M1 with a given value of $Y$, $x_o(Y,Z)$ will repeat exactly after $Z$ has changed by the diameter of a subunit. In M2 for a given value of $Y$, $x_o(Y,Z)$ will repeat exactly when $Z$ has changed by $P/n$, where $P$ is the pitch of the helix and $n$ is the number of helical filaments per pitch. Fortunately there are also planes of symmetry to reduce the ranges of $Y$ and $Z$ in which $x_o(Y,Z)$, and hence $I_N$, need be calculated for both models. It may also be noted that

$x_o(Y,Z)$ will change significantly if each model is compressed in a direction parallel to the electron beam. This is a realistic distortion of the models, since it may be expected that surface tension and other forces that come into play during the preparation of the specimen and the subsequent drying of the stain will distort the structure in this manner. We shall treat such distortions by studying the behavior of $x_o(Y,Z)$ for those cases in which the models have elliptical cross sections with the minor axis parallel to the $X$-axis. The calculation of the normalized intensity $I_N$ over the surface of the image may be carried out in the following stages:

1. Values of $x_o(Y,Z)$ are calculated on the computer for a given model in a particular orientation ($\phi$) at each point of a chosen orthogonal array in the $(Y,Z)$ plane. The grid was chosen to correspond to about the value of the resolution limit of the electron microscope.
2. The values of $x_o(Y_i,Z_i)$, together with the coordinates of the lattice points are fed into another program that will either plot out $x_o(Y_i,Z_i)$ on the teleprinter; or calculate the values of $I_N(Y_i,Z_i)$ using Eq. (6) and then plot these on a similar grid.

To calculate $I_N(Y_i,Z_i)$, it is necessary to assume a functional form for $x_s(Y)$. The general Gaussian curve

$$x_s(Y) = A + Be^{-cY^2} \tag{7}$$

was employed, in which the values of $A$, $B$ and $C$ can be chosen to give plausible shapes for the stain envelope. $x_c(Y)$ can be calculated from the cartesian equation of a circle or an ellipse. The values of the radius, or major and minor axes, can be varied. At this stage we must also assume values for $S$, $\rho_s$, and $\rho_o$. The value of $\rho_o$ was conveniently chosen to be 1 gm cc$^{-1}$ and that of $\rho_s$ as 5 gm cc$^{-1}$. This latter value lies well within the range of negative stain densities examined by Valentine and Horne [51] and is in fact probably slightly higher than that of most practical stains. $S$ was given the value of $3 \times 10^{-5}$ cm$^2$ gm$^{-1}$, which is somewhat larger than either the measured or predicted value. The use of accurate values of $S$, $\rho_o$, and $\rho_s$ is not necessary, since as has been said before, no quantitative comparison of the calculated normalized intensities has been made with experimental values. We are interested only in understanding changes in the distribution of intensity in the $(Y,Z)$ plane, that is, the image plane. In practice $S$ and $\rho_s$ were varied within reasonable ranges until an optimum variation in intensity was obtained.

The values of $I_N(Y_i,Z_i)$ were printed out on the same size grid as $x_o(Y_i,Z_i)$ on the teleprinter, as before.

It may be added that the values of $I_N$ were scaled up by a factor of a thousand in order to elimate decimal points on the teleprinter plot.

3. Contours of equal intensity were then drawn on the $I_N(Y_i,Z_i)$ grid. The spacing of the contours was chosen somewhat arbitrarily and rather smaller than the differences that one can realistically expect to detect by eye in the electron micrographs. Thus the computed images are rather overdetermined.

4. The contoured grid of normalized intensity (or $x_o(Y_i,Z_i)$ ) was photographed and a mirror image of the plot was obtained from the reversed negative. Thus the ideal calculated normalized intensity distribution could be constructed for a section of the tubule by combining several copies of the mirror image and the direct photographic print of the contoured grid.

Stage (i), the determination of $x_o(Y_i,Z_i)$, represents the most elaborate part of the calculation. For model M1 simple analytical expressions can be used to calculate the contributions to $x_o(Y_i,Z_i)$ from different spherical units. The difficulty lies in ensuring that all the contributions to $x_o$ are included systematically. This requires a water-tight search routine. For model M2 the computation of $x_o(Y_i,Z_i)$ is considerably more difficult because there is no analytical expression for the contribution to $x_o$ from each helical filament. Suppose each helical filament is defined as the volume swept out by a circular disk of radius $r_o$ whose center follows a helix of pitch $P$ and radius $R_o$; the plane of the disk remains fixed in the $(x,y)$ plane of a right-handed orthogonal set of axes whose origin coincides with the center of the disk and the $z$-axis is parallel to the tangent to the helix; the $x$-axis lies along the radius of the helix. Let $(X,Y,Z)$ be another set of orthogonal right-handed axes chosen so that $Z$ coincides with the helix axis and that the helix intercepts the $Z=0$ plane at $X=R_o, Y=0$. Consider the cross section of the cylindrical helix in a plane perpendicular to the Z-axis, say at $Z=Z_i$. Then it can be shown that the $(X,Y)$ coordinates of points in the cross-sectional area are given by the two parametric equations:

$$\begin{aligned} X_{Z_i}(r,\theta) &= R(\theta,r)\cos\frac{\omega}{v}\{Z_i - Z_o(r,\theta)\} \\ Y_{Z_i}(r,\theta) &= R(\theta,r)\sin\frac{\omega}{v}\{Z_i - Z_o(r,\theta)\} \end{aligned} \tag{8}$$

where $R(\theta,r)$ is given by

$$R(\theta,r) = [R_o^2 + r^2\cos^2\theta - 2R_o r\cos\theta + r^2\sin^2\theta\sin^2\beta]^{1/2} \tag{9}$$

and $Z_o(r,\theta)$ by

$$Z_o(r,\theta) = -r\sin\theta\cos\beta$$

$\beta$, $\omega$, $v$, $R_o$ are parameters of the helix which is generated from the equation $X = R_o \cos \omega t$, $Y = R_o \sin \omega t$, and $Z = vt$. $\beta$ is the angle of inclination of the tangent to the helix to the $Z = Z_i$ plane ($\beta = \tan^{-1}(v/R_o\omega)$). The other, parametric, variables, $(r,\theta)$ are the polar coordinates of the points in the circular disk with respect to the $(x,y,z)$ axes, i.e., in the $(x,y)$ plane.

If in Eq. (8), $r$ varies continuously in the range 0 to $r_o$ and $\theta$ varies continuously from 0 to $2\pi$, then all points in the cross section will be generated. In order to calculate the contribution to $x_o$ we are interested only in the intersection of the surface of each cylindrical helix with the plane $Z = Z_i$. This locus is obtained from Eq. (8) by putting $r = r_o$ and letting $\theta$ vary in the range 0 to $2\pi$. Of course as $Z_i$ varies, this cross section rotates about the $Z$-axis. Thus for model M2 any plane $Z = Z_i$ will contain six cross sections equally spaced around the circumference of a circle. Again, as $Z$ varies, the circular array of cross sections rotates about the $Z$-axis of the helix. Fig. 17 shows a section through model M2 at the plane $Z = Z_i$. The computational difficulty arises from the fact that, from Eq. (8) we cannot obtain $X = X(Y)$; that is, we cannot eliminate $\theta$. This means there is no analytical expression for contributions to $x_o(Y_i,Z_i)$; for instance $(X_2 - X_1)$ in Fig. 17. A numerical method was devised for solving this problem.

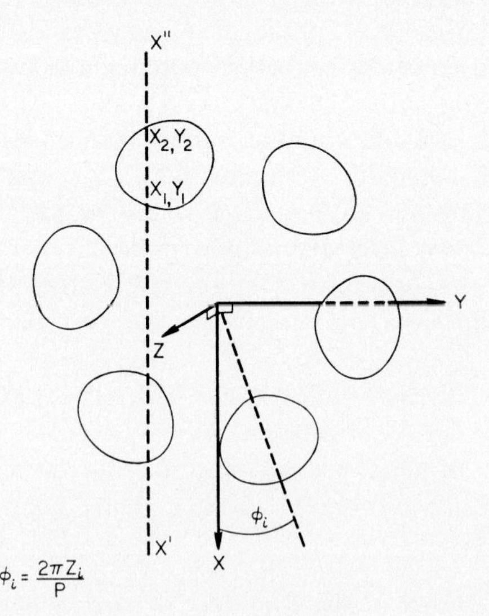


FIG. 17. A transverse cross section through M2 at the plane $Z = Z_i$. Each of the six filament cross sections is obtained by calculating a large number of points on the periphery using Eq. (8). The axes used $(X,Y,Z)$ are the same as those employed in the calculation of $x_o$.

The method is to calculate a large number of points (characterized by different values of $\theta$) on each cross-section circumference and find the nearest values $Y_1(\theta_1)$, $Y_2(\theta_2)$, to a given $Y$, on the cross-section circumference, by an arithmetic comparison search routine. Knowing $\theta_1$ and $\theta_2$ we can obtain $X_1$ and $X_2$ from Eq. (8). Again, careful programming was necessary to include all contributions to $x_o(Y_i,Z_i)$.

The outstanding advantage of this computational approach is flexibility. Once the programs for calculating $x_o(Y_i,Z_i)$ have been written, the dimensions and parameters of each model can be varied at will. Since each set of values, $x_o(Y_i,Z_i)$, is stored on computer tape, the effect of different staining situations can be studied easily and rapidly at stage (2). For instance, the calculation of a set $x_o(Y_i,Z_i)$, may take about 15 minutes on the Elliott 803 computer. The calculation and plotting of $I_N(Y_i,Z_i)$ for a given stain distribution can be obtained in about 5 minutes. The contouring of the grid points and reconstruction of the image takes longer, but, with experience, one can obtain a good idea of the changes in the intensity distribution from the bare, uncontoured grid. Thus, changes due to staining, orientation of the model, and compression distortion can be studied rapidly and easily in different models. The disadvantages of this approach lie in its assumptions, which have already been outlined. It is obviously limited to the study of models that provide some toehold for geometrical analysis. It assumes a regular structure which, as has been pointed out already, may not exist. It is hoped to extend these calculations to cover the case of a more general shape for the subunit; e.g., a spheroid or an ellipsoid. It may be possible to allow for cases in which stain does not completely immerse the upper part of the object, i.e., only fills the grooves in the upper surface. This is more difficult since the shape of stain will be determined by the local geometry of the structure. However, in order to develop the computer program the simpler assumptions already described above were employed.

### *Results of Computations*

The two-dimensional intensity distribution in the image plane is obviously dependent on the values of $x_o(Y,Z)$. It is, therefore, helpful to consider these distributions first. In Fig. 18 are plotted values of $x_o$ for M2 with azimuthal angles 0° and 15°, and with an uncompressed cross section of tubule. The rotation of the model tubule about its long axis through 15° has brought about a striking redistribution of $x_o$. From a hexagonal array (with $\psi \simeq 30°$) of five longitudinal columns of peaks (intensity maxima), with distortion of the outer columns due

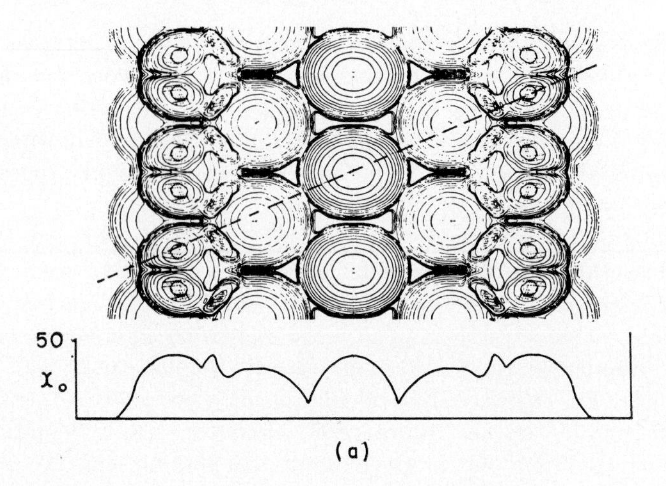


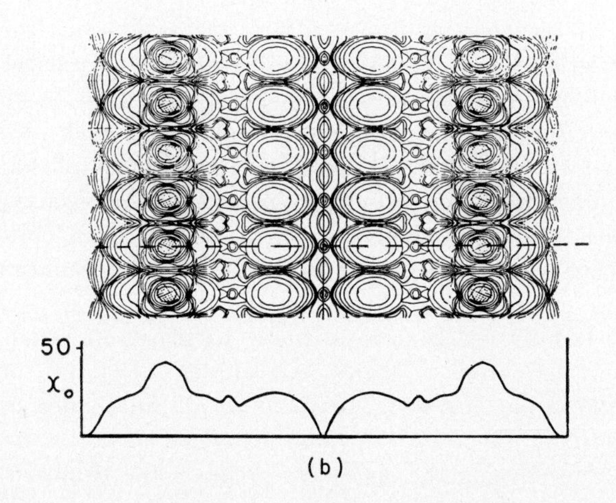


FIG. 18. Plots of the $x_o$ distribution derived from M1 for different values of $\phi$. The trace under each plot shows the variation of $x_o$ along the broken line drawn transversely across the distributions. (a) $\phi = 0°$, (b) $\phi = 15°$.

to edge effects, the pattern changes to a rectangular array of maxima in four columns. The $x_o$ maxima in the two outer columns are more complex in shape and significantly greater than those of the central two columns. The longitudinal extension of each peak has been reduced by half, the horizontal extension by less than a half. Figure 19 displays values of $x_o(Y,Z)$ for a compressed version of M1, with elliptical cross section on major/minor axis ratio of approximately 2 and with the same two orientations. The most striking effect is shown in the 15° plot (Fig. 19b), where each of the outer columns has separated into two columns, with the same peak height as the inner pair. However, we can see from the plot of $x_o$ along the line through the peaks that the trough between these two new columns is not nearly as deep as those on either side of the two central columns. This behavior of $x_o$ for M1 under rotation and compression is one of the most significant results of the computational analysis. A peak or intensity maximum now no longer corresponds to the center of a spherical subunit but to the center of a region of overlap between upper and lower subunits. It will be discussed further with reference to the calculated intensity distributions. We may note that the plot of $x_o$ for $\phi = 30°$ would produce the same distribution as $\phi = 0°$, but translated parallel to the $Z$-axis by a distance equal to the radius of the spherical subunit. Figure

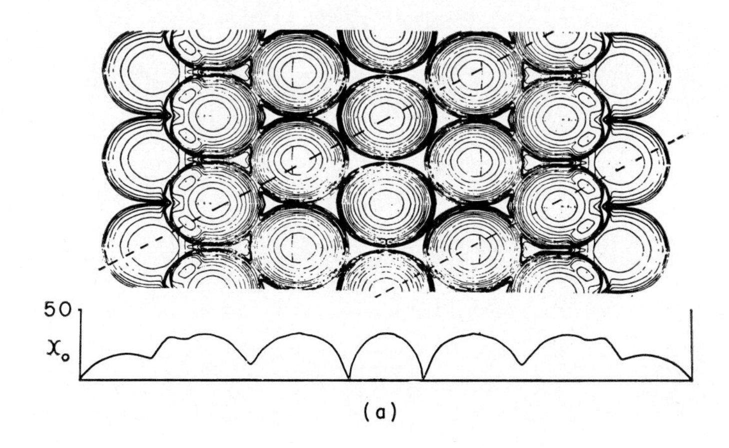


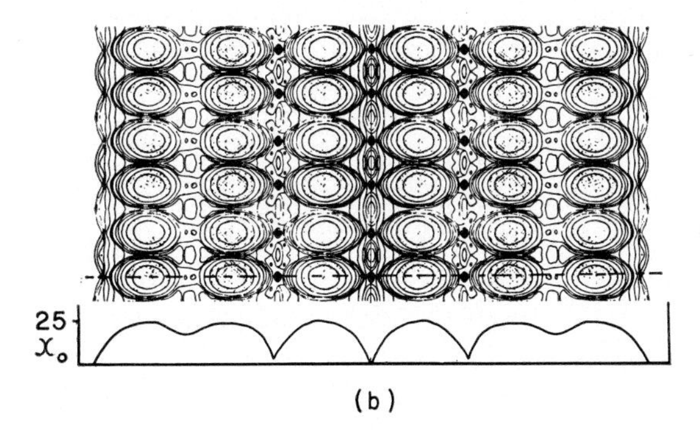


FIG. 19. The distribution of $x_o$ for M1 compressed in a direction parallel to the electron beam, i.e., perpendicular to the plane of the paper. The values of $\phi$ are those of the model structure before compression. (a) $\phi = 0°$, (b) $\phi = 15°$.

20 shows the $x_o$ plots for model M2 which result in a hexagonal array of peaks, the size and shape of the peaks differing from column to column. For this model the $x_o$ contour peaks are not circular in shape but asymmetrically distorted.

The analysis now proceeds to computed normalized intensity plots for different distributions of stain. Figure 21, a–f illustrates different stain envelopes and distributions, labeled as S1, S2, S3, S4, S5, and S6. In Fig. 21, a, c, and e correspond to distributions S1, S3, S5, in which the stain has penetrated the core completely; and in Fig. 21, b, d,

FIG. 20. The distribution of $x_o$ for M2, uncompressed.

and f correspond to those (S2, S4, and S6) in which the stain just surrounds each subunit but is excluded from the core. Examples S3, S4, S5, and S6 have Gaussian-shaped envelopes with S5 and S6 denoting sharper Gaussian distributions than S3 and S4. The Gaussian curves were chosen to have a maximum that approximately coincided with the height of the plane in examples S1, S2 (Fig. 21 a and d). In S3 and S4 the parameters of the Gaussian distribution were chosen so that the curve fell to two-thirds of its maximum at a value of $Y$ just outside the maximum extent of the structure, and at $Y = \infty$ had fallen to half the maximum. The corresponding factors for S5 and S6 are one-half and one-third. In none of the examples considered does the stain form a layer beneath the organelle structure. It will also be noted that the stain envelopes do not extend much above the height of the structures. The tubule projections shown in Fig. 21 a–f correspond to model M1 with an uncompressed cross section. The same stain distributions have been used for calculations on M2. For compressed cross sections the height and shape of the Gaussian envelopes have been altered in order just to cover the flattened structures, in a fashion similar to those of Fig. 21 (see Fig. 22). These will be referred to as S′1, S′2, . . . S′6, using the same factors for the Gaussians as above.

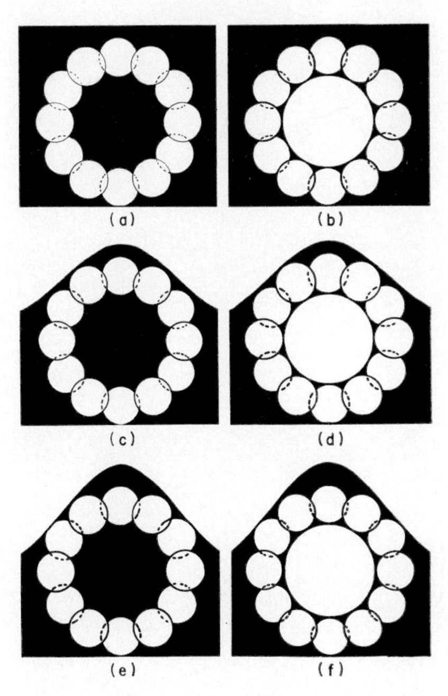


FIG. 21. Various stain distributions used in the calculation of the $I_N$ distribution. The structure embedded in the stain is that of M1 projected onto a transverse plane. The different stain distributions are labeled as follows: (a) S1, (b) S2, (c) S3, (d) S4, (e) S5, (f) S6.

In order to show the effect of these different conditions the intensity plots for M1, uncompressed at $\phi = 15°$ will now be examined. The six appropriate plots are shown in Fig. 23, a–f. For S1 (Fig. 23a) although the same two-dimensional arrangement of intensity follows the plot of $x_o$ (see Fig. 18b) the intensity variation across the peaks is extremely shallow, the peaks themselves being small. For S2 (Fig. 23b) the absence of stain in the core produces an increased intensity for the two central columns of peaks, the outer ones having the same height as for S1. Presumably this would give the appearance of two intense

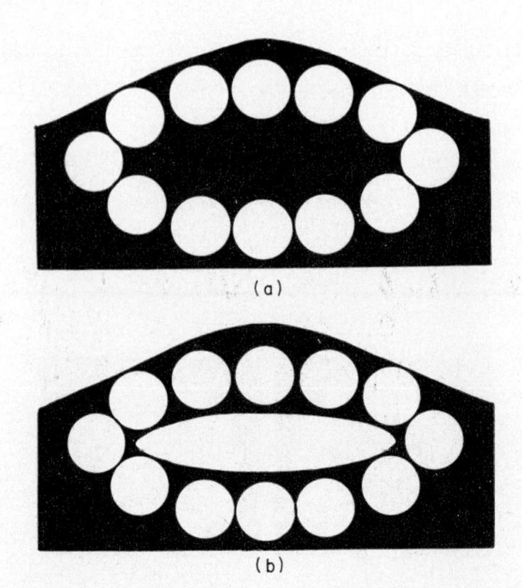


FIG. 22. Stain distributions for a compressed tubule labeled as follows: (a) S′3, (b) S′4.

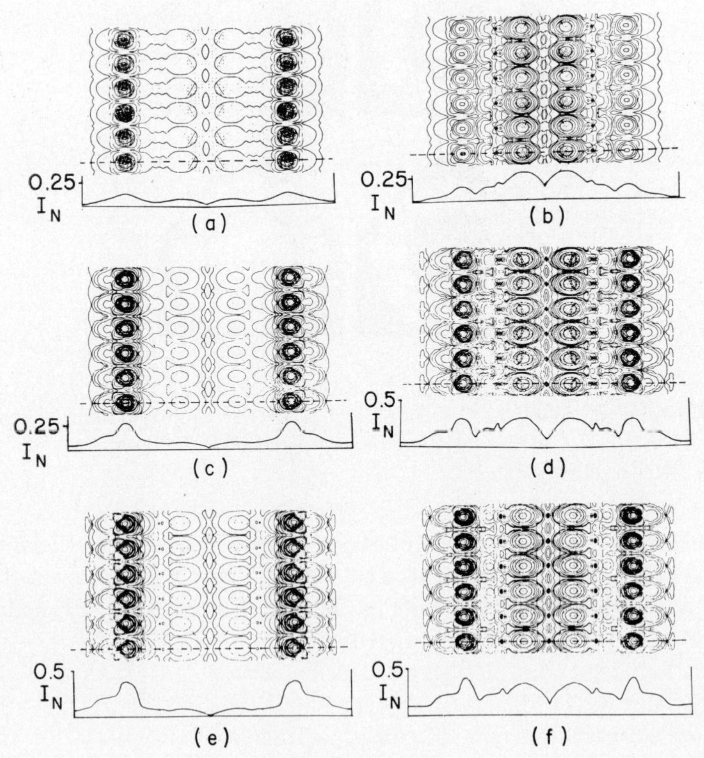


FIG. 23. Six plots of the $I_N$ distribution for M1 at $\phi = 15°$ for the six different stain distributions shown in Fig. 21. (a) S1, (b) S2, (c) S3, (d) S4, (e) S5, (f) S6.

center columns of peaks with two faint outer pairs, which, if they occurred in an actual image, might not be visible. The intensity distribution for stain S3 is shown in Fig. 23c, where the outer pair of columns have a much increased intensity and the inner pair have almost disappeared. The image thus appears as only two columns of sharp peaks separated by a wide region of approximately uniform intensity. It is also to be noted that the intensity falls by only half along a line drawn through the peaks parallel to the columns. It may be expected that when the linear spread of the peaks is small, say ~20 Å, the columns may sometimes, especially in the presence of spherical aberration, appear as continuous lines. (This possibility may be relevant to the two observed forms of bacterial flagella structure.) For S4, the inner two columns of peaks are enhanced due to the absence of stain in the core. We note that these columns have peaks much broader laterally than the outer pair, which have considerably sharper peaks. For stain distribution S5 and S6 (Fig. 23 e and f) are very similar to those of S3 and S4 (Fig. 23, c and d), respectively, the only difference being that in both S5 and S6 the outer peaks are even more intense and sharp. The nature of the calculated images under these different stain conditions can, of course, be understood qualitatively quite simply. When stain is absent from the core (Fig. 23, b, d, and f) there is an obvious increase in the transmitted intensity in the central region of the image; if the stain thickness falls toward the edge of the object, then the intensity will be enhanced toward the edge.

Figure 24 illustrates two plots derived from M1 for $\phi = 0°$ and no compression. Figure 24a (identical with Fig. 18a) shows a plot of $x_o$, and Fig. 24b gives the distribution of intensity for S1. These two diagrams illustrate how, under certain staining conditions, the peaks and troughs in the $x_o$ plot can become smoothed out in the intensity distribution. As can be seen from Fig. 24b the variations in intensity are very small. Figure 25 illustrates three normalized intensity plots all derived from model M1 compressed (as shown in Fig. 22). Figure 25a is for $\phi = 0°$ and stain distribution S′1; Fig. 25b gives $I_N$ for $\phi = 0°$ with stain distribution S′6; and Fig. 25c gives $I_N$ for $\phi = 15°$ and stain S′1. In Fig. 25 a and c there is much greater variation of $I_N$ than for Fig. 24b. This is because the overall thickness of the specimen and stain is less, due to the compression (see Fig. 22). These three plots again show the striking differences that can be produced in the intensity distribution in the image by simply rotating M1 through an angle of 15°. The shape and size of the intensity maxima alter completely.

Figure 26 shows two calculated images derived from M2, uncom-

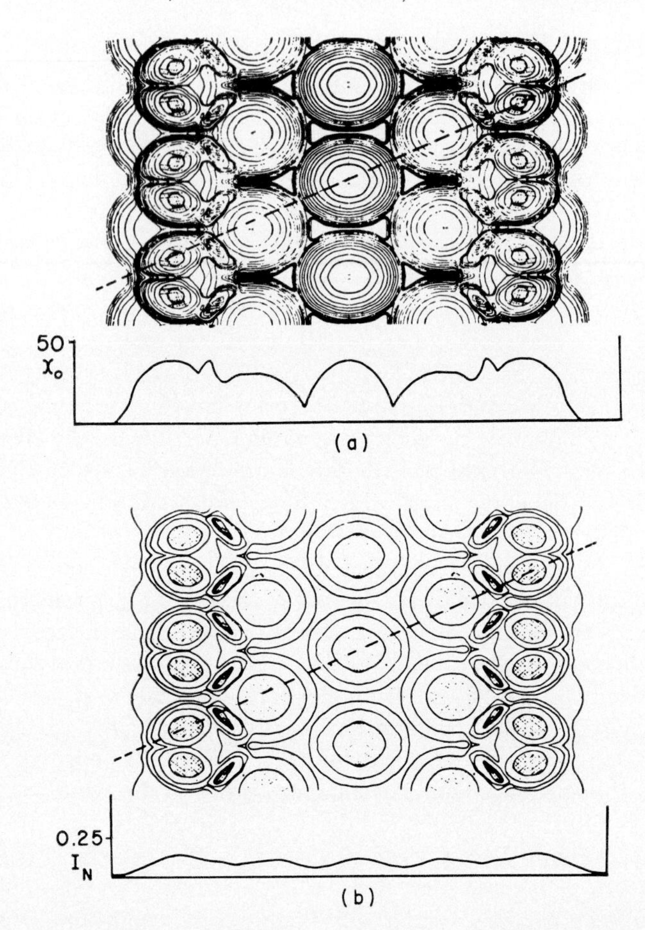


FIG. 24. Distributions obtained from M1 at $\phi = 0°$. (a) $x_o$, (b) $I_N$ for stain distribution S1.

pressed. Figure 26a gives $I_N$ for S1, and Fig. 26b, $I_N$ for S6. For S1 there is again little variation in $I_N$. In Fig. 26b there are considerable variations in intensity, together with an appearance of an approximately hexagonal array of intensity maxima. The peaks of each outer pair of columns are roughly triangular in shape with well-rounded corners. The central columns have roughly elliptically shaped peaks. In each column there are the same number of peaks per pitch of helix as there are component helices in the model. The characteristic size of the peaks is only a little smaller than the peaks given by model M1 for $\phi = 0°$. The peaks in the three central columns are about 3/2 times as intense as the outer pair. Another characteristic feature of Fig. 26b is the appearance of bands of approximately uniform intensity which diagonally cross the central region of three columns.

Figure 27 shows a plot from model M2 with compressed cross section. (It was necessary to make a further approximation in the calculation of $x_o$ for these models under compression. It will not significantly affect the distribution of $x_o$). Figure 27 plots $I_N$ for M2, S′1. The effect of compression is to produce a gradation in overall dimensions of the

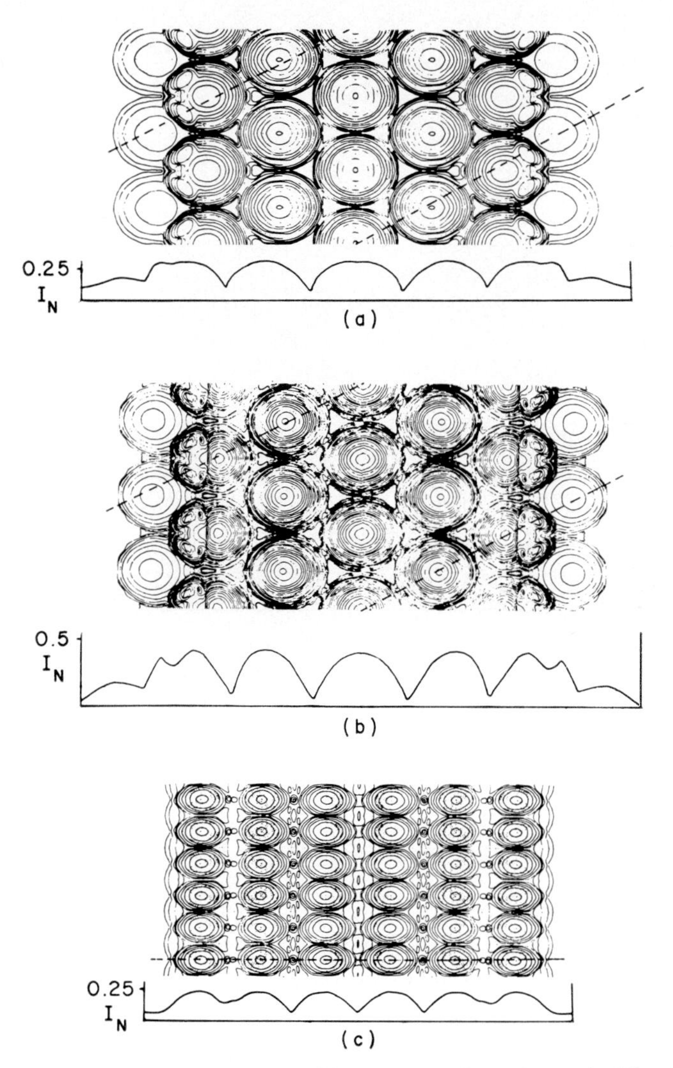


FIG. 25. Three plots obtained from M1 compressed as shown in Fig. 22. (a) $I_N$ for $\phi = 0°$ and S′1. (b) $I_N$ for $\phi = 0°$ and S′6. (c) $I_N$ for $\phi = 15°$ and S′1.

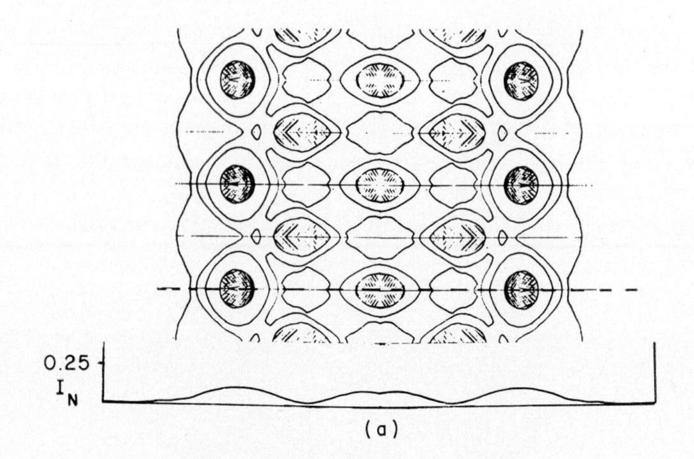


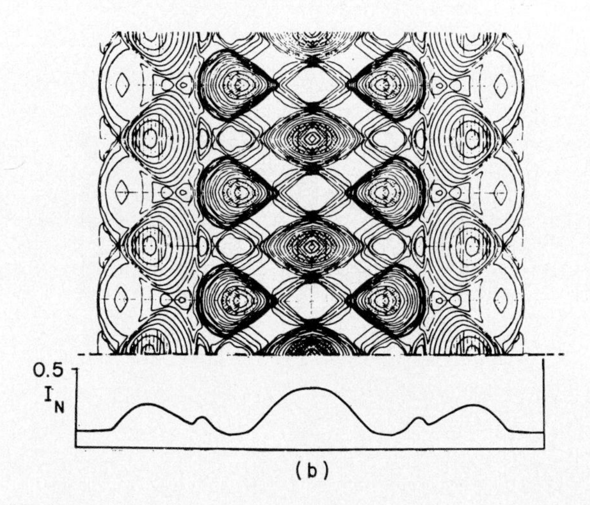


FIG. 26. Two plots of $I_N$ derived from M2, uncompressed. (a) S1, (b) S6.

peaks, decreasing from the main outer column to the center. The central column intensity maxima are particularly small. The bands of uniform intensity have widened considerably and a new, peripheral column has appeared with a somewhat lower intensity than its neighbor the main, most intense, column of maxima.

The main results of the computer calculations can now be summarized:

1. A fundamental change in the pattern of maxima in the distributions of $x_o$ and $I_N$ for M1 can be brought about by a rotation from

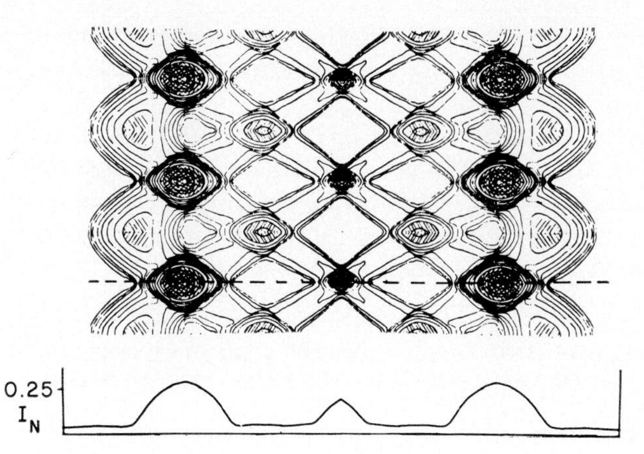


FIG. 27. An intensity plot derived from M2, compressed, with the stain distribution S′1.

$\phi = 0°$ to $\phi = 15°$. From a hexagonal arrangement ($\psi \simeq 30°$) of mostly circularly shaped maxima, the pattern changes to a rectangular arrangement ($\psi = 0°$) with the maxima extended in a transverse direction. The axial separation for $\phi = 0°$ is equal to the diameter of the subunit sphere; for $\phi = 15°$, this separation is equal to the radius of the sphere. For intermediate values of $\phi$ (say 7.5°) the intensity distribution is confused.

2. Very significant redistribution of intensity can be achieved by altering the shape and location of the stain.

3. Model M2 gives a hexagonal ($\psi \simeq 30°$) array of intensity maxima, the central region of the image being crossed diagonally with bands of slowly varying intensity.

### *Description of Preliminary Study of the Column Models*

Superposition effects were studied in some members of the special subset. The value of $n$ was kept the same as in model 1, i.e., 12. This appears, at present, as will be shown in the discussion, to be the most probable value. The values of $m$ studied were 1, 2, 3, 4, and 5. The graphical method used was to project each lattice of points on to the plane containing the cylinder axis, for a particular value of $\phi$. Circles were then drawn around each projected point. It was possible to deduce the pattern of intensity, in central region of each of these models, from the overlap of the circles. The superposition effects near the edges were more complex and will require direct computer calculations to obtain the correct intensity distribution. Two values of $\phi$ were studied, viz. 0° and 15°, as for M1. The most interesting result of this study was that all the models, at both orientations gave either rectangular or hexagonal patterns of intensity in the central region of the image, the hexagonal patterns having $\psi$ values in the range 30°–60°. Even small values of $m$ did *not* give small values of $\psi$, say 10°–20°). This again confirms the conclusion, from the direct computer calculations, that the pattern of intensity in fully contrasted images may be misleading as to the underlying structure of the tubules.

## ANALYSIS AND DISCUSSION OF EXPERIMENTAL RESULTS IN TERMS OF THEORETICAL PREDICTIONS

Before this analysis is carried out it is important to recall that there are two main factors which limit the application of the computer results to electron micrographs of the tubule structures in the ciliary axoneme. The first relates to the preparative procedures and the consequent state of the tubule; and the second factor arises from the necessarily restricted nature of the calculations. It is in fact doubtful whether more extended calculations are warranted until much-improved techniques of specimen preparation are available. The objects of this analysis are to gain some insight into the state of the tubules after preparation for study in the electron microscope and to see in what ways the structures of the tubules differ from those of either M1 or M2. Both of the limiting factors that have been mentioned will necessarily prevent any firm and final conclusions being made at present about the structure of the tubules. The ultimate determination of these structures will almost certainly require the application of other methods such as those of X-ray diffraction, electron diffraction, and optical diffraction of electron micrographs mentioned earlier.

From the previous descriptions of the general features of the electron micrographs and of the computer results it seems more probable, for the following reasons, that the tubules are composed of globular subunits than of filaments:

1. The computer results show that a model composed of globular subunits can give rise to a variety of images according to the orientation of the structure with respect to the electron beam. This is in keeping with the electron microscope results.

2. The apparent globules which occur in the calculated images of a filament model (M2) vary in shape according to the position of the globule in the center or periphery of the image (Fig. 26). No such variation is found in the observed images.

3. Calculated images of the filament model M2 have diagonal bands

of approximately uniform intensity crossing the central region. These are not found in the electron micrographs.

4. If the filament model M2 is compressed, the apparent globules in the calculated image vary in size according to their position; those in the center are smaller than those at the edge. Such variations in size do not appear in electron micrographs of tubules that are obviously compressed.

Since it has not proved possible to find even small areas of agreement between the observed images and those calculated from the filament model M2, it is concluded that the structure is composed of globular subunits rather than helically wound filaments.

There are other reasons why models such as M1 can at present lead only to a broad rather than a precise interpretation of tubule structure. For example, it was subsumed in the section concerning the computer approach that the inner (CP) and outer pairs (OP) of tubules are identical for the purposes of calculation. It has already been indicated in the section on results of electron microscope observations that there are (in *Tetrahymena*) small differences in diameter between the A and B components of an outer pair and between these and the center pair when observed in stained material in transverse section. Since there are difficulties in defining the precise diameters of the stained material, let alone those of the tubules themselves, these differences have not been stressed. Even if neglected, however, there are other factors.

The structure of a particular tubule will depend completely on the shape and size of its subunits and on the number, position, and direction of the external bonds that can be formed. This leads us to ask how many structurally different subunits does each tubule have; and further, are the subunits of the central pair different from those of the outer pair? The more structurally different subunits a given tubule has the more complex its structure is likely to be. Consider the subunits in an outer pair: some form the region of conjunction between A and B (see Fig. 12); some are involved in the binding of the chelae; and some in the attachment of the radial subfibrils. Finally, there is the majority of subunits forming the bulk of the structure and not implicated in any of these specialized features. Thus, from these rather *a priori* arguments, there are four different kinds of subunit in any outer pair of tubules. The distinctions between these four kinds cannot be merely structurally irrelevant chemical differences, since the appropriate subunits must be incorporated at "correct" positions in the tubule. Inferences of a similar nature about the central pair are less certain, but the overall difference in diameter between CP and OP, the existence of a helically wound filament (so called sheath) surrounding CP, and other possible attachments suggested from observations of longitudinal sections, indicate that the central tubules probably also contain more than one type of subunit.

With these differences between the inner and outer tubules in mind, it can now be seen that any of the set of column models described would not be more than an approximation to reality, since in these models each subunit is treated as identical.

### *Comparison of Observed and Calculated Images*

There are three important features of the images on which to base a comparison: (1) the number of columns of intensity maxima; (2) the spacing of intensity maxima within a column; (3) the existence of two-dimensional order or pattern in the images.

The occurrence of 4, 5, and 6 columns of maxima in the observed electron microscope images tends to confirm the choice of the value of parameter $n$ (the number of columns in a tubule) as 12, since it is this parameter that will be the major determinant of the number of columns observed in the micrographs. Figures 28, 29, and 30 each show a calculated image together with a limited selection of electron microscope images. Figure 28 compares observed and calculated 6-column images; Figs. 29 and 30 compare 5- and 4-column images, respectively. The predominant occurrence of 6-column images in the micrographs suggests two possible explanations, the first of which is that the images arise from only partially contrasted objects. The lower side next to the specimen grid is fully contrasted and is thus the portion seen in the image. The second explanation is based on superposition effects in the image of a fully contrasted object, similar to the examples shown for M1 in Fig. 28a (compressed structure, $\phi = 15°$). Each of these explanations supports a value of $n = 12$. However, if the second explanation is correct and one of the special set of column models (with $n = 12$) is a good fit, then regions of rectangular or hexagonal order (with $\psi \lesssim 30°$) would be expected to appear. Detailed examination of the tubules shown in Fig. 28, b and c, shows that this is very rarely the case. The values of $\psi$ lie predominantly in the range of 15°–20° (see regions *1* and *2*, Fig. 28b). Occasionally, but very rarely, small regions of a rectangular pattern of maxima may be observed (Fig. 28c, region *1*). Consequently it must be concluded that either the first explanation is correct (partially contrasted image); or if the second explanation (full contrast and superposition effects) is correct then the tubules must have structures very different from those of the special set of column

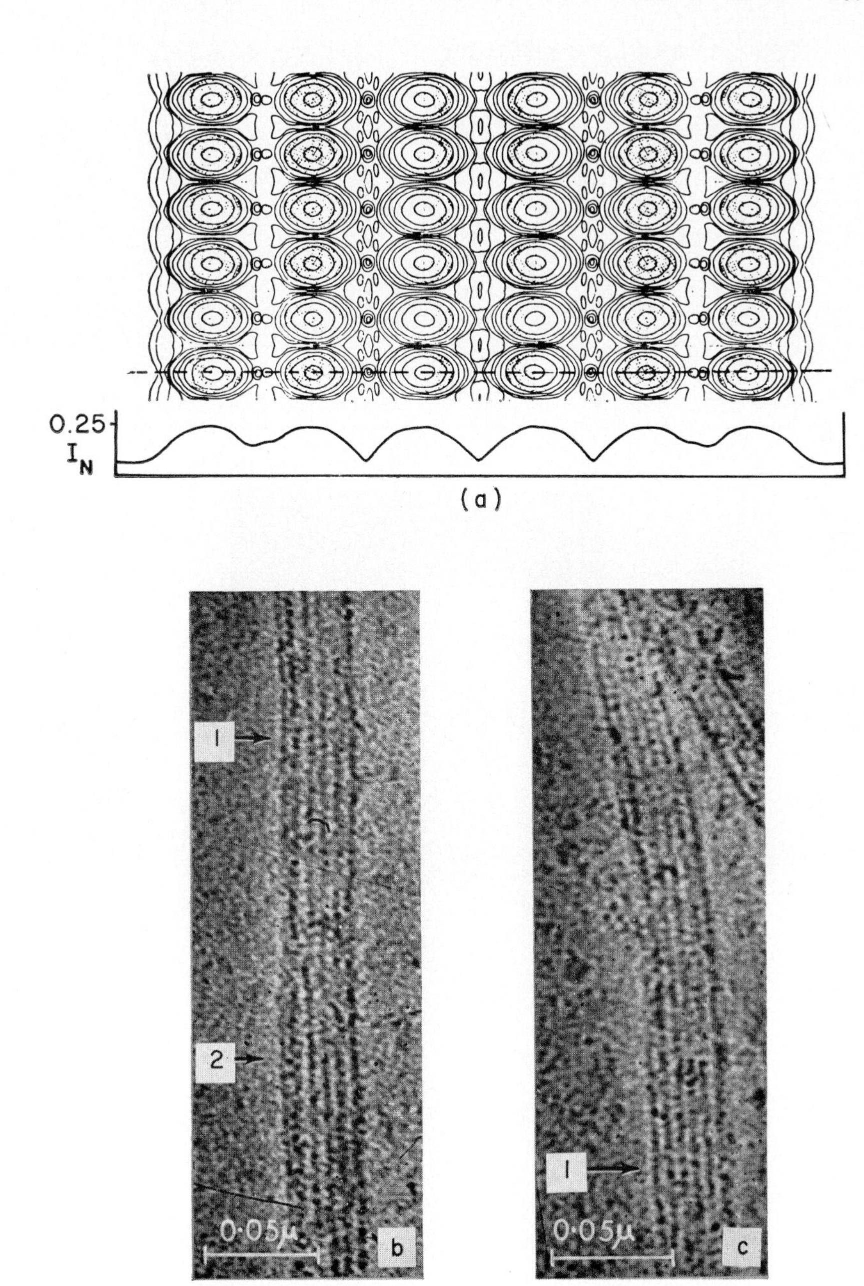


FIG. 28. (a) $I_N$ distribution derived from M1 at $\phi = 15°$ with compressed cross section and stain distribution S′1. (b) and (c) are examples of 6-column images (probably center pair tubules). Electron micrographs. × 330,000.

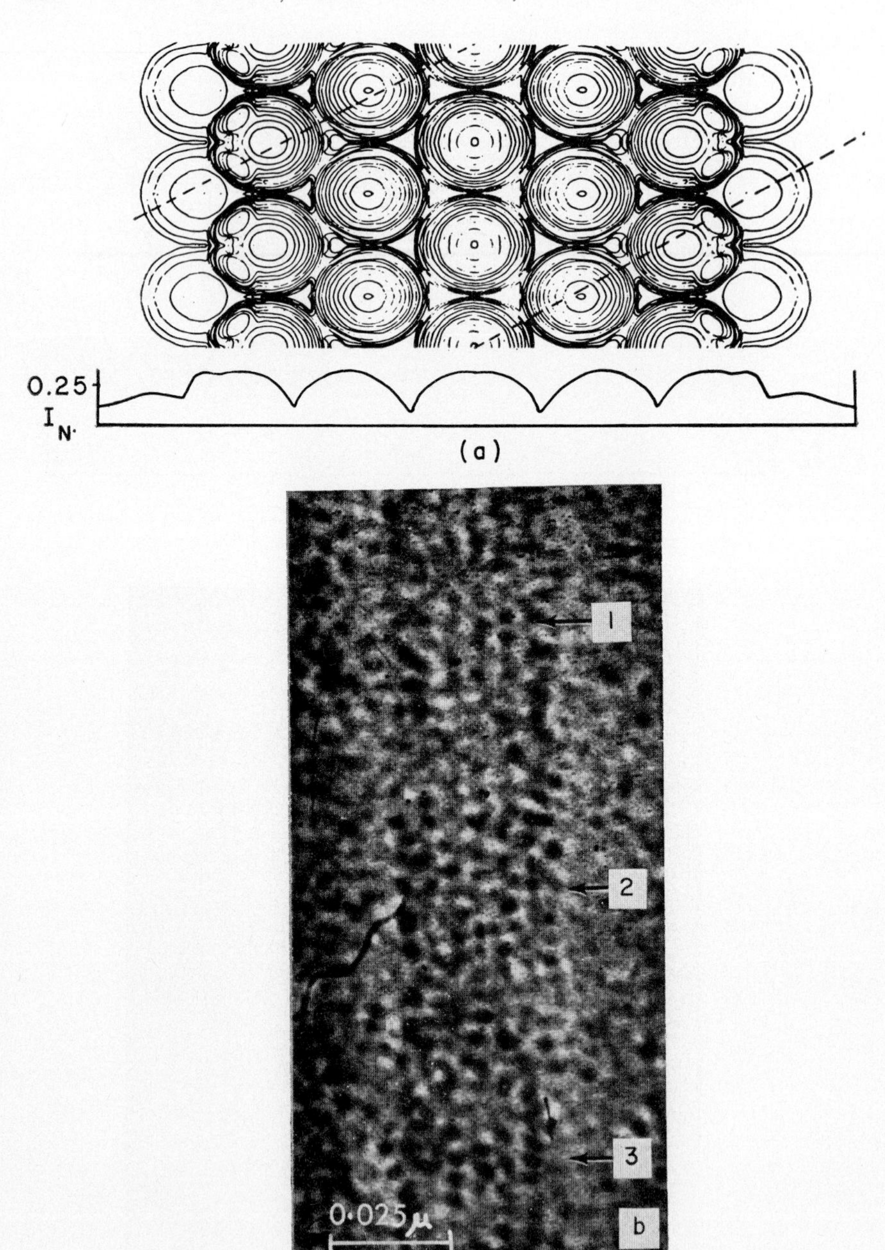


FIG. 29. (a) $I_N$ for M1 at $\phi = 0°$ with compressed cross section and stain distributions S′1. (b) An example of a 5-column image of a tubule somewhat distorted. Electron micrograph. × 650,000.

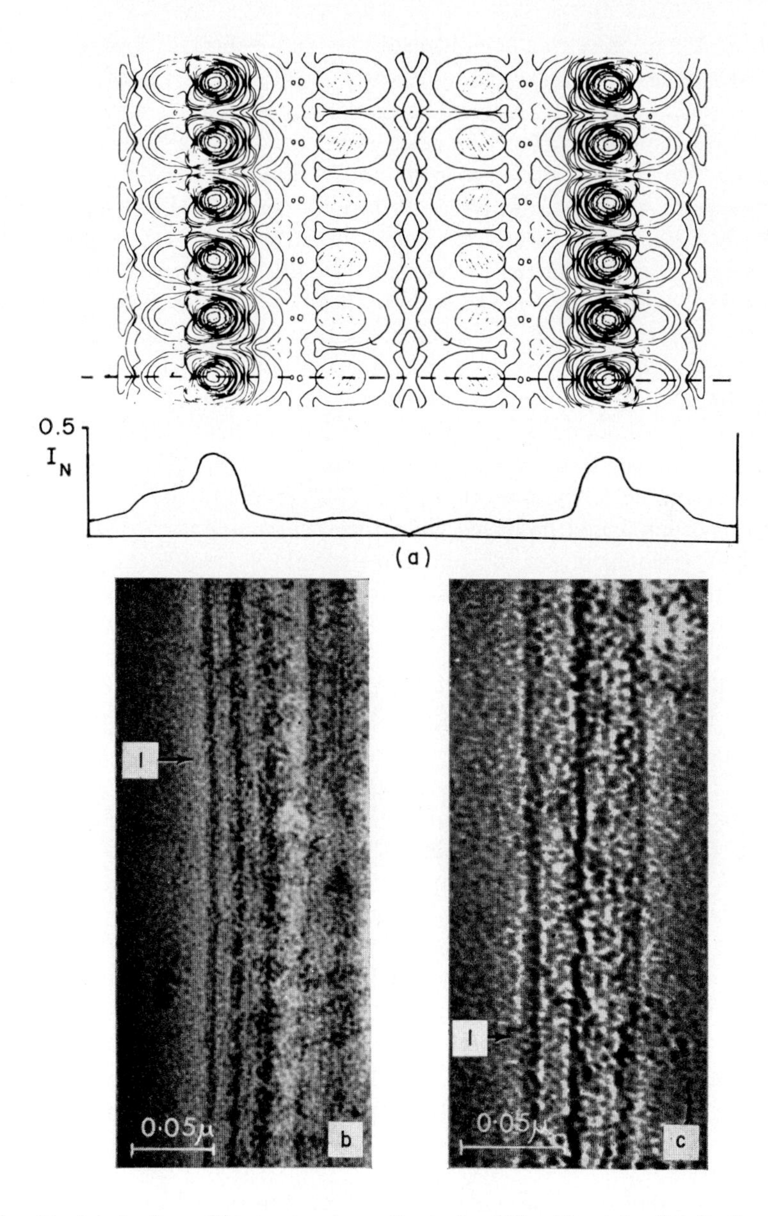


FIG. 30. (a) $I_N$ from M1 (uncompressed) at $\phi = 15°$ with stain distribution S5. (b) and (c) Examples of 4-column images from outer pairs of tubules. The right-hand tubule of each pair in both figures, does not show a 4-column image, whereas both the left-hand tubules have regions containing four columns (see region 1 in b and c). Electron micrographs. b, × 277,000; c, × 350,000.

models. If, in fact, the condition of partial contrast does exist, then the low values of $\psi$ observed (15°–20°) suggest in terms of the column models an $m$ value of 3 or 4. The possibility of partial contrast is supported by the observation of a wide range of spacings between intensity maxima, which would be expected under these conditions to have two predominant values. Additional variations could be brought about by perturbations of the structure during preparation. There is no particular reason why the degree of contrast should not vary within one tubule. While the partial contrast hypothesis seems the most likely basis for the explanation of the results, higher contrast and superposition effects may be necessary to account for the smaller spacings of 25–30 Å often found near the edge of the image.

A similar explanation may be required to account for the features of 5-column images that may be seen in parts of Fig. 29b (regions *1* and *3*), representing a compressed tubule. The 5-column images tend to be more disordered and confused than the 6-column images with both low $\psi$ values (regions *2* and *3*) and high $\psi$ values (region *1*). This image shows no correlation with that calculated for M1 and $\phi = 0°$ (Fig. 29a), and the more complex situation of variable contrast described above may well apply.

Finally, we shall consider the images of outer pair tubules illustrated in Fig. 30, b and c. The important question as to how the two tubules are joined together dominates any attempt to interpret these images. There is some evidence from images of transverse and longitudinal sections that tubules A and B of a pair share a common region (see Fig. 12). If sharing does not take place the wall thickness where the two tubules A and B touch should appear thicker in sectioned material (see Fig. 12, and region 1 of Fig. 30c). The appearance of 4-column regions in at least one member of an outer pair is a frequent phenomenon (see region 1 of Fig. 30, b and c). It is tempting to interpret these 4-column images in terms of a superposition effect similar to that shown in Fig. 30a for M1. It must be realized that if two component tubules of an outer pair share more than two columns of subunits, then they cannot both have circular cross sections. Three or more columns cannot belong to two tubules simultaneously. In terms of column models, there are two possibilities:

1. The tubules of an outer pair share more than two columns of subunits and one tubule has a circular cross section. This would frequently give rise to a 4-column image of one tubule because of some constraint on its azimuthal orientation imposed perhaps by its relationship to the other member of the pair.

2. The tubules of an outer pair share two columns of subunits and

both tubules have an approximately circular cross section. If the plane containing their axes is constrained to be perpendicular to the electron beam (which will be the case if both tubules lie on the carbon surface covering the specimen grid), then both tubules will be positioned with $\phi = 15°$. In order to explain the usual occurrence of 4-column images in one tubule only of a pair, it must be further supposed (under this hypothesis) that one tubule is more compressible than the other. This is in keeping with the probability of structural differences between the two components. Two further observations tend to support hypothesis 2: occasionally each tubule of a pair gives rise to a 4-column image, and spacings between maxima in the intense outer columns tend to be small (30–40 Å). Therefore, in terms of the column models, present evidence is in favor of hypothesis 2.

There remains the problem of the shape and size of the subunit. It has been stated that the mean intracolumn spacing of intensity maxima is about 40 Å. Coupled with this, however, is the wide range of spacings. Consequently the mean value of ~40 Å could be misleading. Although it is too early to identify units inferred from measurements in solution, it can be observed that a protein "sphere" of 40 Å diameter would have a much smaller molecular weight than that of approximately 60,000 reported by Gibbons [21a]. This discrepancy may well arise from such effects as the distortion of the structure in the electron microscope and to irregularities of stain distribution. (Out-of-focus images probably affect the size of a maximum, but not the mean spacing between maxima.) Such effects could produce confused images and superposition effects difficult to interpret. At this stage the apparent size of the subunit of ~40 Å diameter should be treated with reserve.

From this comparison of observed and calculated results the following provisional conclusions are put forward:

1. The ciliary tubules are composed of globular subunits.
2. Most of the tubules examined so far are only partially contrasted with stain.
3. Some distortion of the tubules is present, but it is not as great as might be superficially concluded from observations of confusion on the image.
4. In terms of column models the most probable value of $n$, the number of columns per tubule, is 12.
5. The value of $m$ is more uncertain; if 2 above is correct then a value of $m = 3$ or 4 is likely.
6. The shape and size of the subunit is uncertain. The mean intra column spacing between maxima is 40 Å. On the other hand, the observed mean diameter of the cylinder on which the subunits lie suggests

an equivalent diameter of ~50 Å when calculated for the special subset of column models with $m = 3$ or 4 and $n = 12$.

7. The region of conjunction between the two component tubules of an outer pair may contain two shared columns of subunits. However this is the least certain of the conclusions.

## References

1. Beermann, W., *in* "Genetics Today," 11th Intern. Conf. Genetics, 1963, Vol. 2, p. 309. Macmillan (Pergamon), New York, 1965.
2. Beisson, J., and Sonneborn, T. M., *Proc. Natl. Acad. Sci. U.S.* **53,** 275 (1965).
3. Burge, R. E., Ph.D. Thesis, Univ. of London, 1957.
4. Burge, R. E., and Silvester, N. R., *J. Biophys. Biochem. Cytol.* **8,** 1 (1960).
5. Burge, R. E., and Smith, G. H., *Proc. Phys. Soc. (London)* **79,** 673 (1962).
6. Capecchi, M. R., *Proc. Natl. Acad. Sci. U.S.* **55,** 1517 (1966).
7. Caspar, D. L. D., *J. Mol. Biol.* **15,** 365. (1966).
8. Caspar, D. L. D., and Klug, A., *Cold Spring Harbor Symp. Quant. Biol.* **27,** 1 (1962).
9. Chatton, E., and Lwoff, A., *Arch. Zool. Exptl. Gen.* **77,** Fasc. 1, 1 (1935).
10. Cosslett, V. E., *J. Roy. Microscop. Soc.* **78,** 1 (1958).
11. Cosslett, V. E., *Lab. Invest.* **14,** No. 6 (1965).
12. Coupland, J. H., *Proc. Phys. Soc. (London)* **B69,** 648 (1956).
13. Dirksen, E. R., and Crocker, T. T., *J. Microscopie* **5,** 629 (1966).
14. Edgar, R. S., and Wood, W. B., *Proc. Natl. Acad. Sci. U.S.* **55,** 498 (1966).
15. Epstein, R. H., Bolle, A., Steinberg, C. M., Kellenberger, E., Boy de la Tour, E., Chevalley, R., Edgar, R. S., Susman, M., Denhardt, G. H., and Lielausis, A., *Cold Spring Harbor Symp. Quant. Biol.* **28,** 375 (1963).
16. Favre, R., Boy de la Tour, E., Segrè, N., and Kellenberger, E., *J. Ultrastruct. Res.* **13,** 318 (1965).
17. Fawcett, D. W., *in* "The Cell" (J. Brachet and A. E. Mirsky, eds.), Vol. 2, p. 217. Academic Press, New York, 1961.
18. Finch, J. T., and Klug, A., *J. Mol. Biol.* **13,** 1 (1965).
19. Finch, J. T., and Klug, A., *J. Mol. Biol.* **15,** 1 (1966).
20. Fraenkel-Conrat, H., "Design and Function at the Threshold of Life." Academic Press, New York, 1962.
21. Gibbons, I. R., *Arch. Biol. (Liege)* **76,** 317 (1965).
21a. Gibbons, I. R., this symposium, p. 99.
22. Gibbons, I. R., and Grimstone, A. V., *J. Biophys. Biochem. Cytol.* **7,** 697 (1960).
23. Grassé, P. P., *Compt. Rend.* **252,** 3917 (1961).
24. Gurdon, J., *Advan. Morphogenesis* **4,** 1 (1964).
25. Hall, C. E., *J. Appl. Phys.* **22,** 655 (1951).
26. Hall, C. E., *J. Biophys. Biochem. Cytol.* **1,** 1 (1955).
27. Hall, C. E., and Inoué, T., *J. Appl. Phys.* **28,** 1346 (1957).
28. Jacob, F., and Monod, J., *in* "Cytodifferentiation and Macromolecular Synthesis" (M. Locke, ed.). Academic Press, New York, 1963.
29. Kellenberger, E., *in* "New Perspectives in Biology," Vol. 4, 234. Elsevier, Amsterdam, 1964.
30. Kellenberger, E., and Boy de la Tour, E., *J. Ultrastruct. Res.* **13,** 343 (1965).
31. Klug, A., and Berger, J. E., *J. Mol. Biol.* **10,** 565 (1964).

32. Klug, A., and Finch, J. T., *J. Mol. Biol.* **11,** 403 (1965).
33. Ledbetter, M. C., and Porter, K. R., *J. Cell Biol.* **19,** 239 (1963).
34. Lenz, F., *Z. Naturforsch.* **9a,** 185 (1954).
35. Lwoff, A., "Problems of Morphogenesis of Ciliates." Chapman & Hall, London, 1950.
36. Mazia, D., and Child, F. M., *Experientia* **12,** 161 (1956).
37. Randall, Sir J., and Jackson, S. F., *J. Biophys. Biochem. Cytol.* **7,** 1 (1958).
38. Randall, Sir J., and Disbrey, C., *Proc. Roy. Soc.* (*London*) **B162,** 473 (1965).
39. Randall, Sir J., Hopkins, J. M., Eadie, J. M., and Butcher, R. W., *Proc. Linnean Soc. N.S. Wales* **174,** Session 1961–1962, Pt. 1, 31 (1963).
40. Randall, Sir J., Warr, J. R., Hopkins, J. M., and McVittie, A., *Nature* **203,** 912 (1964).
41. Reimer, L., *Z. Naturforsch.* **14b,** 566 (1959).
42. Renaud, F. L., and Swift, H., *J. Cell Biol.* **23,** 339 (1964).
43. Ringo, D. L., Ph.D. Thesis, Univ. of Texas, Austin, Texas, 1966.
44. Schuster, F., *J. Protozool.* **10,** 297 (1963).
45. Silvester, N. R., Ph.D. Thesis, Univ. of London, 1960.
46. Smith, G. H., and Burge, R. E., *Proc. Phys. Soc.* (*London*) **81,** 522 (1963).
47. Smith-Sonneborn, J., and Plaut, W., *J. Cell Sci.* **2,** 225 (1967).
48. Stubblefield, E., and Brinkley, B. R., this symposium, p. 175.
49. Tartar, V., "The Biology of *Stentor*." Macmillan (Pergamon), New York, 1961.
50. Tilney, L. G., and Porter, K. R., *Protoplasma* **60,** 317 (1965).
51. Valentine, R. C., and Horne, R. W., *Symp. Intern. Soc. Cell Biol.* **1,** 263 (1961).
52. Warr, J. R., McVittie, A., Randall, Sir J., and Hopkins, J. M., *Genet. Res. Cambridge* **1,** 335 (1966).
53. Watson, M. R., and Hopkins, J. M., *Exptl. Cell Res.* **28,** 280 (1962).
54. Watson, M. R., and Hynes, R. D., *Exptl. Cell Res.* **42,** 348 (1966).
55. Watson, M. R., Hopkins, J. M., and Randall, J. T., *Exptl. Cell Res.* **23,** 629 (1961).
56. Watson, M. R , Alexander, J. B., and Silvester, N. R., *Exptl. Cell Res.* **33,** 112 (1964).

# ARCHITECTURE AND FUNCTION OF THE MAMMALIAN CENTRIOLE

**ELTON STUBBLEFIELD AND B. R. BRINKLEY**

*Section of Cytology, Department of Biology, The University of Texas M. D. Anderson Hospital and Tumor Institute, Houston, Texas*

## INTRODUCTION

For many years cytologists have known of the existence of the centriole from observations with the light microscope. Other than the fact that these tiny granules were associated with the process of mitosis in animal and some plant cells, very little progress was made in understanding either their structure or function until the electron microscope became available for biological studies. Light microscopists concluded that the centriole somehow organized the poles of the mitotic apparatus, but until recent years even the existence of the mitotic apparatus as a cell structure was doubted by many biologists. A second cellular organelle, the basal body, which has long been thought to be a structure derived from the centriole in ciliated cells [13, 21], has in recent years been clearly demonstrated to be the centriole in a second type of activity.

In 1956, DeHarven and Bernhard [5] were the first to publish ultrastructure studies of the centriole using electron microscopy. The ultrastructure of the basal body was quickly demonstrated to be identical with that of the centriole, and the homology of the two structures was later established by several groups of workers [9, 30, 36].

The activities of the centriole in both mitosis and ciliogenesis have recently been studied extensively in our laboratory [2, 3, 37]. Since both centriole functions are present in a fibroblast cell line *in vitro,* it has now become possible to study transitional stages in centriole behavior and to time centriole replication and ciliogenesis in terms of the cell reproductive cycle. In addition, new information on centriole ultrastructure is available as a result of improved preservation techniques for thin-sectioned material and photographic methods that reduce image noise in electron micrographs. The architecture of the centriole will be considered first, and experiments and discussions of centriole function will follow.

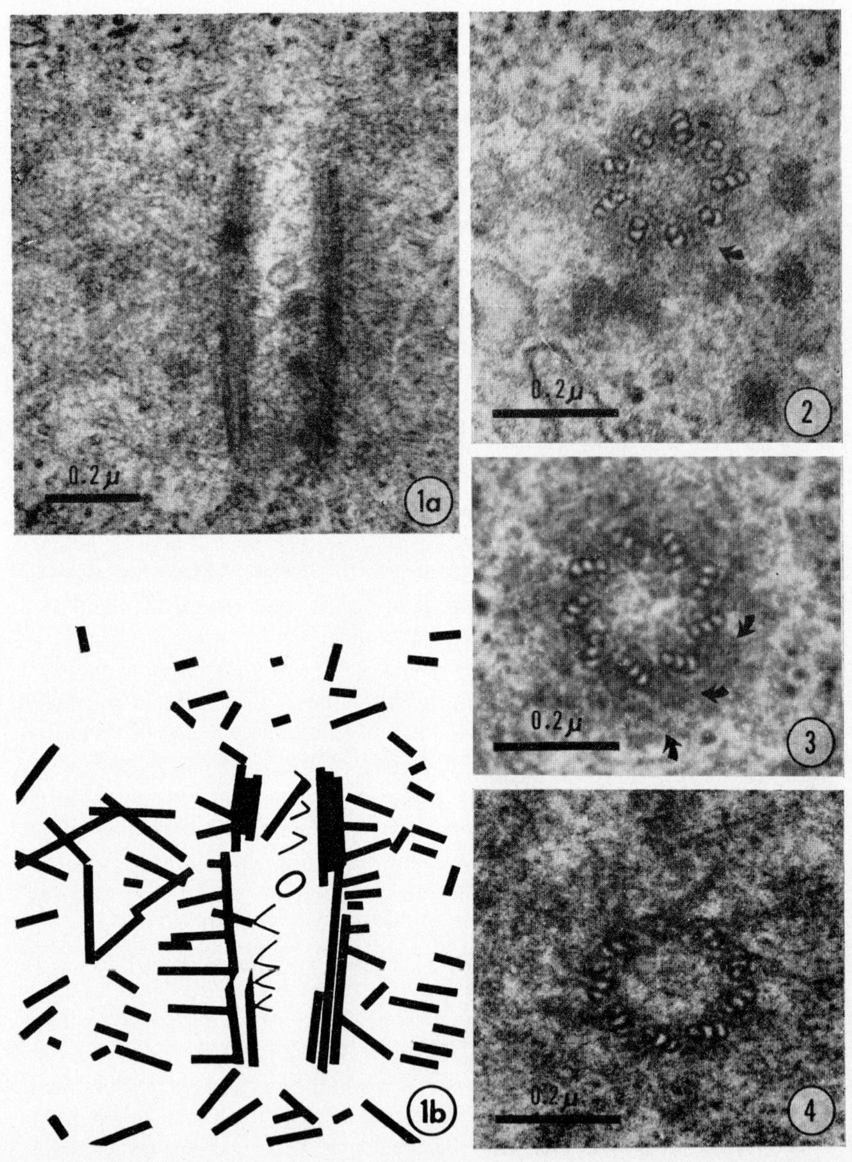


All material prepared for electron microscopy was fixed in 3% glutaraldehyde (pH = 7.4), postfixed in 1% osmium tetroxide, dehydrated in ethanol, and embedded in Epon 812 (for more details see [2, 3, 37]).

FIG. 1. (a) Longitudinal section of Chinese hamster centriole in a mitotic cell. (b) Diagram showing the positions of the microtubules seen in (a) and also indicating the *central vesicle* and segments of the *internal helix* (see text).

## ARCHITECTURE OF THE CENTRIOLE OF THE CHINESE HAMSTER

Almost all available information about centriole ultrastructure comes from electron microscopic studies. Only tentative biochemical studies on basal body composition have yet been made [14], simply because of the difficulty in isolating sufficient material; however, this deficiency will no doubt eventually be overcome. Cilia and the mitotic apparatus have been purified and analyzed biochemically [11, 23, 24, 39], so some chemical information is also available from the studies of these related structures.

One or more centrioles are visible in about 6% of the cell thin sections studied of the Chinese hamster cell strain Don-C [26]. Most sections through centrioles are oblique, but occasionally good longitudinal and cross sections are obtained as in Figs. 1–4. From these four micrographs alone we can extract much of what is known about centriole architecture. Basically, the centriole is a hollow cylinder. Although the overall shape is easy to demonstrate, a number of structures are contained within which are more difficult to see; these will be considered in sequence.

### *The Outer Wall*

The outer wall of the centriole is composed of 27 microtubules that appear to run the length of the centriole parallel to its long axis. These microtubules are separated into 9 groups of 3, with the members of each triplet fused together to form a blade. The blades form an angle with the cylinder surface in such a way that when viewed in cross section (Figs. 2–4), they have a "pinwheel" arrangement. The microtubules of one triplet do not touch members of the adjacent triplets, but they appear to be embedded in a common osmiophilic matrix in the centriole wall. The microtubules are each about 200–250 Å in outer diameter; the overall centriole structure is about 0.25 μ in diameter

---

FIG. 2. Cross section of centriole at the proximal end showing the *cartwheel.* Note the microtubule (arrow) connecting the centriole wall and a *pericentriolar satellite;* several of the latter structures are present. Microtubule C is shorter than A or B in 6 of the triplets at this end of the centriole.

FIG. 3. Cross section of a centriole. One-third of a turn of the *internal helix* is visible at the left side of the centriole lumen. In 3 cases (arrows) microtubules are seen that attach to the centriole wall between triplets on the *triplet base* (see text).

FIG. 4. Cross section of a centriole at the distal end showing external fibrous appendages (*transition fibers*) and a dense structure in the centriole lumen which differs from the cartwheel. The section is oval in shape because of compression during sectioning (compare with Fig. 5).

and 0.5–0.7 μ in length. The microtubules of a triplet are designated by convention as A, B, and C, with A the innermost microtubule [12].

### *The Central Vesicle*

Within the lumen of the centriole is a small vesicle about 600 Å across (Fig. 1). This structure is thought to be a vesicle rather than a ring, since it is always circular in profile when seen in cross, oblique, or longitudinal sections of centrioles. It has been observed in centrioles and basal bodies of other organisms [6, 29, 35] and may be present in all species. Its function is completely unknown.

### *The Internal Helix*

A second component of the centriole lumen, and much more difficult to demonstrate, is a large helix which spirals just under the triplets for the full length of the centriole cylinder. In Fig. 1 the longitudinal sections of this helix are faintly visible as diagrammed below the figure. The element of the helix appears to be about 50–75 Å in diameter. About 8–10 turns can be counted with a spacing of about 750 Å per turn. In cross sections of the centriole only part of one turn of the helix is visible; in Fig. 3 it appears as a semicircle on the left side of the centriole lumen. Measurements indicate that the distance across the helix is about 1300 Å and the distance between turns about 750 Å; this would indicate a total stretched length of about 4–5 μ for this structure.

Perfect cross sections of centrioles provide a figure with 9-fold rotational symmetry. Such structures may be studied with a special photographic method devised by Markham *et al.* [22] for probing the structure of virus capsids. The method is based on the principle that an object viewed along an axis of $n$-fold symmetry may be seen with improved resolution (because of a decrease in "noise") if it is viewed in all its $n$ different but equivalent rotational positions simultaneously. This effect can be achieved photographically, in the case of a centriole, for example, by making a 9-fold multiple exposure photograph with the centriole cross section rotated 40 degrees around its central axis between each exposure. The result is a kind of photographic "average," with enhancement of structure that repeats every 40 degrees and simultaneous reduction of the randomly distributed noise in the photograph. At the same time structures that repeat at other intervals (i. e., 60 degrees) will also be reduced in contrast.

Before the method could be applied to centriole cross sections, however, an additional problem had to be solved. We intuitively felt that the centriole cross section should be symmetrical, i.e., circular, but it was always more or less oval in appearance. This was probably the result of compression of the structure as it was sectioned. If this was indeed the case, we reasoned, it should be possible to "recircularize" the image photographically. A photograph can be "stretched" in a given direction by projecting it on a slanting surface. That this process could be successfully applied to centrioles is seen in Figs. 4 and 5.

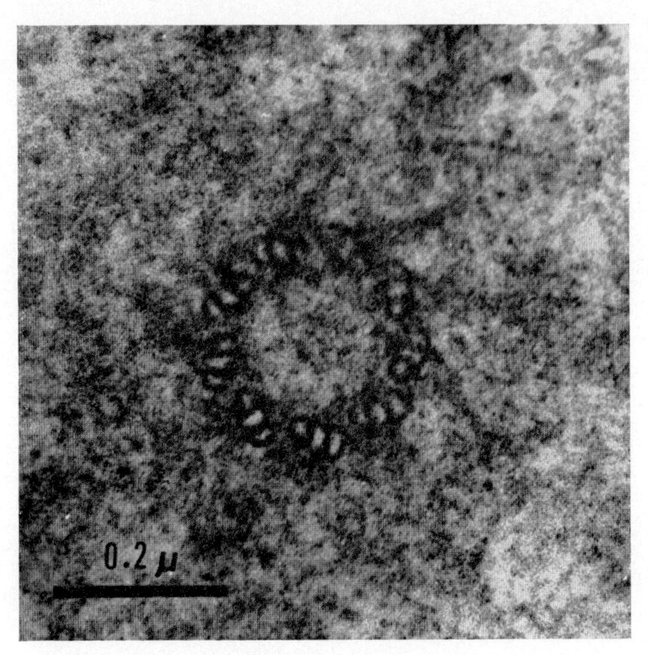


FIG. 5. Same section as Fig. 4 after "recircularizing" the image by printing the negative on a slanting surface to restore rotational symmetry.

Restoring overall circularity to the image also restored rotational symmetry, in that triplet angles and positions all became equivalent. The high-resolution photographs resulting from applying the Markham technique to the recircularized images of Figs. 2, 3, and 4 are presented in Figs. 6, 7, and 8, respectively.

### *The Triplet Base*

With improved resolution, each triplet blade of the centriole was seen to rest on a faint double structure between it and the centriole lumen (Fig. 7). This *triplet base* appeared to be bounded peripherally by the dense matrix of the centriole wall and in the lumen of the centri-

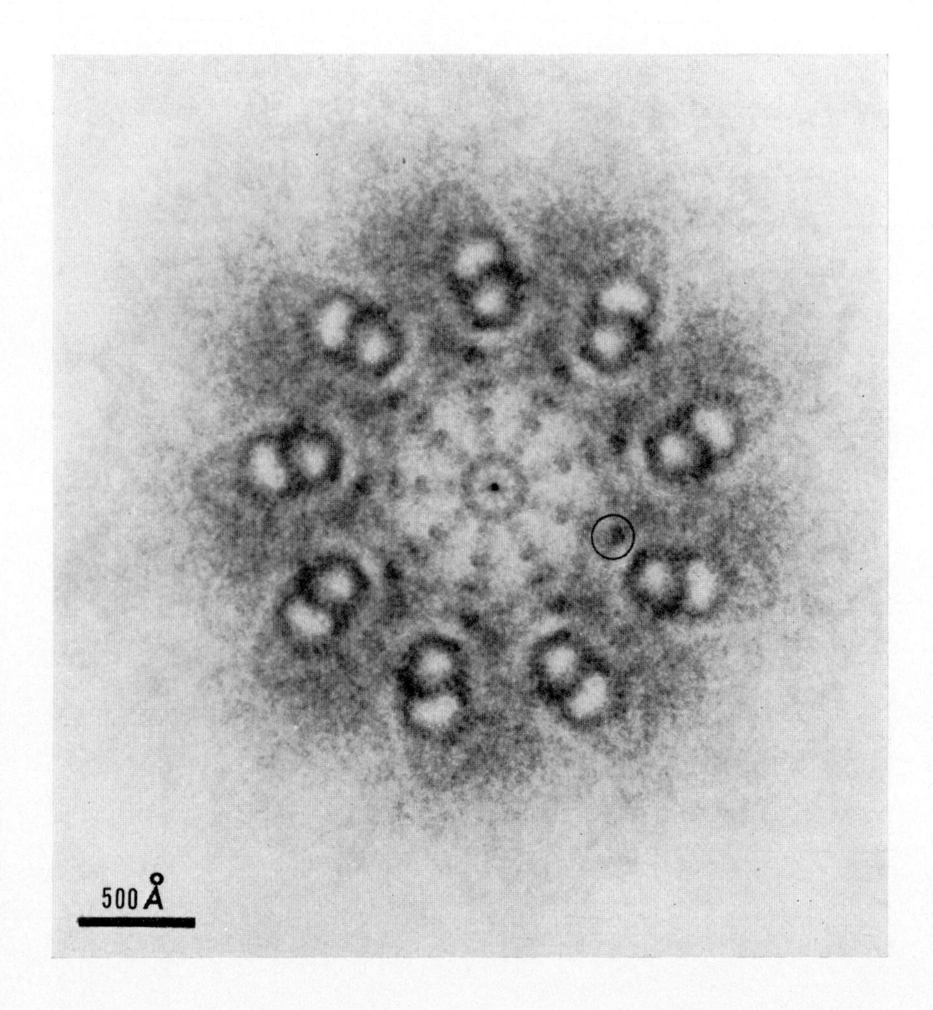


FIG. 6. High-resolution photograph produced with the technique of Markham *et al.* [22] using the recircularized image of Fig. 2. The *cartwheel* is now seen with greater clarity; note the segmentation of the spokes and the fine detail of the central hub. Microtubule C occurs but 3 times in this section and so does not reinforce well in this photograph. The dense particle of stain (at 1 o'clock in Fig. 2) is also eliminated by this procedure since it does not occur repeatedly at 40-degree intervals; all random stain (noise) is likewise reduced, while repeating detail is enhanced. Substructure is also now visible in the walls of the triplet microtubules, which indicates that the triplet architecture is rigidly related to the whole centriole structure. The dense structure in the circle is found in all cross sections (see Figs. 7–10) and is thought to contain RNA (compare with Fig. 15).

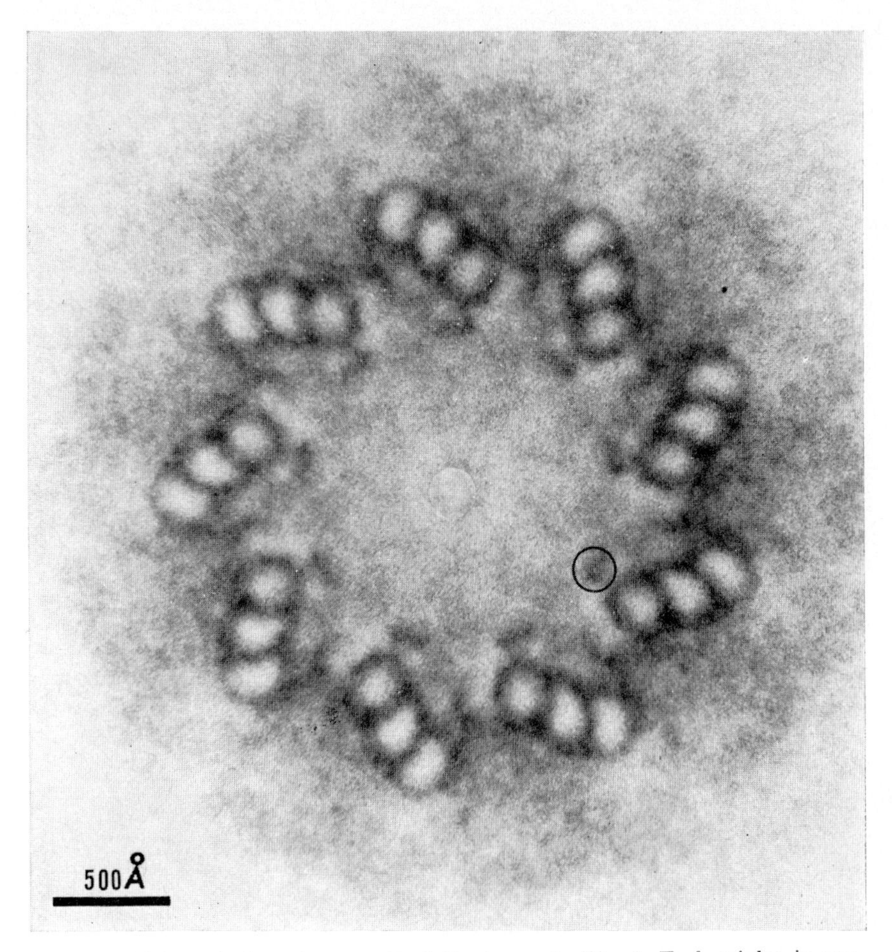


FIG. 7. High-resolution photograph of the image in Fig. 3. Each triplet is associated with a double structure, the *triplet base*. In the centriole lumen, each triplet base is bounded by a dense "foot" (circle). Peripherally, the triplet base attaches to microtubules oriented vertically to the long axis of the centriole, as can be seen in Figs. 2 and 3.

ole by a dense footlike appendage extending from microtubule A (circled in Figs. 6–8). All the "feet" appeared to rest on a circle about 1300 Å in diameter. The spiral structure described earlier also had a helical diameter of about 1300 Å, so we believed that the spiral was centered in the lumen on these feet. In Fig. 3 the spiral extends for only a short distance on one side of the centriole and did not reinforce adequately to appear in the high resolution photograph. However, in a thicker section a sufficient part of the spiral was present, and the rota-

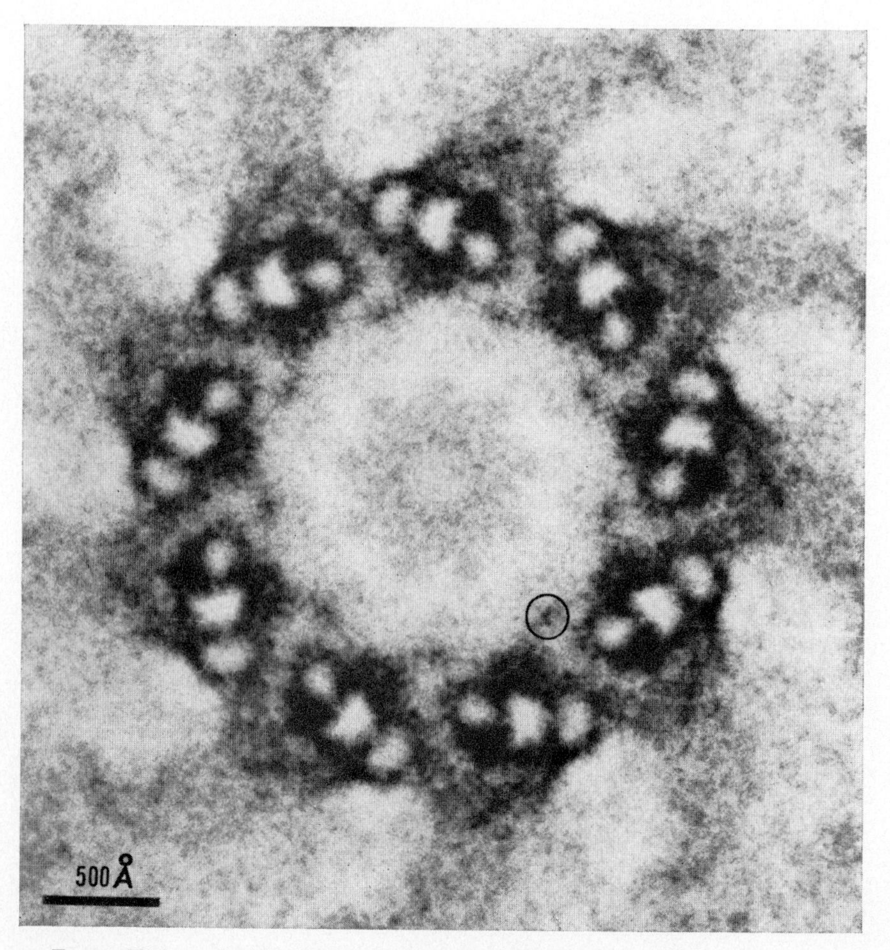


FIG. 8. High-resolution photograph of the image in Fig. 4. The *triplet base* is reduced to a single dense structure adjacent to microtubule B in this end of the centriole; the dense "foot" (circle) is still present, however. The central structure does not appear to have 9-fold symmetry since it does not reinforce well (compare with Fig. 13).

tional photograph shown in Fig. 9 demonstrates that the spiral does contact the feet projecting from microtubule A.

Peripherally, the triplet base sometimes appears to be continuous with microtubules extending away from the centriole. This is best demonstrated in cross sections, and several examples are shown in Figs. 2 and 3 (arrows). As in the case of Fig. 3, microtubules are present on only 2 or 3 triplet bases. This suggested the possibility that triplet bases may appear different in different positions around the centriole if they are arranged in some sort of spiral pattern. Accordingly, a 3-fold rotational photograph of the centriole cross section in Fig. 3 was prepared in which only 3 adjacent triplets were superimposed in each case (Fig. 10). The triplet bases looked much the same in each position; this was interpreted to indicate that they are linear structures running continuously from end to end parallel to the triplet microtubules. In Figs. 6 and 9 the triplet base is also present and appears to maintain a fixed relationship to the triplet plane. However, in Fig. 8, at the opposite end of the centriole from Fig. 6, the triplet base is apparently reduced to a single dense structure adjacent to microtubule B.

The concept of direct attachment of microtubules in the cell cytoplasm to the wall of the centriole has been suggested by other investigators [9, 34] but never clearly demonstrated. Although at first glance the microtubule populations in Figs. 1 and 11 (diagrammed in 1b) seems to bear no obvious relationship to the centriole, a more careful analysis reveals that many of the segments are roughly perpendicular to the long axis of the centriole. Those microtubules nearest to the centriole are more difficult to detect because of the increase in density of the surrounding material near the centriole wall. Two additional factors add to the confusion: (a) in mitosis some microtubules appear to be produced by the kinetochores of the chromosomes, and these probably do not actually attach to the centriole wall, and (b) in interphase there is more than one active centriole; thus many microtubules may be oriented with respect to a second unseen centriole and confuse the overall picture. Nevertheless, the only point of attachment seen consistently in mammalian centriole sections is to the triplet base as in Figs. 2 and 3. The attachment to several consecutive triplet bases suggests that the microtubules are attached in a spiral arrangement corresponding roughly to the helix spiral within the centriole lumen.

### *The Cartwheel*

Cross sections at one end of the centriole reveal an additional structure (Fig. 2). Connecting the microtubules of the centriole wall to a central circle are 9 radial spokes. This "cartwheel" structure has been seen in both centrioles and basal bodies of several organisms [9, 12]; in some protozoan species it appears to be a multiple structure, occurring in successive cross sections for some distance through the length of the basal body [12]. However, in Chinese hamster centrioles it occurs in only the end section and only at one end.

In Fig. 6, the high resolution photograph of the end section shown in Fig. 2, the "cartwheel" structure can be seen with greater clarity.

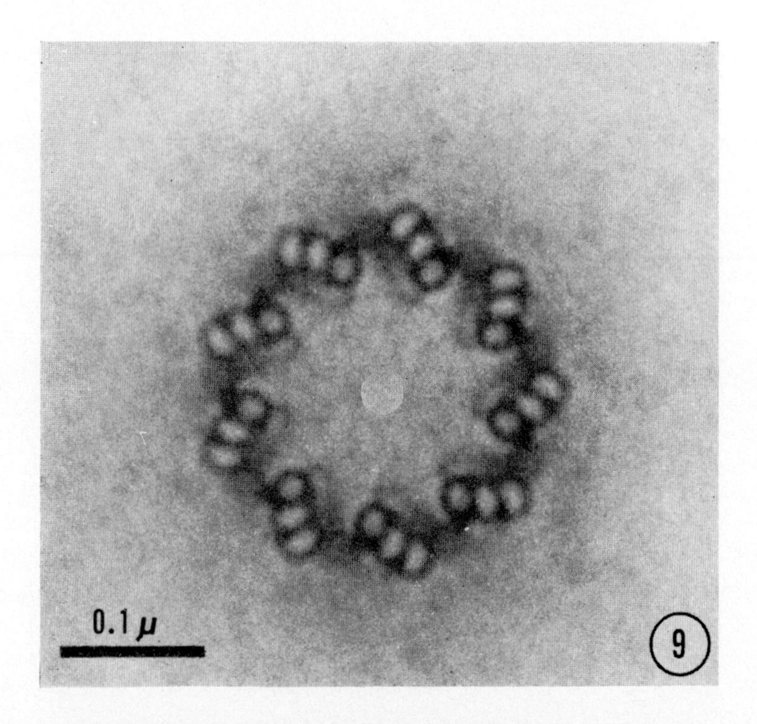
0.1 μ
9

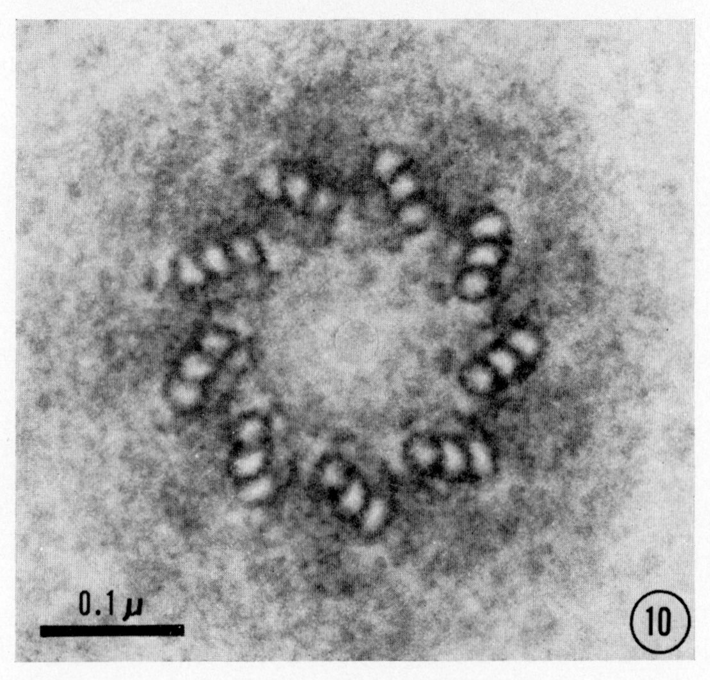
0.1 μ
10

TABLE I. *Relationship between Triplet Angle and Internal and External Centriole Diameters*[a]

| Case | Triplet angle | Internal centriole diameter (Å) | External centriole diameter (Å) |
|---|---|---|---|
| Fig. 2 (end section) | 10° | 1600 | 2350 |
| Fig. 9 | 33° | 1860 | 2320 |
| Fig. 3 | 38° | 1840 | 2420 |
| Fig. 14e (3rd section) | 40° | 1840 | 2360 |
| Fig. 14d (2nd section) | 46° | 1880 | 2310 |
| Fig. 14c (end section) | 50° | 1900 | 2320 |
| Fig. 4 (end section) | 57° | 1950 | 2320 |

[a] In all cases the average distance within all triplets from the center of microtubule A to the center of microtubule is taken as 400 Å. The *triplet angle* is defined as the smaller angle at the intersection of a line through the triplet microtubule centers and the centriole radius through the center of microtubule C. The *internal centriole diameter* is defined as the average distance from the center of a microtubule A to the centers of the two microtubules A opposite it (across the lumen). The *external centriole diameter* is the analogous measurement between microtubules C. All measurements were made using high-resolution rotation photographs (see text).

The "spokes" are definitely segmented and appear to attach at one end of each triplet base. The microtubules in this centriole end-section were so well positioned that their subunits reinforced in Fig. 6, and the microtubule structure appears to be similar to that of other species as described elsewhere in this volume [20].

In Fig. 2 we also note that only 3 of the triplets are complete; in 6 cases the outer microtubule is missing. We interpret this to indicate that the outer microtubule is usually shorter than the other two in the end of the centriole containing the cartwheel. This was also the conclusion of Gibbons and Grimstone in their studies of flagellate basal bodies ([12], see their Fig. 13).

It is also of interest that in Fig. 2 the angle of the triplet plane with respect to the radius of the centriole is much smaller than in Fig. 3. Careful measurements show that the outer diameters of the centrioles in Figs. 2 and 3 are almost identical; the decrease in the triplet angle in Fig. 2 resulted in a decrease of the lumen diameter (Table I). It

FIG. 9. High-resolution photograph of a centriole cross section similar to Fig. 3. However, since the section was thicker, the *internal helix* was visible over an arc large enough to reinforce as a circle in the final photograph. The helix contacts the *triplet base* where the "foot" appears.

FIG. 10. Rotation photograph of Fig. 3 in which only 3 adjacent 40-degree sections were superimposed. The *triplet base* is visible at each position, suggesting that it is also a linear structure like the triplet.

thus appears that the triplet structures are hinged along microtubule C and that the triplet angle is decreased by moving microtubule A toward the central axis of the centriole.

The triplet angle does not appear to be the same at different points along the centriole axis. Data given in Table I for three consecutive sections at one end of a centriole indicate a change of 10 degrees in 3 sections. Complete serials are not yet available, however. At the point of transition from the basal body to the cilium such a twist in the triplet plane is evident in ciliates [12]. Figure 11 seems to indicate that the triplets are twisted from one end of the centriole to the other, but the serial sections of a centriole shown in Fig. 12 show all microtubules lying parallel with no evidence of any twist in the triplet plane. Therefore, we have concluded that the triplets are mounted in such a way that the triplet twist can be introduced during different centriole activity states. We will return to this point later.

### *External Fibrous Appendages*

Figures 4 and 8 reveal another centriole structure, a series of fibrous appendages, which radiate away from the centriole near one end approximately vertical to each triplet plane. Similar structures occur near the junction of basal bodies to cilia, and these have been named "transition fibers" [12, 29]. That these fibrous structures also are found near one end in Chinese hamster centrioles is evident in the serial sections presented in Fig. 14. The fact that they do not reinforce well in the rotation photograph (Fig. 8) suggests that they are not rigidly fixed in space or are of a rather irregular composition.

### *Octagonal End Structure*

A somewhat blurred structure is present in the lumen of the end section shown in Fig. 4. It does not appear to be the same as the cartwheel seen before (Figs. 2 and 6), since it is rather dense and does not possess the obvious 9-fold rotational symmetry of the cartwheel. In Fig. 8, the structure does not reinforce well, indicating a lack of 9-fold symmetry. The symmetry of this structure may have been partially destroyed by fixation and embedding, but we felt that a test of other possible symmetries was indicated. Accordingly, rotations of 120, 90, 72, 60, 51.4, and 45 degrees between exposures were tried (corresponding to tests for 3-, 4-, 5-, 6-, 7-, and 8-fold symmetry, respectively). The resulting rotation photographs are shown in Fig. 13. Surprisingly, the best reinforcement was obtained at 90- and 45-degree intervals, indi-

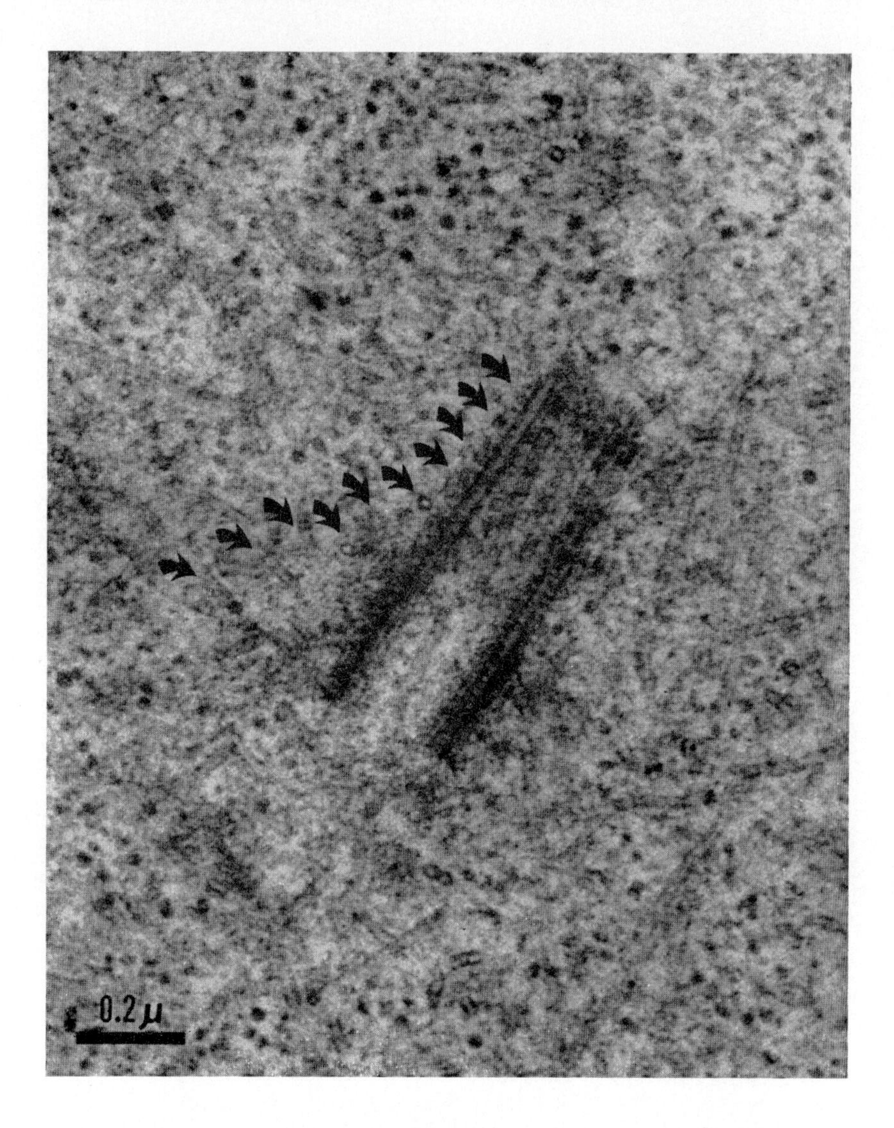


FIG. 11. Longitudinal section of a centriole in a mitotic cell. The twist of the triplets from one end to the other is apparent in the slant of the microtubules A in the lumen at the upper end. The arrows indicate the regular spacing of microtubules near the centriole wall; careful analysis reveals a similar spacing on the opposite side. Since the microtubules leave the surface of the centriole parallel to the triplet plane, instead of exactly vertical to the wall, only short segments are visible near the centriole. At the proximal end (bottom) the microtubules are more nearly vertical because of the triplet twist, thus longer segments are seen closer to the centriole surface.

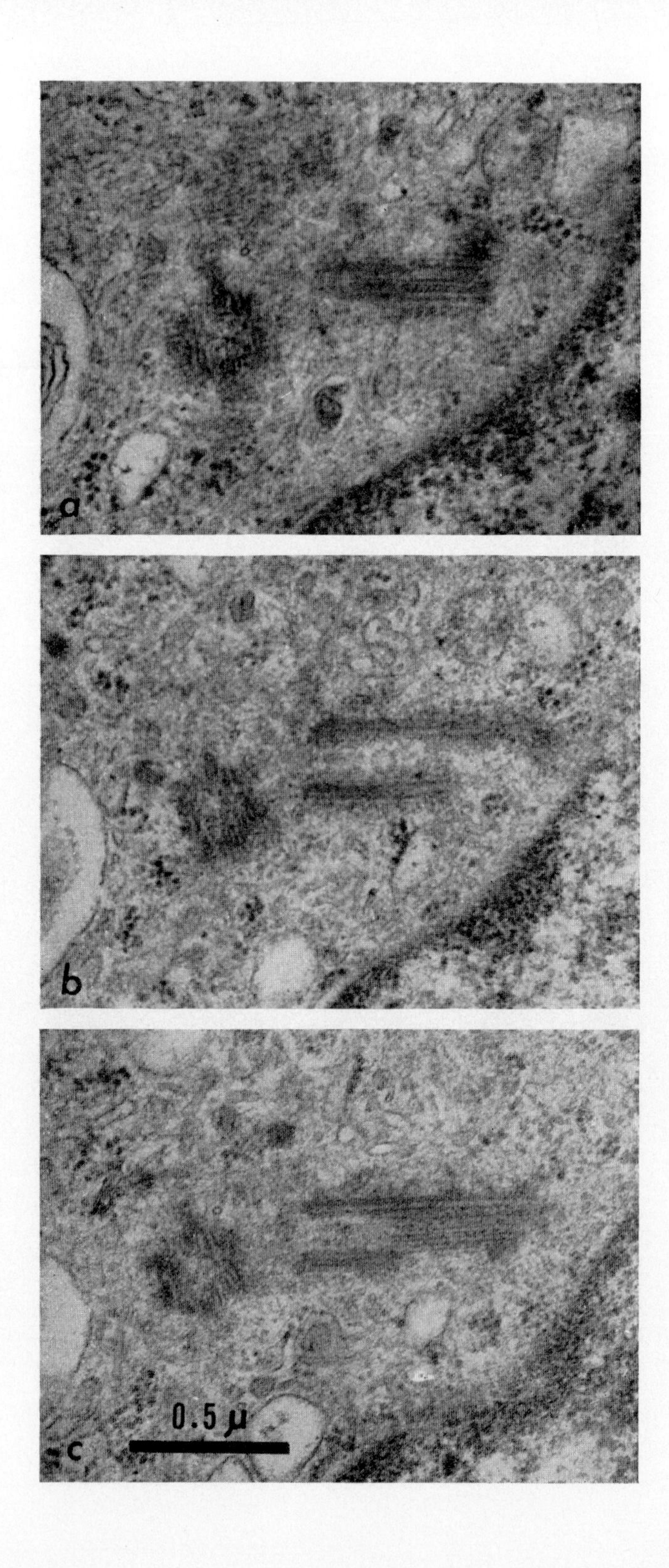
a
b
c
0.5 μ

cating that this end structure has 8-fold rotational symmetry. The implications of this discovery will be discussed later.

### *Pericentriolar Satellites*

Many centriole sections reveal numbers of irregular dense masses near the centriole as in Fig. 2. These are termed *pericentriolar satellites* and their function remains unknown. Occasionally, they appear to be associated with microtubules near the centriole, and some authors have proposed that these are the point of microtubule attachment [5, 27]. This point will be raised again in a later section.

### *Structures Peculiar to Basal Bodies*

In order to complete this section, we must include descriptions of several structures that were first described in basal bodies and therefore acquired different names, although some may be modifications of structures we have already considered. The *basal plate* [7, 12] is a flat structure across the end of the basal body nearest the cilium. It may be homologous with the octagonal end structure which we have already considered. The *basal foot* (see Fig. 23) may be a modified procentriole, although some find greater similarity to the pericentriolar satellites [6]. The basal foot always lies in the plane of ciliary beating [10], but its function is unknown. *Rootlets* appear at the end of the basal body opposite the cilium in some species [6, 7]. These are usually cross-striated and may extend several microns into the cytoplasm; they may serve in anchoring the basal body in the cytoplasm.

## CHEMICAL COMPOSITION OF CENTRIOLE COMPONENTS

The biochemical composition of the architectural components of the free centriole remains open to investigation. Some information about the nucleic acids occurring in the basal body is available, however, and the composition of free centrioles will probably prove to be similar. Hoffman [14] found evidence for the presence of a small amount of RNA (about 2% of the total preparation) in isolated basal bodies

---

FIG. 12. Longitudinal serial sections through an immature procentriole. The parent centriole is obliquely sectioned on the left. In the procentriole, note that all the microtubules in the wall are parallel, i.e., there is no indication of a twist in the triplet plane (compare with Fig. 11). This suggests that the twist in the triplet plane is induced later as the centriole begins to function in microtubule production. The *central vesicle* is shown in (b).

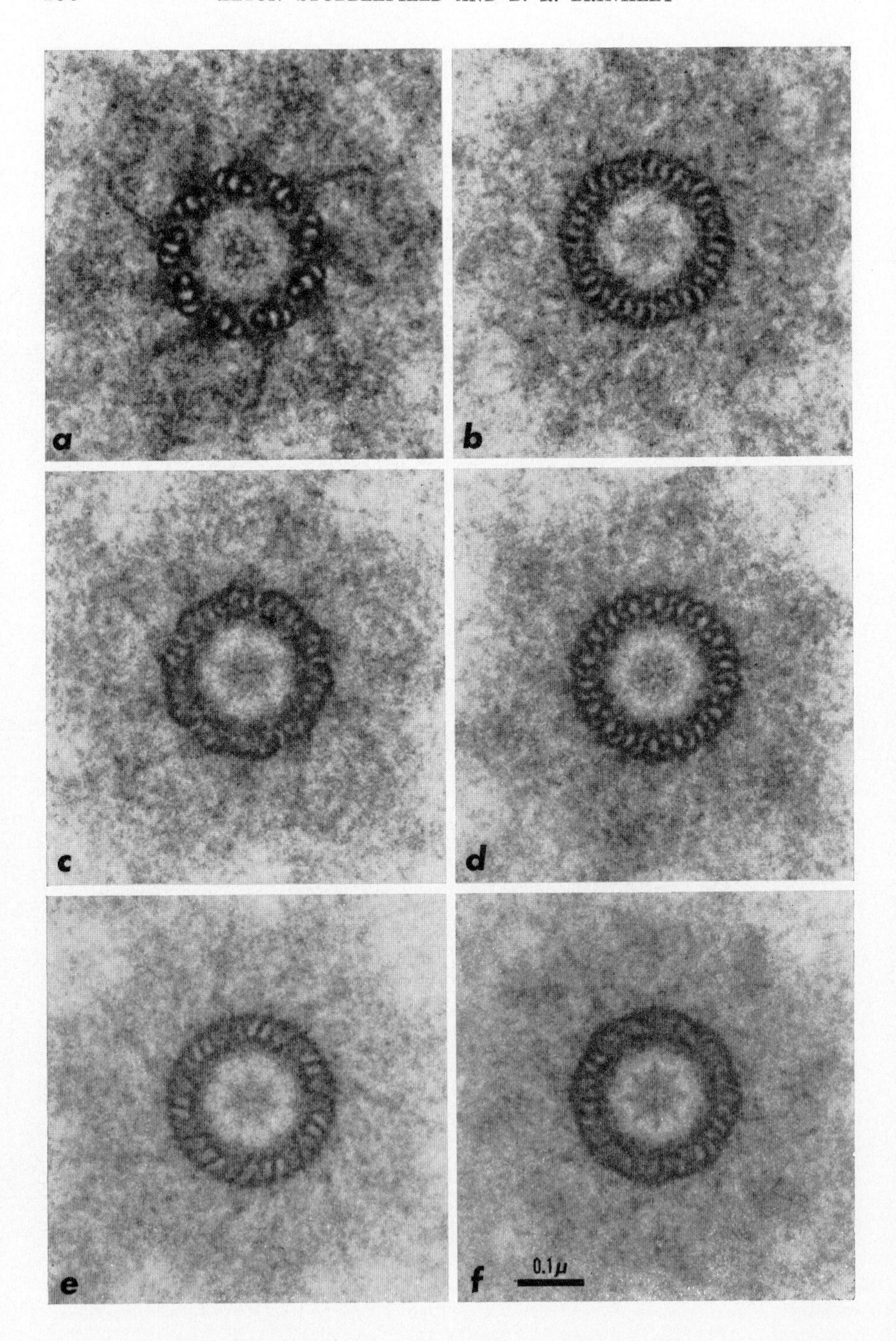
a
b
c
d
e
f
0.1μ

from *Tetrahymena*. DNA was less than 1% of the total preparation, and this amount was too low to be detected with certainty. Randall and Disbrey [28] found evidence for the presence of both DNA and RNA in the basal bodies of *Tetrahymena* using the fluorescent dye acridine orange. We can infer that the triplet microtubules of the centriole are protein (perhaps ribonucleoprotein), since this is the case for the microtubules of cilia, which are direct extensions of microtubules A and B of the basal body [11, 12].

We can now present indirect evidence concerning the chemical composition of a specific centriole component, the dense footlike appendage on the inner surface of the triplet base. Digestion of the glutaraldehyde-fixed cells with ribonuclease[1] before the osmium postfixing step resulted in the disappearance of this structure. This is demonstrated in Fig. 14, which shows a series of 5 consecutive sections through one end of a centriole in a ribonuclease-treated preparation. Ribosomes are now missing in the cytoplasm of this cell, so we can reasonably be sure that the enzyme was effective. The rotation photographs of the 3 centriole sections in Fig. 14 are presented in Fig. 15. Microtubules, fibrous appendages, and the octagonal end structure in the end section are all present. However, the dense foot normally attached to the triplet base and microtubule A is not visible. Fig. 8 demonstrates that the foot should be present at this end of the centriole, even though the triplet base is reduced in size. The presence or absence of the spiral structure after ribonuclease digestion is difficult to determine, but we believe it is still present in Fig. 14e as an irregular filament in the lumen of the centriole, as diagrammed in Fig. 16. This filament seems still to be attached to one triplet base (arrow) where the footlike structure normally appears.

The analogous treatments with deoxyribonuclease have been inconclusive, and the chemical composition of all other centriole components remains unknown.

[1] Ribonuclease digestion: RNase (Worthington) (2 mg/ml) dissolved in 0.1 *M* phosphate buffer pH 5.0; 60 minutes at 37°C. The enzyme solution was heated to 100°C for 5 minutes prior to use to destroy any contaminating enzymes.

---

FIG. 13. Symmetry tests of the image shown in Fig. 5. The lack of 9-fold rotational symmetry in the central structure in the distal end of the centriole (Fig. 8) prompted a search for other possible symmetries of this component. In (a) the triplets and external fibers reinforce well when tested for 3-fold symmetry (see text), but the central structure does not. In (b) through (f) are presented test images for 4-, 5-, 6-, 7-, and 8-fold rotational symmetry, respectively. The best reinforcement is seen in (b) and (f), where a central image with 8-fold symmetry is seen in both cases. We have termed this the *octagonal end structure*.

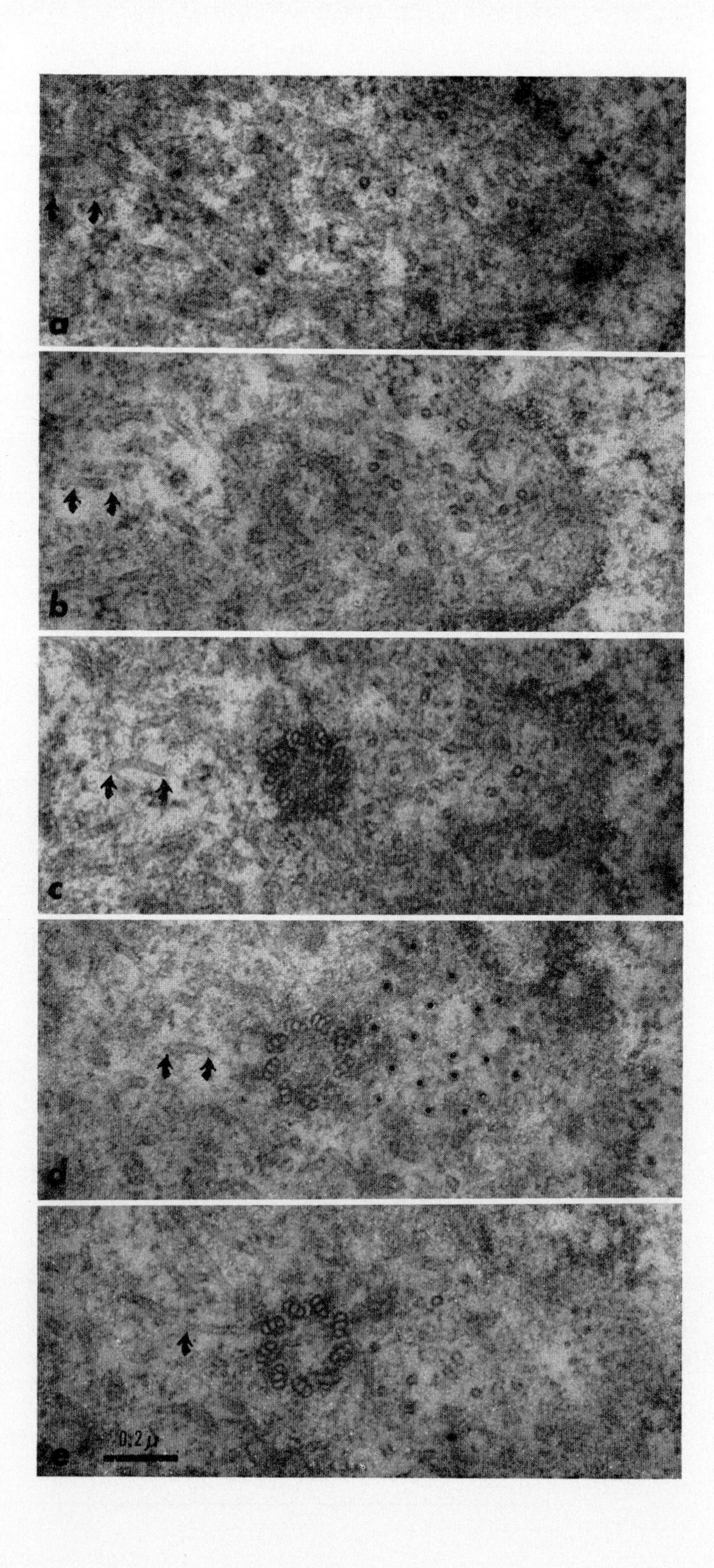
a
b
c
d
e
0.2 μ

## REPLICATION OF THE CENTRIOLE

In favorable cases where centrioles could be seen and counted by light microscopy [23], the centriole population in most animal cells numbers 2 at the start of a cell reproductive cycle (immediately after mitosis). Sometime before the cell divides again, the centrioles must be duplicated, just like all other cell components. When and how does this occur?

In living Chinese hamster fibroblasts the centrioles are not visible because of the multitude of small granules and vesicles around them in the cytoplasm of these cells. However, there are dilute alcohol solutions designed for direct isolation of the mitotic apparatus [17] that stabilize cell organelles containing microtubules while allowing most other cellular components to dissolve. For reasons unknown, the addition of digitonin to cells fixed briefly in Kane's fixative [17] brings out centrioles and other microtubular organelles in sharp definition when viewed with a phase contrast microscope [3]. Examples are given in Fig. 17. Since Chinese hamster fibroblasts can be easily synchronized by selection of mitotic cells [38], the question of when centriole reproduction occurs can be answered, in part, by direct counting in synchronous cultures.

Figure 18 shows a series of photographs of daughter cells prepared as described above at various intervals after mitosis. The generation time for cells is about 12 hours. Two centrioles are present in each cell during the first 8 hours of the cell cycle; by 10 hours almost all of the cells have 4 visible centrioles. This experiment does not tell us anything about the early stages of centriole replication; we learn only about the growth of the procentrioles into structures large enough to resolve with a light microscope. We must rely on electron microscopy to provide the details of the early events.

Centrioles reproduce by a *generative* mechanism, i.e., each centriole is the maturation product of a centriole *germ*, termed the *procentriole*, somehow produced by a parent centriole. This mechanism is in contrast to the process of fission, where a parent structure divides into two equiv-

---

FIG. 14. Serial cross sections through the distal end of a centriole. After glutaraldehyde fixation the preparation was treated with RNase before osmium fixation and embedding. Ribosomes and nucleoli were effectively removed by this procedure. Most of the centriole structure was left undisturbed; however, the octagonal end structure in (c) is displaced to one side. There were other more subtle effects shown in Figs. 15 and 16. To the right of the centriole is seen a large bundle of microtubules in cross section; their spacing is indicated in (d). This bundle is probably directed toward the second unseen parent centriole nearby. The consecutive positions of a single microtubule in each section is indicated by arrows. The cell nucleus is at the extreme right.

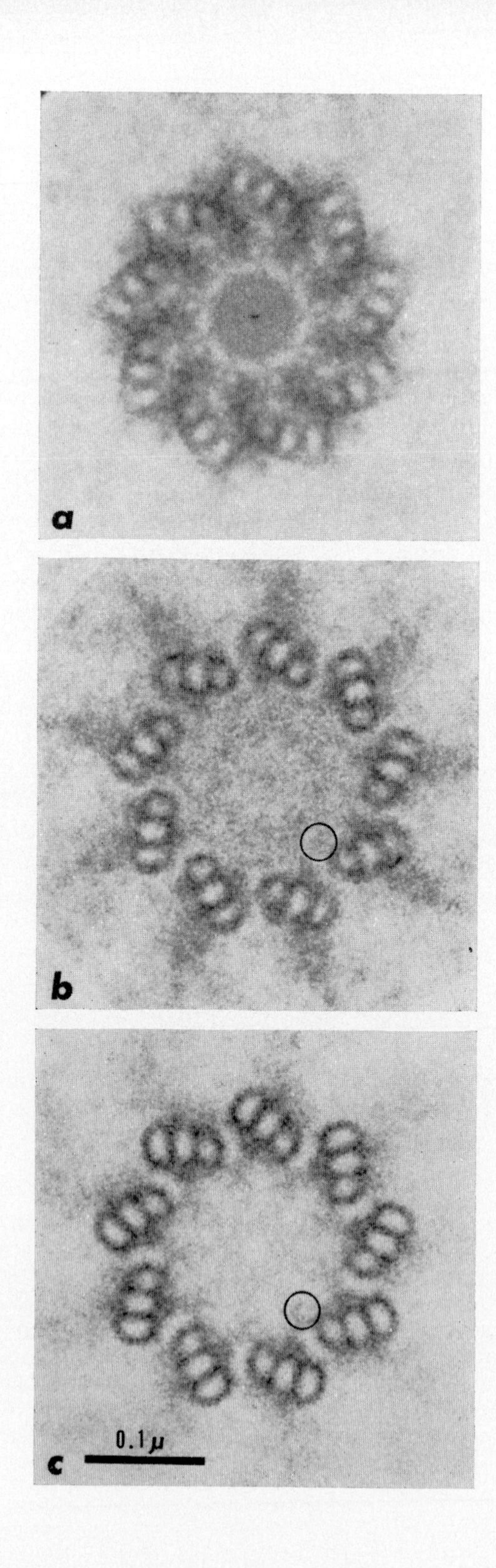
a
b
c
0.1μ

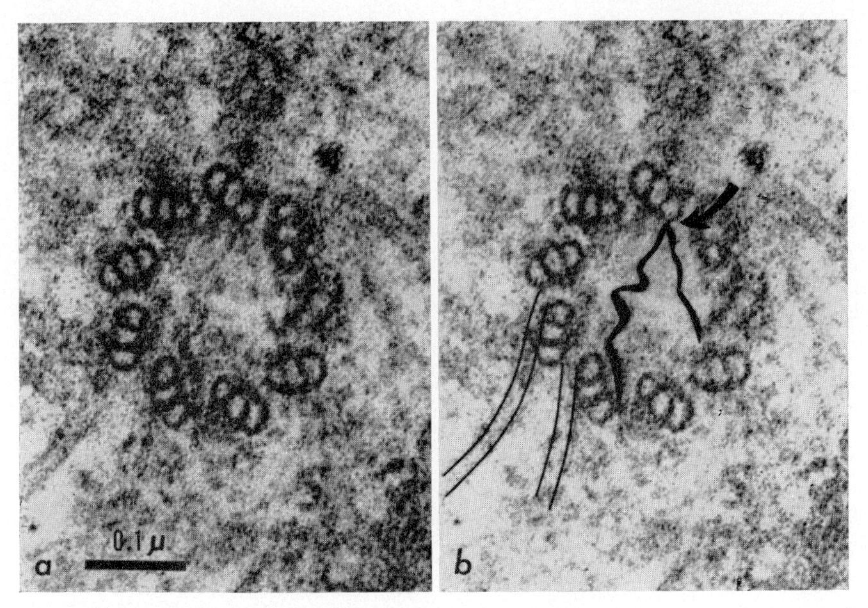


Fig. 16. (a) Enlargement of Fig. 14e. What is thought to represent a segment of the *internal helix* is seen in the lumen of the centriole, diagrammed in (b); at the point indicated by the arrow, the filament seems still attached where a "foot" structure usually appears.

alent daughter units [see Mazia, 23]. From our knowledge about the architecture of centrioles, considerable information can be inferred about the reproductive process.

The earliest event so far detected in centriole reproduction is the appearance of the procentriole near one end of the parent, as shown in Fig. 19. Viewed in its various aspects, the procentriole appears to be a very short cylinder of almost the parental diameter and possessing many of the cross-sectional features of the parent [9]. As it matures, the cylinder simply lengthens at right angles to the parent until it reaches full size. Figure 20a shows 2 parent centrioles in cross section, each with a maturing procentriole perpendicular to it. In the usual case each parent forms only one daughter, but potentially many procentioles can be produced simultaneously by a single parent. In the snail *Viviparus,* Gall [9] demonstrated that mulberry-like clusters of procentrioles around the parent are possible, each daughter roughly at right angles to the parent.

---

Fig. 15. High-resolution rotation photographs of the cross sections of the centriole in Fig. 14. Note the absence of a dense "foot" near microtubule A (circle) in (b) and (c) (compare with Figs. 6–8).

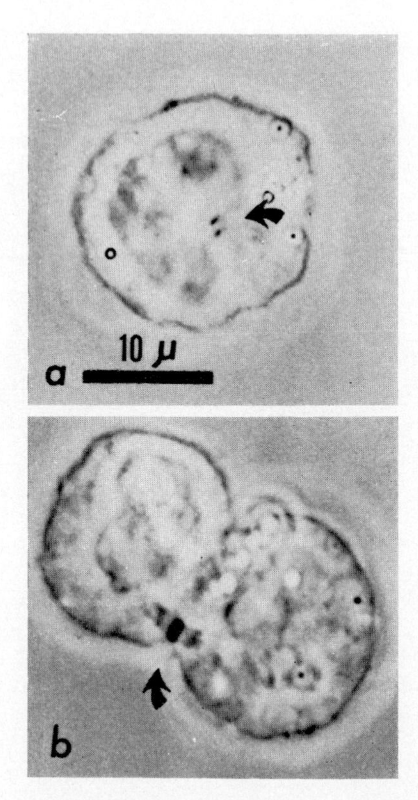


FIG. 17. Phase-contrast micrographs of centrioles (a) and Flemming body (b) (arrows) in cells treated with Kane's fixative and digitonin (see text).

Why does the procentriole develop perpendicular to the parent? A rather simple explanation is suggested by an understanding of the relationship of a mature centriole to the microtubules around it. As we demonstrated earlier, microtubules are attached to the triplet base and project approximately at right angles to the centriole wall. If the centriole actually functions as a producer of microtubules, which is almost certainly the case [23], then the microtubules projecting vertically around the centriole wall are the ones made by the centriole. If the procentriole relies on the parent as a source of microtubules for its outer wall, it will develop perpendicular to the parent, for its building blocks are already oriented that way. The observation that some of the microtubules of the developing procentriole actually terminate in the parent wall as in Fig. 20 (lower centriole pair) adds considerable support to this view.

What organizes the parent microtubules into the precise structure

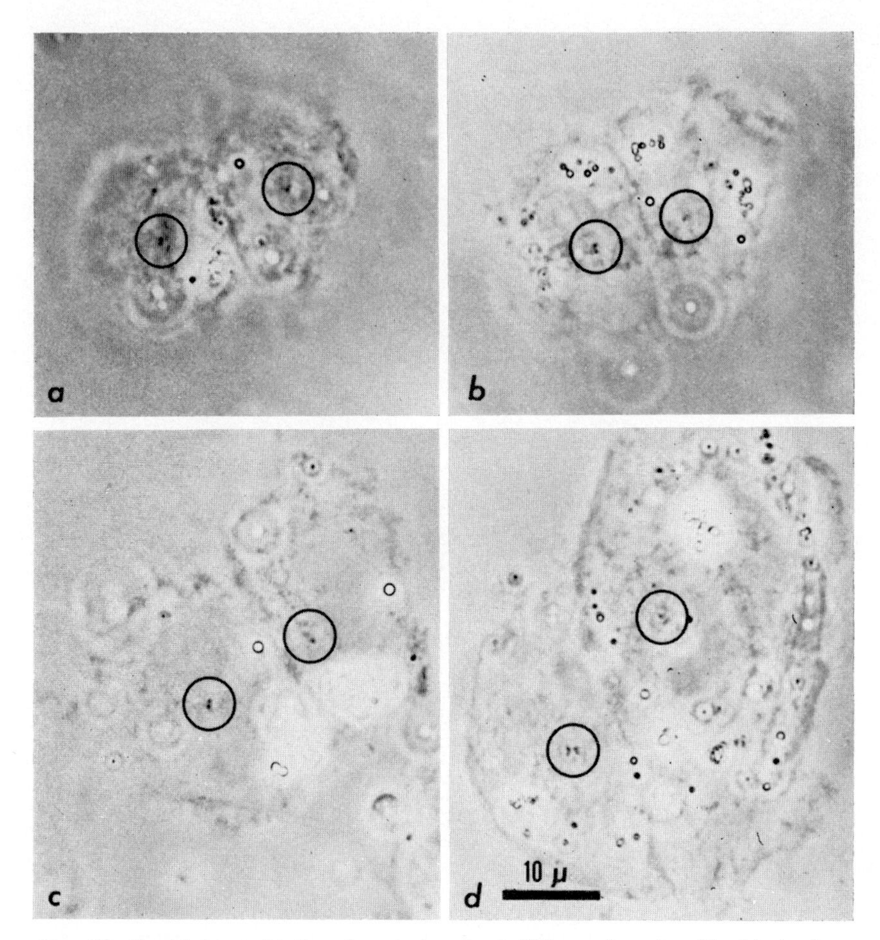


FIG. 18. Centriole replication in synchronized Chinese hamster fibroblasts (prepared as in Fig. 17). In each case the centrioles in the cell on the left are in focus; the other cell centrioles are somewhat out of focus. In (a) through (d) the cells were 3, 5, 8, and 10.5 hours old, respectively. The average generation time for these cells is 12 hours. Daughter centrioles become resolvable in the light microscope after 8 hours, although earlier replication stages are seen as early as 6 hours in electron micrographs (Fig. 19).

of 9-fold symmetry seen in the procentriole? Theoretically, the component best suited for this activity seems to be the cartwheel. It has several properties that are requisite for a procentriole organizer. (a) It has 9-fold symmetry and can thus serve as an appropriate attachment template for microtubules. (b) Whereas it seems to be a single structure in the centrioles of Chinese hamster fibroblasts, occurring only in one end section, it is clearly a multiple structure in basal bodies of

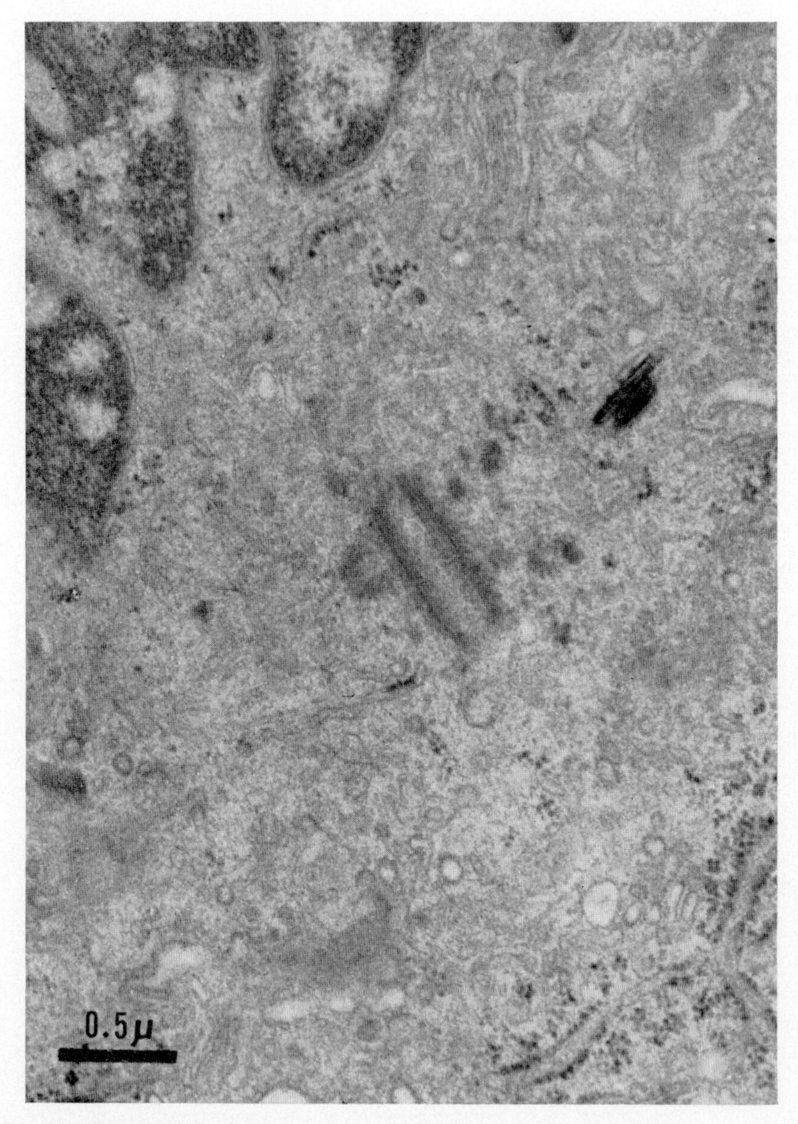


FIG. 19. Procentriole formation in a cell 6 hours old. The daughter forms at right angles to the parent as a short cylinder with most of the cross-sectional features of the parent.

---

FIG. 20. (a) Two parent centrioles (both in cross section) with immature daughter procentrioles sectioned longitudinally. The positions of all the microtubules in the photograph are emphasized in (b), and most of them are oriented toward the two parents, rather than the daughters. Note the apparent penetration of microtubules through the nuclear membrane (large arrow) and the numerous microtubules directed toward the field of nuclear pores at the extreme right. A fine fibril

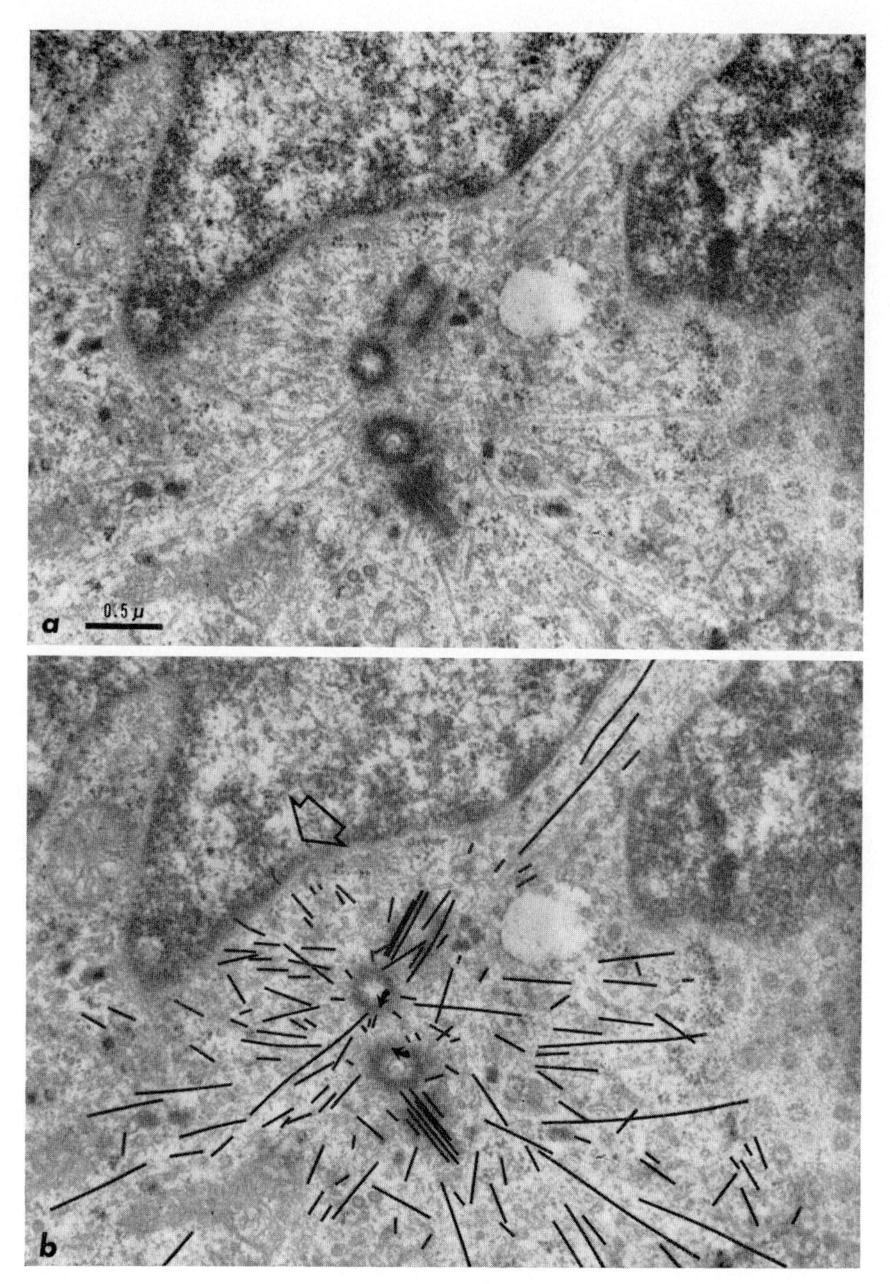


connecting the two parent centrioles is indicated by the small arrows in (b). In the lower parent centriole, the oblique section has included a large part of a turn of the *internal helix*. Also note that two of the microtubules in the wall of the lower procentriole are still connected to the wall of the parent. See Fig. 12 for a later stage of procentriole maturation.

ciliates [12] occurring repeatedly throughout much of the length of the centriole lumen. This fact suggests that the cartwheel may be able to reproduce itself. (c) It is present in procentrioles [9].

Of course, the details of centriole reproduction are largely beyond our current understanding. To suggest that the cartwheel is the procentriole organizer only converts the question, "How do centrioles reproduce?" into "How do cartwheels reproduce?" However, in the cartwheel we at least have a structure of molecular dimensions where replication is theoretically simpler, if technically more difficult to study.

## CILIOGENESIS

Recently we reported that Chinese hamster fibroblasts treated briefly during interphase with Colcemid were stimulated to generate cilia [37]. The details of ciliogenesis were already known from studies in other species [30, 32, 36], but the opportunity to study the process *in vitro* at timed intervals after stimulation was informative.

The first visible step in ciliogenesis is the approach of a small cytoplasmic vesicle (possibly of Golgi origin) to one end of a centriole. The vesicle flattens and invaginates to form a double-membraned cap over the end of the centriole (Fig. 21a). Microtubules A and B of each triplet extend into the cap and the ciliary bud is formed (Fig. 21b). At this stage the mechanism generating the axial doublet of the cilium is also formed. Ciliary bud formation is somehow promoted in Chinese hamster fibroblasts by treatment during interphase with Colcemid [37]. The mechanism is unknown. However, the growth of the ciliary bud into a cilium does not occur so long as the Colcemid is present. One hour of Colcemid treatment is long enough to complete bud formation. If the cells are then transferred to medium lacking Colcemid, the shaft of the cilium rapidly develops. In the span of 2 hours cilia up to 15 $\mu$ long can be produced. Each cell is potentially able to make 2 cilia, but actually only about half of the cells make cilia, and many of these are short and immature. However, some cilia seem to be structurally complete (Fig. 22) and in a few instances have been seen to beat erratically.

Whether or not cells that form cilia are capable of further division is not known. We have never seen any cilia in dividing Chinese hamster fibroblasts [37]. The experiment shown in Table II indicates that ciliogenesis occurs only during the latter half of the cell cycle, after centriole duplication has occurred. Since only the functioning parent centrioles participate in ciliogenesis [30, 37], the cells may retain the daughter centrioles for mitotic activity at a later time. The fate of either the cilia or the cells producing them has not yet been examined.

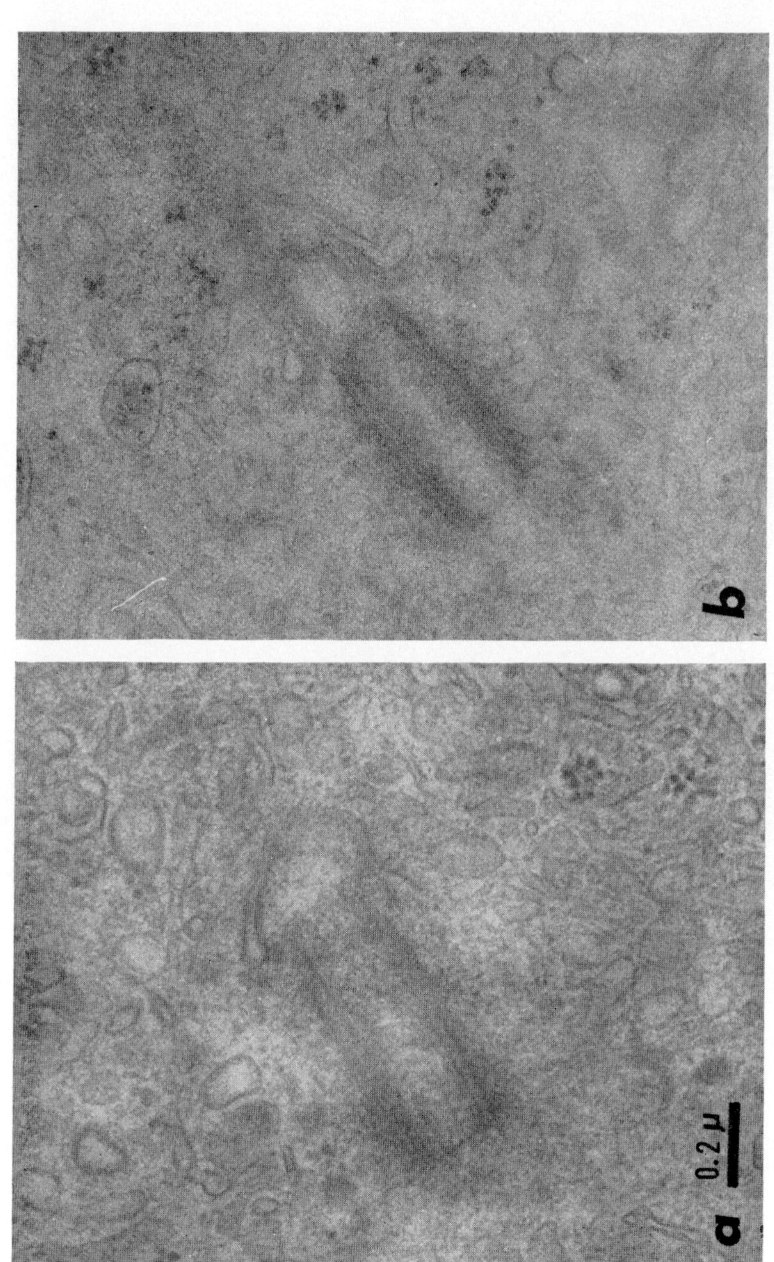


FIG. 21. Early stages of ciliogenesis in Chinese hamster centrioles. In (a) a vesicle has flattened across the distal end of the centriole to form a double membrane cap. In (b) centriole microtubules extend into the cap to complete the formation of a ciliary bud. Ciliary buds are formed in the presence of the drug Colcemid during the latter half of the cell reproductive cycle. After bud formation is complete, removal of the drug allows the cilium to develop as in Fig. 22.

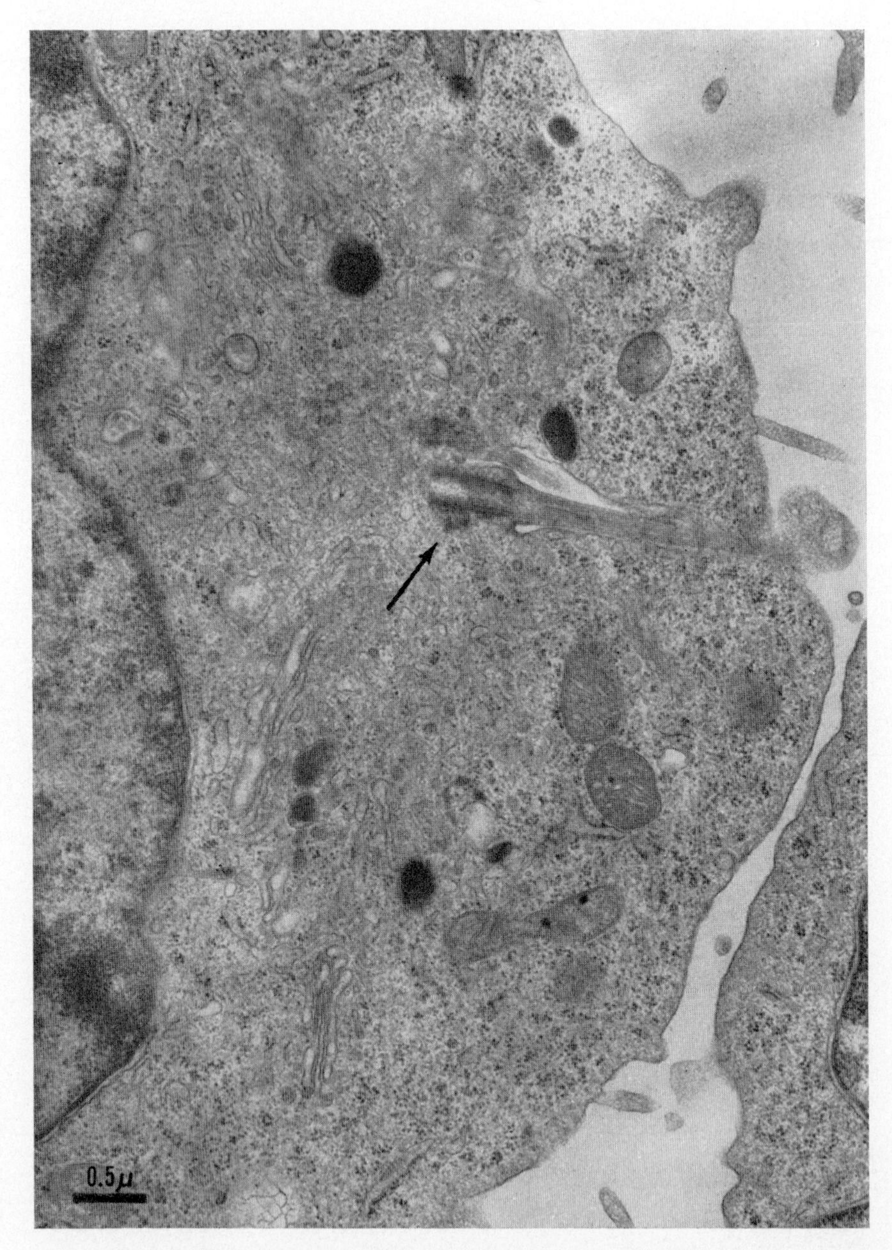


FIG. 22. Complete cilium formed in a Chinese hamster fibroblast by treatment with Colcemid (0.06 μg/ml) for 2 hours and further incubation for 1 hour in the absence of the drug. A second cilium is also present (obliquely sectioned). Arrow indicates the basal foot. Note the extensive development of the Golgi system in this cell.

TABLE II. *Colcemid Induction of Ciliogenesis in Synchronized Chinese Hamster Cells at Different Times in the Cell Cycle*[a]

| Culture No. | Colcemid treatment for synchronization (hours) | Age of synchronized culture at beginning of 2nd Colcemid treatment for 1 hour (hours) | Post-Colcemid incubation (hours) | Percent of cells containing cilia |
|---|---|---|---|---|
| 1 | None | — | — | 0 |
| 2 | 3 | (Control) | (0 to 7) | 16 |
| 3 | 3 | 4 | 5 to 6 | 10 |
| 4 | 3 | 6 | 7 to 8 | 11 |
| 5 | 3 | 7 | 8 to 9 | 46 |
| 6 | 3 | 8 | 9 to 10 | 35 |

[a] Colcemid-treated metaphase cells were incubated in medium lacking Colcemid to initiate the synchronous cultures [38]. At various times after mitosis the cells were again treated with Colcemid (0.06 μg/ml) for 1 hour to establish ciliary buds. This was followed by an additional hour of incubation in medium lacking Colcemid to allow the cilia to grow (post-Colcemid incubation). The cells (attached to coverslips) were then fixed and "stained" with digitonin as in Fig. 17 to reveal cilia, an example of which is shown in Fig. 23. Contaminating interphase cells in the original mitotic cell population produce cilia in response to the first Colcemid treatment (Culture No. 2), whereas untreated exponential cultures make no cilia (Culture No. 1).

Which end of the centriole is involved in ciliogenesis? By definition Gibbons and Grimstone [12] and Gall [9] termed the end attached to the cilium the *distal* end of the centriole and the opposite end the *proximal* end. Gibbons and Grimstone [12] also established that in ciliates the cartwheels are situated in the proximal end of the basal body. Although the centriole has radial symmetry, it does not possess bilateral symmetry. It necessarily follows that there are two possible centriole enantiomorphs, i.e., right- and left-handed forms. Ciliate basal bodies all appear to be of one type; possibly all species have centrioles of the same enantiomorph. When viewed along the major axis with the distal end (bearing the cilium) away from the viewer, the triplet planes slant away to the left at the top (counterclockwise), as in Figs. 2–10. Gall [9] cites an unpublished study indicating that parent and daughter centrioles are oriented with their proximal ends together in Lepidoptera, the distal ends later producing flagella without the pair first separating. We also showed that in the Chinese hamster centriole ciliogenesis and procentriole formation occur at opposite ends [37].

## THE CENTRIOLE IN MITOSIS

The role of the centriole in mitosis was reviewed extensively by Mazia [23]. More recent results have in general supported some older

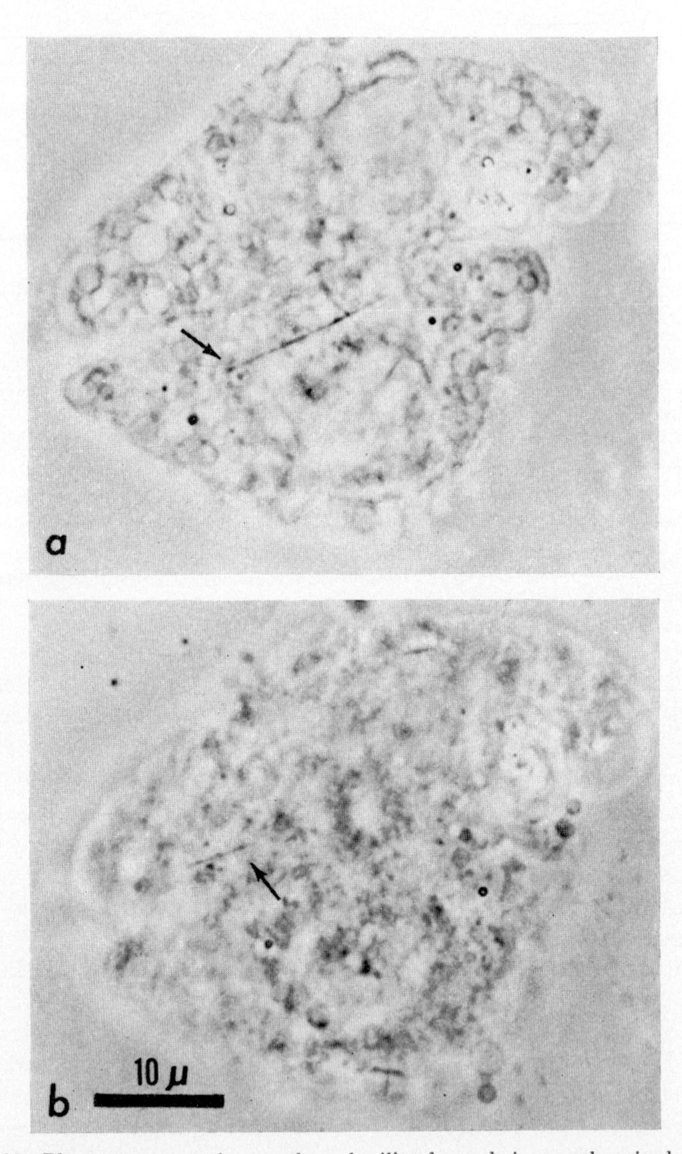


FIG. 23. Phase-contrast micrographs of cilia formed in synchronized Chinese hamster fibroblasts (treated as in Fig. 17). Each daughter contains a single cilium (arrows). Cells were 7 hours old when treated with Colcemid for 1 hour to initiate ciliary bud formation, and the culture was incubated an additional hour in the absence of the drug to allow the cilia to develop.

views of centriole activity in mitosis, and a simple updating of the information given by Mazia is all that is necessary in this section.

The mechanism which is responsible for the partitioning of daughter chromosomes into 2 daughter cells is called the mitotic apparatus. In animal cells it consists of 3 main components: the chromosomes, the centrioles, and a superstructure made of microtubules traditionally referred to as the spindle. Some higher plants, however, seem to lack centrioles entirely [23], so we may tentatively conclude that the centrioles are not an essential element of mitosis in certain cases. As a generalization, with certain special exceptions [33], we may state that centrioles are probably essential in cells which do not have a rigid wall.

Our recent ultrastructure studies of the effects of Colcemid on the mitotic process in Chinese hamster fibroblasts [2, 3] have shed some light on the role of centrioles in mitosis. Cells entering mitosis in the presence of Colcemid seem normal in all respects with one exception. Whereas normally the 2 centriole pairs separate to establish the poles of the mitotic spindle, in the presence of Colcemid the centrioles do not separate. Instead, the 4 remain clustered in the center of the cell, and chromosomes distribute peripherally around them. Careful examination revealed the presence of microtubules in the cytoplasm, but most of them were attached to chromosomal kinetochores, while near the centrioles most microtubules had disappeared. Furthermore, in each chromosome microtubules were attached to, and presumably made by, only one of the two daughter kinetochores. The one with microtubules was always oriented toward the cell center where the centrioles were.

Upon removal of Colcemid, the cells were able to divide normally in about 30 minutes. Within 5 minutes the centriole pairs began to separate, with an abundance of microtubules appearing all around them. Then the chromosomes began to move to the equator of the spindle between the centrioles, and microtubule attachment to both kinetochores on each chromosome became apparent. Within 20 minutes after removal of Colcemid most cells had established a metaphase configuration that was normal in all respects, so far as we were able to discern.

The structural defects observed in the Colcemid-blocked mitosis can be explained as the result of the effect of the drug on a single process, the formation of microtubules by the centriole. That Colcemid does not destroy all microtubules in the spindle, at least at the dosage which we employed, is seen in the fact that those microtubules attached to chromosomal kinetochores persisted. The studies of Inoué and coworkers [8, 16] very strongly suggest that the spindle is a structure in dynamic equilibrium, with simultaneous synthesis and degradation of its component microtubules. In plants, where mitotic spindles are produced in the absence of centrioles, there is good reason to believe that the microtubules are somehow organized by the kinetochore region of each chromosome; at least they are attached there. The most coherent picture emerging from these observations is consistent with the assumption that the microtubules of the mitotic apparatus are made in two places in animal cells, i.e., at the ends where they attach either in the kinetochore or to the centriole. Whether the "making" of microtubules means synthesis from amino acids or assembly of preexisting proteins is not clear, but the latter seems more probable [23]. If microtubules are in equilibrium, they are probably being disorganized at their free end.

In the presence of Colcemid, if one kinetochore of each chromosome is able to make microtubules, why is its sister unable to do so? The sequential activation of kinetochores may be a phenomenon of normal mitosis (see discussion by Mazia [23]). The rationale is that if the parent (older sister) kinetochore orients toward one pole first, the daughter (younger sister) kinetochore will be "aimed" at the opposite pole when it become active. The impression gained from Fig. 20 is that there are microtubules penetrating the nuclear membrane (large arrow) during late interphase which would perhaps substantiate the idea of a prior orientation of one kinetochore (the older sister) to the centrioles.

It seems equally apparent, however, that both active parent centrioles in Fig. 20 have such connecting microtubules, and since the two parents normally separate at prophase, each accompanied by its immature daughter, we must discount the notion that the microtubule associations in interphase are simply those left over from the previous mitosis. Figure 24, which depicts an anaphase spindle at one pole, clearly demonstrates orientation of the microtubules toward only one member of the centriole pair at the pole. Figures 1 and 11, which are longitudinal sections of centrioles in mitosis, prove that microtubules are oriented along the full length of the centriole cylinder, so we cannot suppose that the second member of the centriole pair in Fig. 23 is simply sectioned at the wrong level to reveal the microtubules around it. That only one centriole at each pole is active was also concluded by Murray *et al.* [27]. The inactive daughter centriole matures early in the following interphase to become the second parent (as in Fig. 20), and subsequently it may also secure connection to some kinetochore sites.

The clearest demonstration that centrioles make part of the microtubules of the mitotic apparatus is seen in the phenomenon of centriole separation during prophase [23]. That this phenomenon requires microtubule formation has been inferred from observations using phase contrast microscopy [4]. Our own observation of the appearance of microtubules around the separating centriole pairs following removal of Colcemid inhibition [2, 3] also strengthens this conclusion.

How is it that the production of microtubules causes centrioles to

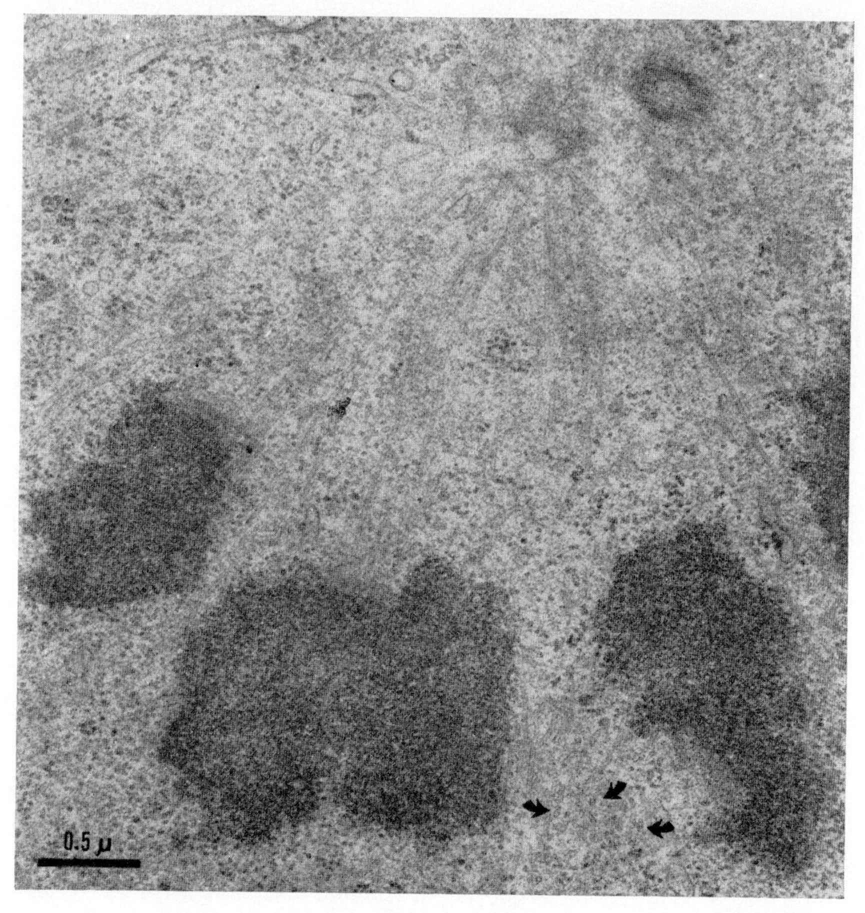


FIG. 24. Anaphase in a Chinese hamster fibroblast. The microtubules of the spindle are seen to converge on one centriole which was just grazed in this section; the second centriole, obliquely sectioned, is the immature daughter procentriole. The spindle microtubules are gathered into bundles; those reaching from pole to pole appear to be doublets in cross section (arrows).

separate? Individual microtubules appear to be relatively stiff structures, but they can be bent over short distances. In the mitotic spindle they occur most frequently in *bundles,* rather than single units, and such aggregate structures are probably much more rigid. The bundles are large enough to be seen with light microscopy as the spindle *fibers.* In cross section, spindle fibers are seen to contain about 10 microtubules (Fig. 25). Such bundles of microtubules are also seen in interphase (Figs. 14 and 20) in the region of the centrioles. If we think of the bundle of microtubules as the active unit of the spindle mechanism, then it is possible to formulate some tentative explanations for centriole and chromosome movement during mitosis.

The reason why microtubules associate into bundles can readily be ascribed to their molecular structure. They are likely to be highly polarized structures, because of their great length and their attachment at one end to relatively massive structures, the centriole or the chromosome. Microtubules growing from one centriole would be expected to diverge in many directions, since the like charges at their free ends would repel each other. One such microtubule, upon encountering a microtubule from another centriole, would readily align with it if the two could align with opposite polarity. Charge forces would immediately draw the free ends of each microtubule toward the opposing centriole. In like fashion other microtubule pairs would align and join together to form bundles of microtubules connecting the two centrioles. As each centriole lengthens the microtubules attached to it, the two will effectively propel themselves to the opposite poles.

The microtubules of the mitotic spindle form two kinds of bundles [23], those which run from one parent centriole to the other (pole-to-pole fibers) and those which connect chromosomal kinetochores to the poles (chromosomal fibers). The former are not easily distinguished from the latter until anaphase, when the pole-to-pole fibers are seen at the spindle equator between the two sets of separating chromosomes. We may suppose that the same forces act to form bundles of microtubules between kinetochore and centriole as we have suggested for centriole-to-centriole fibers. From our earlier description we recall that the kinetochore microtubules appear to retain their association with the centrioles throughout interphase, although some kinetochores may shift their association to the newly active daughter centriole. Such permanent orientation to the centrioles should effectively prevent the mistake of forming kinetochore-to-kinetochore bundles. The additional feature of delayed activation of the "new" kinetochores, as discussed earlier, also will help to prevent misalignments.

Additionally, it may be supposed that the charge distribution is different in kinetochore and centriole microtubules. If the centriole microtubules are more highly polar than the kinetochore microtubules, so that one centriole microtubule could effectively align with several kinetochore microtubules, then centriole microtubules could perhaps disrupt any misaligned kinetochore-to-kinetochore bundles by competing more strongly for the kinetochore microtubules.

The idea of chromosomal fibers containing mostly kinetochore microtubules becomes even more probable when we take into account the number of microtubule attachment sites on the centriole. If microtubules attach to the triplet bases in a spiral pattern that follows the large helix in the centriole lumen, and if the helix makes 10 turns in the total centriole length, then there are 90 attachment sites per centriole. Normally, there is only one active centriole at each pole, so we must construct the entire spindle structure using only 180 centriole-attached microtubules (90 at each pole) plus an undetermined number of kinetochore microtubules. If we allot half of the microtubules to make 9 pole-to-pole bundles of 10 microtubules each (5 from each pole in each bundle) then we are left with only 2 microtubules to form chromosomal bundles to each of the 22 kinetochores in the Chinese hamster cell. The problem is more acute in cells with higher chromosome numbers, where we must consider alternative ideas. The attachment sites on the centriole may bear no direct relationship to the internal helix, in which case there may be more sites. However, the upper limit is about 300 sites if the microtubules were packed tightly along the 9 triplet bases. On the other hand the centrioles may make microtubules on the triplet bases and then transfer them to an auxiliary attachment site, such as the pericentriolar satellite bodies, in order to make room for more microtubule production. The actual solution may be simpler than any of the foregoing. In species with many chromosomes, more microtubules may be provided by an earlier maturation of the daughter centriole, so that it also can function during mitosis. This would provide about 180 centriole-attached microtubules at each pole, which would be plenty for most species. In polyploid cells it might be anticipated that there is more than one centriole pair at each pole, since such cells frequently are multipolar. Such an assumption would account for the 500–600 pole-to-pole microtubules counted by Krishan and Buck [19] in L cells.

In connection with the foregoing speculation about microtubule bundle formation, it is interesting to note that there appear to be two types of bundles of microtubules in electron micrographs of dividing cells. These are shown in Figs. 25 and 26. In the one type (Fig. 25) all the microtubules are single, and there are usually 2 or 3 in the center of the bundle with 8 or 9 others peripherally distributed around them. In the second type (Fig. 26) some of the microtubules occur as *paired doublets*. Krishan and Buck [19] have already determined that the pole-to-pole fibers contain doublet microtubules, and this is in complete agreement with the concepts that we have presented here. The microtubules from one centriole, in forming pole-to-pole fibers, interact with the microtubules from a second centriole that have equal charge but opposite polarity (free ends point in opposite directions) to form doublets. The single fibers seen in such bundles may actually be double at a point nearer the spindle equator. In the case of the bundles running from the kinetochore to the poles, the microtubules of

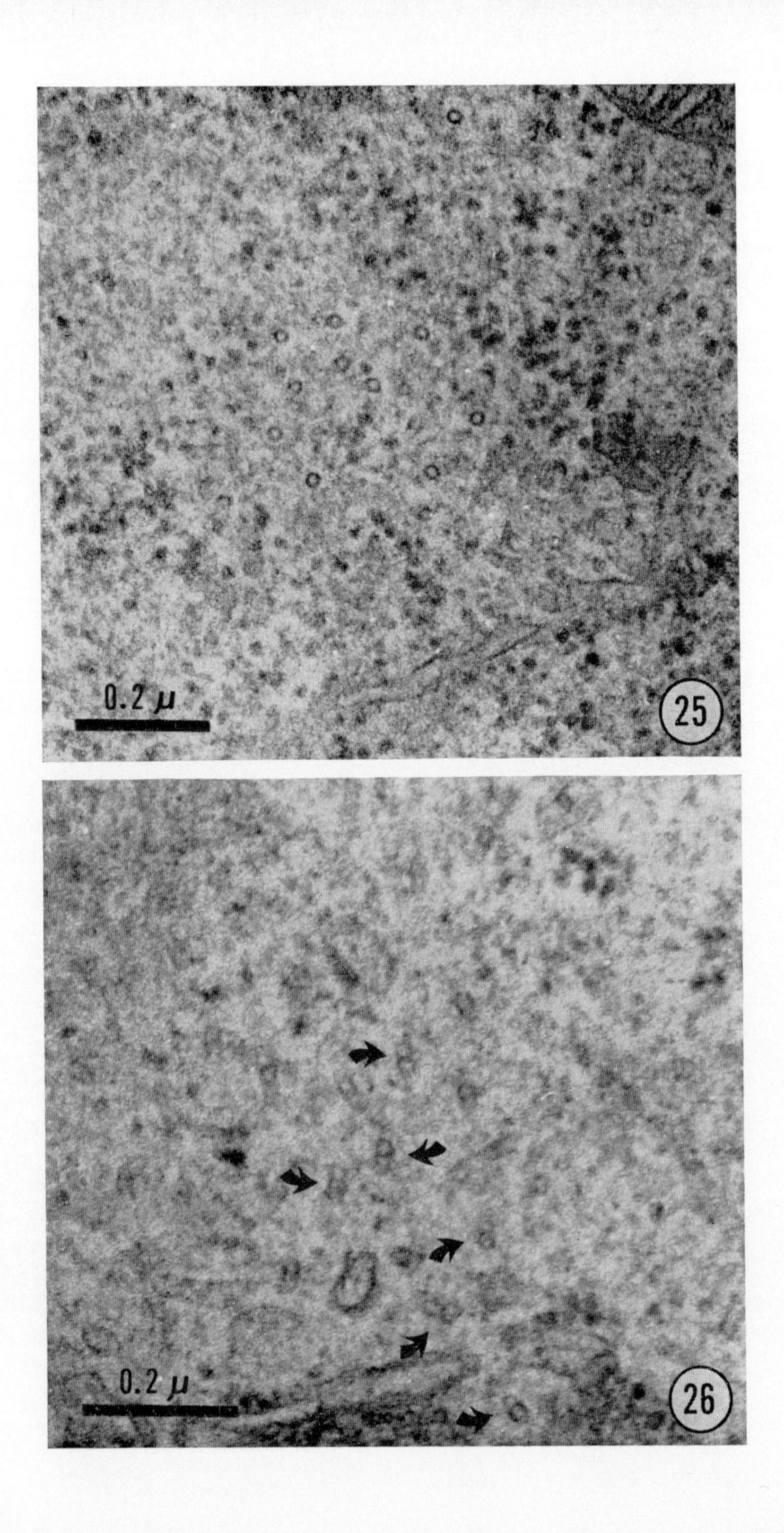
0.2 μ
25
0.2 μ
26

opposite polarity do not form doublets, but instead, a distinct space is maintained between them. Such an arrangement would possibly allow the kinetochore microtubules to slide along the centriole microtubules, which is important to our concept of an anaphase mechanism.

During anaphase the chromosomal fibers are shortened, and this can be accomplished by any process, probably enzymatic, which degrades the free ends of the kinetochore microtubules in the vicinity of the centrioles. This is another possible function of the pericentriolar satellites. Actually the chromosomal microtubules are probably in an equilibrium state all the time, with simultaneous synthesis at the kinetochore and breakdown at the centrioles [8]. Anaphase could be initiated by either accelerating degradation or decreasing the rate of synthesis. If there are only one or two centriole microtubules in the chromosomal fibers, then we need not specify their equilibrium state. At anaphase the free ends of the centriole microtubules could pass directly through the metaphase chromosome with no ill effects, and such micrographs are occasionally seen (see Robbins and Gonatas [31], their Fig. 16). If the centriole microtubules are more polar, we might postulate that their free ends do not degrade as rapidly as those of the kinetochore microtubules. The proposed model would then account for the simultaneous decrease in chromosomal fiber length at anaphase, while the pole-to-pole fibers actually lengthen, if we speculate that production of centriole microtubules increases in anaphase, perhaps as a response to increased availability of precursors arising in the degradation of kinetochore microtubules.

The finding that plant cell walls have microtubules embedded in them [20] allows us to generalize our picture of a mitotic mechanism to include plant cells. All that is required is to specify that the end walls of a plant cell are able to produce microtubules, and the need for a centriole in plant cell mitosis is eliminated. The interaction of kinetochore microtubules with microtubules anchored in the end walls to make bundles of microtubules and the digestion of the free ends of the less polar kinetochore microtubules will account for metaphase alignment and anaphase movement analogous to the situation in animal cells. Although there are exceptions [33] mentioned earlier in this section, it appears that cells without rigid walls must manufacture a temporary scaffolding of some sort to anchor the poles toward which

---

FIG. 25. Cross section of a bundle of microtubules probably connecting a kinetochore region of a chromosome with a centriole at one pole of a spindle.

FIG. 26. Bundle of pole-to-pole microtubules in cross section. Note that many appear to be doublets (arrows).

the chromosomes move in mitosis. This seems to be the major function of the centriole in mitosis.

## DISCUSSION

Centrioles appear to be capable of activities associated with 3 major cellular events: mitosis, ciliogenesis, and centriole self-replication. Although each of these processes is distinctly different, all have certain aspects in common. We will attempt a correlation of what we know about centriole architecture with a hypothetical description of centriole activity in a molecular sense.

Basically, the centriole appears to be a specialized machine for microtubule production, since it produces microtubules for the mitotic apparatus, for the outer wall of the daughter centriole, and for the shaft of the cilium (or flagellum). However, the microtubules are made by two distinctly different processes. The microtubules of the mitotic apparatus are produced at right angles to the long axis of the centriole; these are attached to the structure we have termed the *triplet base*. On the other hand, the microtubules in the wall of the cilium are direct extensions of microtubules A and B in the centriole triplet [12], and are therefore produced parallel to the long axis of the centriole. A third type of microtubule, the central pair of the cilium, appear to arise from the basal plate and also extend parallel to the centriole long axis. Thus, although the microtubules all look alike, and may even be made of identical precursors, there are obviously several separate manufacturing processes, insofar as site on the centriole and direction of synthesis are concerned. In addition we can consider the microtubules which arise in the kinetochore and the plant cell wall as further examples of similar microtubles made in other sites and directions. Of these, perhaps the simplest in structure is the chromosomal kinetochore. Although we do not yet know the chemical makeup of this structure, we can make some reasonable conjectures about how it may function in microtubule production.

The kinetochore seems to be an integral part of the chromosome, and it has the same general "lampbrush" structure of the meiotic chromosome [2]. Thus it is composed of two major components, a pair of cohelical axial filaments and many fine filamentous loops extending vertically from the axial filaments to form the "brush." There are two such structures running parallel through the centromere region of each chromatid. Although the chemistry of the kinetochore is unknown, we may reasonably suppose that it is similar to the synaptinemal complex in meiosis, where some tentative conclusions have been drawn about the chemistry by Moses and Coleman [25]. In short, it is probably safe to say that the kinetochore is composed of nucleoproteins, but it is not yet known which component contains DNA or RNA.

The very nature of the kinetochore suggests that it is an open, active segment of the metaphase chromosome, while the remainder of the chromosome appears to be compacted into an inert mass (except for nucleolus organizer sites, see Hsu *et al.* [15]). If the kinetochore is doing anything in mitosis, the most logical assumption is that it is making RNA. What has RNA synthesis to do with the formation of microtubules?

Chemical analysis of the mitotic apparatus from sea urchin eggs have consistently indicated that about 5% of the structure is RNA, and this suggests that the microtubules may actually be made of ribonucleoproteins [23, 39]. It is perhaps only a coincidence that there is a well-known case of a ribonucleoprotein structure that is a hollow microtubule of about the same diameter as spindle microtubules, and tobacco mosaic virus (TMV) also contains 5% RNA. The association of the RNA and protein in TMV is now well documented [18], with the protein subunits fitted in between the turns of the long RNA helix. The TMV structure is apparently self-assembled in the tobacco cell cytoplasm, and the RNA appears to determine the length of the TMV cylinder [1].

If the kinetochore produces a specific RNA capable of organizing microtubule subunits in a fashion analogous to TMV assembly, then we would expect the same process to occur at other sites of microtubule production, particularly in the centriole.

That the centriole contains RNA seems relatively certain; it probably also contains DNA. If there is DNA in the centriole, then we would anticipate its presence in the large helix in the centriole lumen. The analogous steps could then be postulated for microtubule formation in the centriole as in the kinetochore; i.e., the DNA helix would deposit RNA on the inner surface of the triplet base, which would then organize the microtubules extending vertically from the centriole wall.

The formation of the cilium microtubules is more difficult to imagine. Here we must consider the extension of existing microtubules A and B in the centriole wall, and a tentative explanation will require that we return briefly to the problem of daughter centriole formation.

The formation of *triplets* in the centriole wall can be understood on the same basis as the association of the pole-to-pole microtubules of the mitotic spindle into *doublets*. We recall that doublet formation would require only the contact of two microtubules aligned with *opposite* polarity; the doublets then form spontaneously. In the procentriole we would then logically expect triplet formation to proceed by several similar steps: (1) A microtubule of the parent (A) is first fused to a second microtubule (B) of opposite polarity (and thus from a separate source, such as the kinetochore). (2) The end of the doublet would then be disconnected distally, but left attached to the parent centriole to reestablish overall polarity of the doublet. (3) The third microtubule (C), of "foreign" origin, would be added and stabilized in position to form the triplet. (4) The A and C microtubules would then be disconnected at each end to complete the formation of the short triplet of the procentriole. The fact that microtubule A appears slightly smaller than B and C ([12] and our Figs. 6–8) also supports its different origin and polarity.

The walls of the procentriole appear to lengthen distally as the procentriole matures [9]. We are thus faced with the problem of explaining how the triplet lengthens and how it stops growing when the centriole attains the proper length.

A related question, which must be considered first, is whether a short microtubule segment is structurally alike at either end. Since it contains protein subunits, which are asymmetric, we might anticipate that the ends are asymmetrical; however, if the basic building block is actually a symmetrical protein dimer, then the ends could be alike. If the microtubule has an RNA component, however, then the two ends of the nucleic acid would be different, unless the RNA were double-stranded, which is improbable.

Returning to the TMV analogy, we would postulate the splicing of RNA molecules of a specific length into the distal ends of the procentriole triplet microtubules and the subsequent self-extension of the microtubules until the length of RNA is used up. However, the distal end of microtubule A may differ from those of B and C, if the polarity is opposite. The analogous process, with much longer RNA strands spliced into microtubules A and B, would account for the formation of the doublets of the cilium. In such a process, however, the growth point of the microtubules would always be at their distal tips; subunits would have to be shipped out to the end of the cilium before they became incorporated into the microtubule. We thus have reasons for specifying two different types of microtubule formation: in one case (spindle microtubules) the microtubule moves away from its fixed point of synthesis, whereas in the other case (cilium doublets) the point of synthesis moves away from the site of primary attachment on the centriole (basal body). The central pair of microtubules in the cilium may be of the former type, i.e., synthesized at the basal plate.

There are several features about the architecture of the centriole as a whole that suggest that it contains a rotation mechanism. Upon building a centriole model, we discovered that a few assumptions about the kinetic relationships between the parts turned the model into a very clever machine. The model is shown in Fig. 27 and is constructed

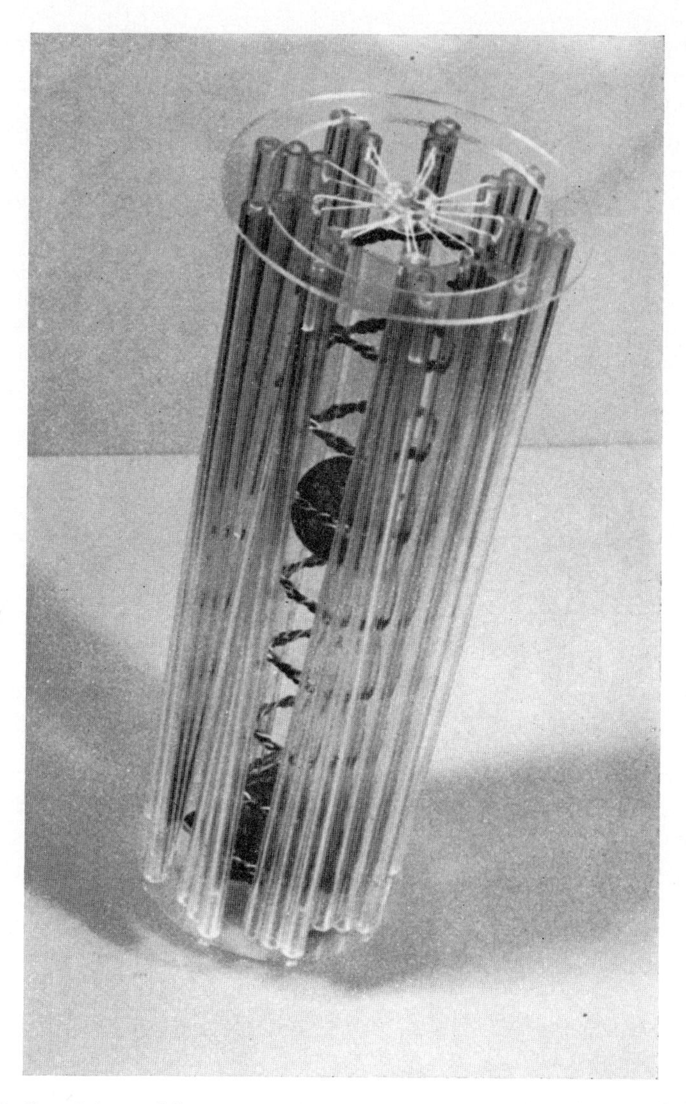

FIG. 27. Centriole model constructed to demonstrate possible kinetic relationship of its parts (for details see discussion in text).

with the triplets hinged along microtubule C. The cartwheel, in the proximal end, attaches to microtubule A and the triplet base of each triplet. The octagonal end structure in the distal end is attached to the spiral in the central lumen of the centriole. Assuming the cartwheel to be a contractile structure that pulls the triplets inward, one after the other in a counterclockwise sequence (viewed from the proximal end), the octagonal end structure becomes a ratchet that is driven clockwise and can thus be used to turn the central helix. If the helix were DNA, then this would be a very ingenious mechanism for moving the information source (template) past the fixed attachment sites of microtubule formation in the triplet bases.

The fact that the cartwheel is a multiple structure in basal bodies suggests that some such motion in the triplet blades may be related to ciliary beating; the added mass of the cilium would require the extra cartwheels to twist the doublets. We are currently studying this possibility.

Admittedly, much of what we have presented in this report is speculative and rather abstract. However tenuous the arguments may be, they have been presented with the aim of stimulating more investigation into the nature and function of a very fascinating, though complex, cell organelle, the centriole.

## SUMMARY

Ultrastructure studies presented in this report have indicated that the centriole of the Chinese hamster is a complex structure made of a number of separate components:

1. The *outer wall* is a hollow cylinder composed of 27 microtubules separated into 9 groups of 3, with the members of each triplet fused together to form a blade. The blades form an angle with the cylinder surface that varies continuously from one end of the centriole to the other in some centrioles, while no such twist is found in other cases. The twisting of the triplet blades may be characteristic of certain centriole functions.

2. A small spherical vesicle, termed the *central vesicle,* is found in the centriole lumen.

3. A filamentous structure spirals just inside the outer wall of the centriole; this *internal helix* makes about 8–10 turns through the entire length of the cylinder.

4. Each triplet blade rests on a diffuse *triplet base,* located inside the centriole wall and running parallel to the microtubules of the wall. The triplet base appears to be the site of assembly and attachment of *spindle* microtubules oriented perpendicular to the long axis of the centriole.

5. In one end of the mammalian centriole a *cartwheel* structure is found. Radial spokes connect the triplets of the centriole wall with a small central circle. The cartwheel is a multiple structure in basal bodies. We propose that it is a contractile structure capable of twisting the triplet blades in the centriole wall.

6. *External fibrous appendages,* projecting vertically from each triplet blade, are found near one end of both centrioles and basal bodies; in the latter case they are termed "transition fibers."

7. In the end opposite the cartwheel is a flat stellate structure with 8-fold symmetry, which we have called the *octagonal end structure.* It may be homologous to the basal plate of basal bodies.

8. *Pericentriolar satellites,* dense masses of irregular composition, are frequently found near the centriole; their function remains unknown.

9. A *procentriole* is frequently seen near a mature centriole. It possesses most, if not all, of the cross-sectional features of the mature centriole, but it is much shorter. It matures into a centriole by lengthening at right angles to the parent centriole to which it is attached.

These components, along with certain specialized structures found in basal bodies, such as the *basal foot* and *rootlets,* are the known parts of the centriole.

Chemically, the microtubules of the centriole wall are probably ribonucleoprotein, and RNA is also present on the inner surface of the triplet base. The chemical composition of the other centriole components remains undetermined.

The probable functions of the centriole in mitosis, ciliogenesis, and procentriole formation are discussed in detail, and a hypothetical model for microtubule formation is presented.

## Acknowledgments

The authors wish to thank Dr. T. C. Hsu for his helpful criticism and advice and also Dr. Arthur Cole for the use of his electron microscope. The competent technical assistance of Miss Patricia Murphy and Mr. John Carnes is also greatly appreciated.

This work was supported in part by research grants DRG-269 from the Damon Runyon Memorial Fund for Cancer Research, E-286 from the American Cancer Society, Inc., and HD-2590 from the United States Public Health Service, Institute for Child Health and Human Development.

## References

1. Anderer, F. A., *Advan. Protein Chem.* **18,** 1, (1963).
2. Brinkley, B. R., and Stubblefield, E., *Chromosoma* **19,** 28 (1966).
3. Brinkley, B. R., Stubblefield, E., and Hsu, T. C., *J. Ultrastruct. Res.* **19,** 1 (1967).
4. Cleveland, L. R., *J. Morphol.* **97,** 511 (1955).
5. DeHarven, E., and Bernhard, W., *Z. Zellforsch. Mikroskop. Anat.* **45,** 378 (1956).
6. Doolin, P. F., and Birge, W. J., *J. Cell Biol.* **29,** 333 (1966).
7. Fawcett, D., *in* "The Cell" (J. Brachet and A. E. Mirsky, eds.), Vol. II, pp. 217–297. Academic Press, New York, 1961.

8. Forer, A., *J. Cell Biol.* **25,** 95 (1965).
9. Gall, J. G., *J. Biophys. Biochem. Cytol.* **10,** 163 (1961).
10. Gibbons, I. R., *J. Biophys. Biochem. Cytol.* **11,** 179 (1961).
11. Gibbons, I. R., this symposium, p. 99.
12. Gibbons, I. R., and Grimstone, A. V., *J. Biophys. Biochem. Cytol.* **7,** 697 (1960).
13. Henneguy, L. F., *Arch. Anat. Microscop.* **1,** 481 (1897).
14. Hoffman, E. J., *J. Cell Biol.* **25,** 217 (1965).
15. Hsu, T. C., Brinkley, B. R., and Arrighi, F., *Chromosoma,* in press (1968).
16. Inoué, S., *in* "Primitive Motile Systems in Cell Biology" (R. Allen and N. Kamiya, eds.), Academic Press, New York, 1964.
17. Kane, R. E., *J. Cell Biol.* **25,** 137 (1965).
18. Klug, A., and Caspar, D. L. D., *Advan. Virus Res.* **7,** 225 (1960).
19. Krishan, A., and Buck, R. C., *J. Cell Biol.* **24,** 433 (1965).
20. Ledbetter, M. C., this symposium, p. 55.
21. Lenhossek, M., *Verhandl. Deut. Anat. Ges. Jena* **12,** 106 (1898).
22. Markham, R., Frey, S., and Hills, G. J., *Virology* **20,** 88 (1963).
23. Mazia, D., *in* "The Cell" (J. Brachet and A. E. Mirsky, eds.), Vol. III, pp. 77–412. Academic Press, New York, 1961.
24. Mazia, D., this symposium, p. 39.
25. Moses, M. J., and Coleman, J. R., *in* "The Role of Chromosomes in Development" (M. Locke, ed.), pp. 11–49. Academic Press, New York, 1964.
26. Mullins, R., and Wette, R., *J. Cell Biol.* **30,** 652 (1966).
27. Murray, R. G., Murray, A. S., and Pizzo, A., *J. Cell Biol.* **26,** 601 (1965).
28. Randall, J., and Disbrey, C., *Proc. Royal Soc.* (*London*) **B162,** 473 (1965).
29. Reese, T. S., *J. Cell Biol.* **25,** 209 (1965).
30. Renaud, F. L., and Swift, H., *J. Cell Biol.* **23,** 339 (1964).
31. Robbins, E., and Gonatas, N. K., *J. Cell Biol.* **21,** 429 (1964).
32. Roth, L. E., and Shigenaka, Y., *J. Cell Biol.* **20,** 249 (1964).
33. Roth, L. E., Obetz, S. W., and Daniels, E. W., *J. Biophys. Biochem. Cytol.* **8,** 207 (1960).
34. Ruthmann, A., *J. Biophys. Biochem. Cytol.* **5,** 177 (1958).
35. Schuster, F. C., *Anat. Record* **150,** 417 (1964).
36. Sorokin, S., *J. Cell Biol.* **15,** 363 (1962).
37. Stubblefield, E., and Brinkley, B. R., *J. Cell Biol.* **30,** 645 (1966).
38. Stubblefield, E., and Klevecz, R. R., *Exptl. Cell Res.* **40,** 660 (1965).
39. Zimmerman, A. M., *Exptl. Cell Res.* **20,** 529 (1960).

# STRUCTURE AND FORMATION OF SOME FIBRILLAR ORGANELLES IN PROTOZOA

A. V. GRIMSTONE

*Department of Zoology, University of Cambridge, Cambridge, England*

## INTRODUCTION

A description of organelle formation, if it is to be at all comprehensive, must ultimately be couched in molecular terms. This requirement obviously precludes for the time being a description of any degree of completeness, since in no case is the molecular structure of an organelle known in any but the most general terms. It is hardly possible to describe organelle formation in any detail without knowing what it is that forms. From this it follows that for the most part the fundamental questions of organelle formation, regarding this as a problem of controlled molecular synthesis and assembly, must at present be set aside, and attention directed instead toward elucidating organelle structure in greater, that is to say, molecular, detail. In this vein, the first section of this paper presents some new information on the molecular architecture of cilia and flagella.

Yet, while the ultimate goal must be explanation in molecular terms, there is obviously much first to be learned about structure and events at a supramolecular level of organization. Organelles may be of such formidable complexity that a direct approach to the problem of their formation in molecular terms is unlikely to be profitable. It is necessary first to break down events to a series of part processes simple enough to permit molecular description. The second part of this paper therefore presents some data about the more complex events in organelle formation as seen in certain ciliates and flagellates.

## SUBSTRUCTURE OF FLAGELLAR FIBERS

The structure of flagella has been explored in sectioned material down to the limits of resolution at present attainable by this method [4], with results that stop some way short of the macromolecular level. Further analysis at present depends on the one hand on the type of chemical dissection described by Gibbons [3] and on the other on a return to the study of fragmented material, such as was used in the early days of biological electron microscopy, before the advent of sectioning techniques. The new technical developments that have again made the latter line of investigation profitable, and extended its resolution down to the molecular level, are first, the use of negative staining, and second, the development of optical methods for analysis of electron micrographs [11]. The application of these methods to flagella will be described here. Most of the observations have been presented more fully elsewhere [9].

The flagella which have been examined in the present work belong to various species of the flagellate *Trichonympha*. This material has previously been studied by means of sections [4]. The flagella of these and other large flagellates from termites seem to be particularly suitable for study by negative staining, since they readily fall apart into their components without recourse to mechanical or chemical methods of fragmentation. The flagellar membranes are highly fragile and are always lost in negative staining unless steps are taken to preserve them by fixation, and this leaves the fibers freely exposed. Most of the observations to be described here were made on unfixed material prepared by mixing living flagellates with sodium phosphotungstate solution at neutral pH and allowing droplets of the suspension to dry on carbon-coated grids. In general, the more rapidly the material is prepared and dried down, the more nearly intact are the fibers. Material that is allowed to remain in the stain for some minutes, or to dry slowly, usually provides examples of fibers that are to some extent disintegrated. The observations to be described here will be restricted to the outer and central fibers, and no attempt will be made to deal with the more fragile secondary fibers or the radial links. Basal bodies will also not be considered.

As is well known, the *outer fibers* are doublets, composed of two subfibers that appear hollow in section. Their dimensions are known fairly accurately from sections, and this knowledge provides a useful basis for assessing the degree of intactness of the fibers after negative staining. In fibers similar in dimensions to those seen in sections (about 360 Å wide overall) the two subfibers usually have dark, stain-filled centers (Fig. 1); this suggests that their seeming hollowness is

---

FIG. 1. A group of three negatively stained outer fibers, preserved more or less intact and with dark, stain-filled cores. Note that little evidence of substructure is visible. In the two outermost fibers the central partition is beginning to split. × 250,000. From [9].

FIG. 2. A partly collapsed outer fiber, showing longitudinal filaments made up of globular subunits. At right one of the filaments is fraying out. Note the patches of regular, apparently helical packing of subunits. × 250,000. From [9].

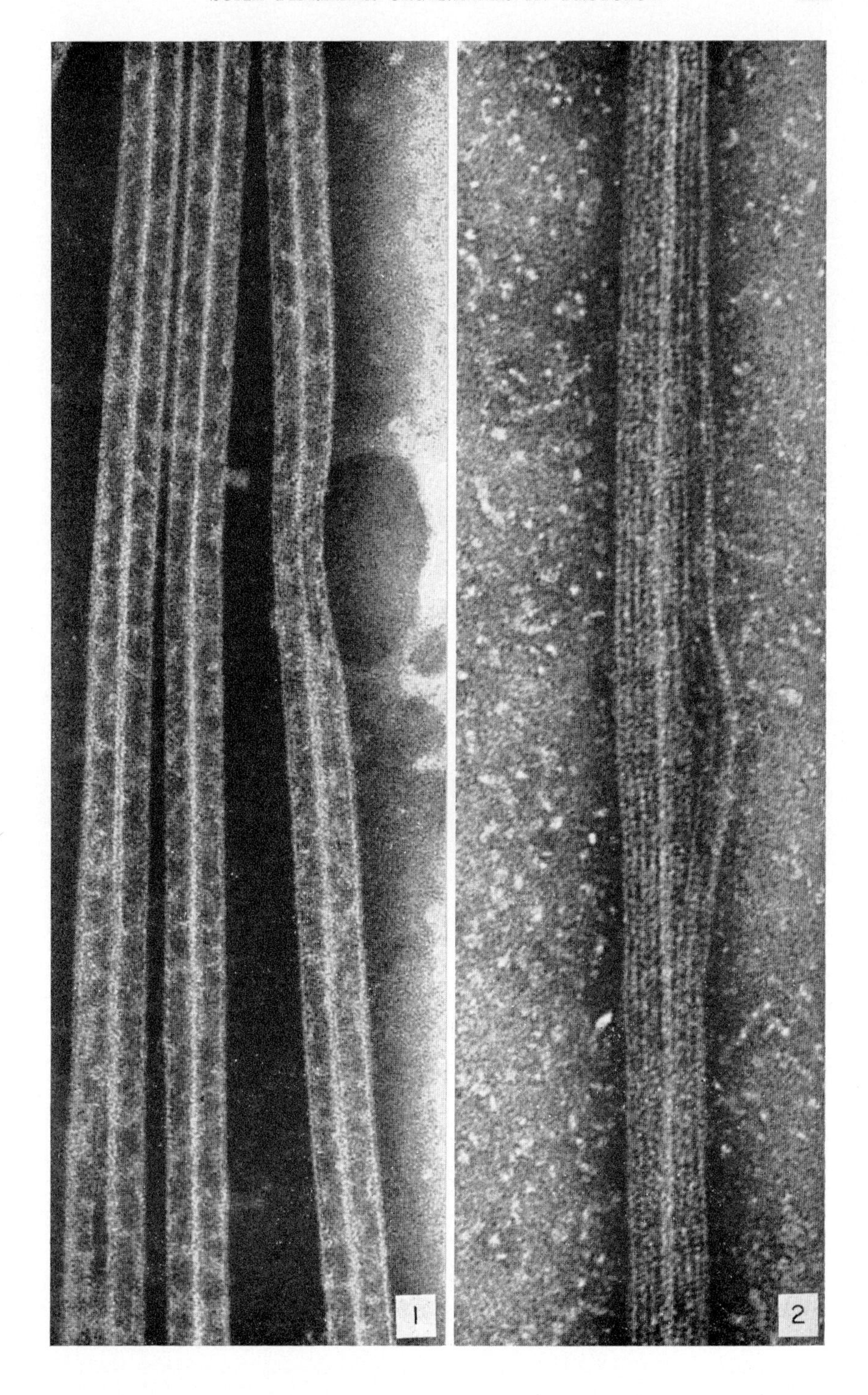
1
2

real. The thickness of their walls is 35–40 Å. There are indications that the central partition between the two sufibers is initially of about the same thickness, but in imperfectly preserved fibers it very readily broadens and splits into two. There are indications that each subfiber may have a complete wall and that these perhaps interdigitate along their area of contact, but this point requires further study.

The substructure of the wall of the subfibers is revealed in negatively stained material as a variety of longitudinal and transverse periodicities. The extent to which these are visible in any particular fiber varies considerably with the staining, and with the extent to which the fiber has remained intact. The most nearly intact outer fibers, judged by their dimensions and the absence of any marked separation of the subfibers, usually display the least striking evidence of periodic substructure (Fig. 1). This suggests that the structure has to "open up" to some extent, allowing penetration of stain between components, before substructure becomes visible. The most readily perceived and interpreted substructure is found in collapsed fibers or fiber fragments, and it is convenient to begin the description with these.

In almost all such subfibers it can be seen that the walls are made up of longitudinal beaded filaments, spaced 40–50 Å apart, center to center, and apparently running straight along the length of the subfiber (Fig. 2). These filaments are the "protofibrils" described by André and Thiéry [2]. Each is made up of a single row of approximately globular subunits, spaced about 40 Å apart, center to center, along the filament. This dimension probably corresponds to the diameter of the subunits, though in the micrographs the diameter of the globules usually appears slightly smaller than this, presumably because their boundaries are obscured or to some extent penetrated by the stain. The overall thickness of the filaments agrees with the thickness of the walls as seen in section, and it is reasonable to suppose that the walls are simply made up of a single layer of filaments. The number of filaments forming each subfiber cannot be unequivocally determined in negatively stained material, but is about a dozen and could well be thirteen, as found in the microtubules described by Ledbetter and Porter [12]. The filaments seem to be real structures with some integrity of their own, since they persist singly at the frayed ends of fibers (Fig. 3) or even lying free around disintegrating fibers.

Direct inspection of micrographs of collapsed fibers or fiber fragments commonly reveals areas in which the subunits form patterns, commonly in the form of an apparent helical arrangement. In other words, the subunits of adjacent filaments are not exactly in register horizontally but are staggered, forming rows running helically around the subfiber (Figs. 2 and 4). The areas in which this appearance can be clearly seen are, however, relatively small in extent and in many cases the images are difficult to interpret. This is a not uncommon state of affairs in micrographs of negatively stained material. The confusion arises from a variety of sources, of which the two principal ones are variation in the depth of penetration of the stain in different regions of the specimen, and the fact that often both sides of the specimen are imaged simultaneously, superimposed on each other in the micrograph. The result is that a precise and unequivocal description of substructure is almost impossible to achieve simply by inspection. A more objective method of analyzing the micrographs is needed, and this has been provided recently by the optical diffraction technique [11, 14]. The basis of this method is that selected areas of the electron micrographs are examined in an optical diffractometer, using visible light. A diffraction pattern is generated, the form of which is related to the distribution of densities in the micrograph. Repeating structures will reinforce and lead to the formation of distinct spots, and from the form and distribution of these many features of the original structure can be deduced. The method has several advantages. First, in "two-sided" micrographs each side of the structure will generate a separate family of spots in the diffraction pattern, and the two sets can usually be distinguished without difficulty on the basis of differences in intensity. Second, since the method involves averaging over a considerable area, rather than concentration on small regions in which structure is supposedly well preserved and clearly revealed, it provides an accurate and objective measure of spacings and angles. Third, it reveals or draws attention to structural regularities that may escape notice on direct examination of the micrographs.

An optical diffraction pattern from a partly collapsed subfiber is shown in Fig. 6. The dominant features are two sets of spots on layer lines corresponding to a spacing of 40 Å, and equatorial spots at a spacing of 40–50 Å. The former correspond to the longitudinal spacing of the subunits, the latter to the side-by-side separation of the filaments. Each side of the fiber gives rise to two spots on the 40-Å layer line, and these, together with the equatorial spots, define the reciprocal cell of the basic surface lattice (see [9] for further description). Figure 5 shows a reconstruction of this surface lattice. The postulated arrangement of subunits agrees with that seen in small areas of the original micrographs (Figs. 2 and 4), but considerably greater confidence can be attached to the reconstruction than to the appearances seen by direct examination.

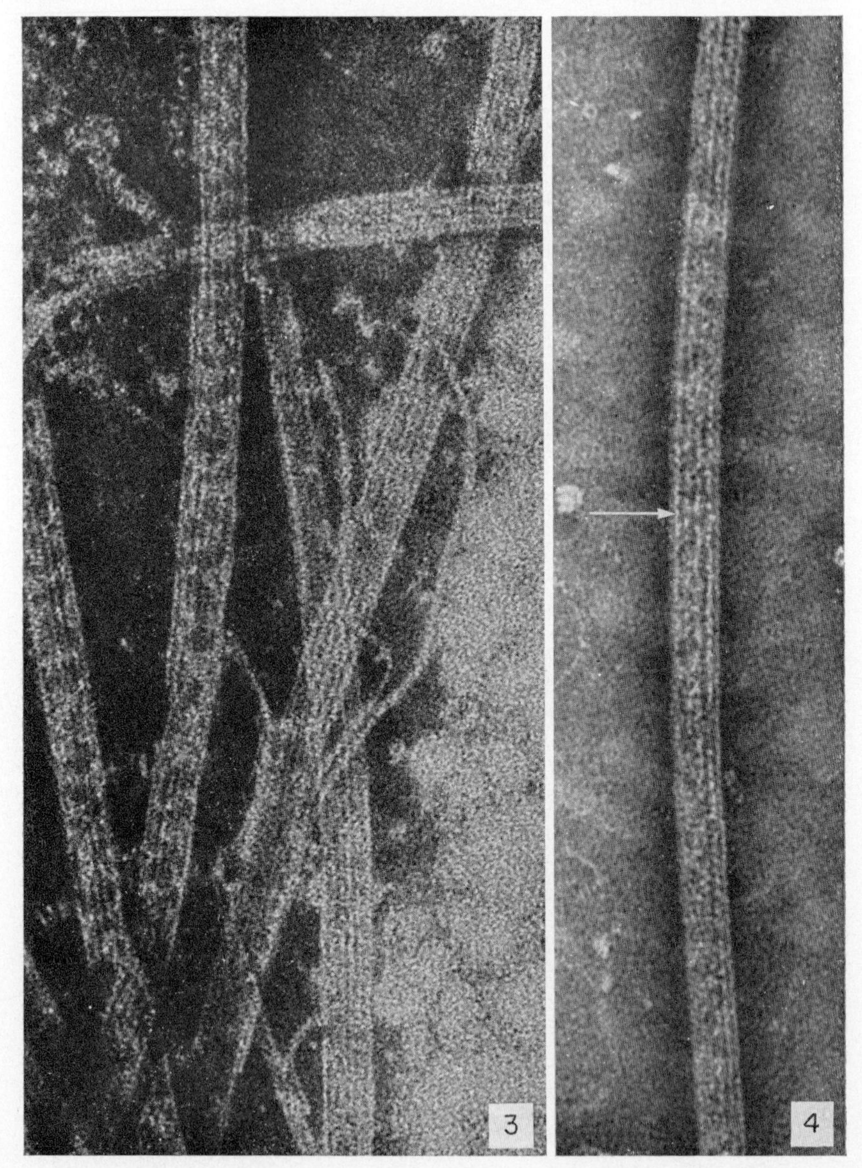


FIG. 3. A group of outer fibers, showing fraying into the constituent filaments at the broken ends. × 250,000.

FIG. 4. A nearly intact outer subfiber, showing in places an apparent zigzag arrangement of subunits along the filaments (arrow). × 250,000. From [9].

The basic surface lattice depicted in Fig. 5, while it is undoubtedly a close approximation to the structure seen in some of the micrographs, is almost certainly not present in simple form in the intact fibers. The diffraction pattern from which this lattice was derived came from a partly collapsed fiber; the patterns obtained from more nearly intact fibers are considerably more complex and reveal the presence of several longer spacings. The degree of prominence of these longer spacings varies a good deal in different examples, but spots on layer lines at 80 and 160 Å are almost always present (Fig. 7). The origin of these spacings has not yet been determined with certainty, but it seems

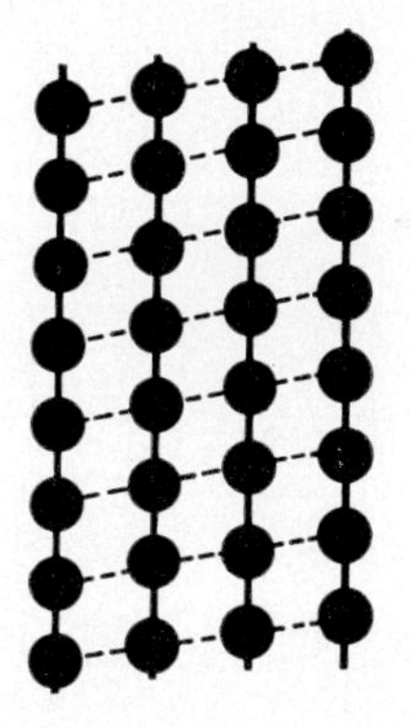

FIG. 5. Diagrammatic reconstruction of the basic surface lattice as seen in a collapsed fiber. The diameter of the circles does not correspond to the real dimensions of the subunits. From [9].

highly probable that the 80-Å periodicity arises from the fact that the subunits do not normally lie in exactly straight rows, as shown in Fig. 5, but are alternately displaced slightly in different directions off the main filament axis. In other words, the subunits are arranged in zigzag fashion, with the repeating unit two subunits long. This appearance can be seen in some of the micrographs (Fig. 4). It seems likely that the displacement of subunits is not simply in the plane of the surface of the fiber, but occurs in a radial direction as well. The fact that the 80-Å periodicity is not seen in either micrographs or diffraction patterns obtained from collapsed fibers suggests that its maintenance depends on the preservation of relatively labile bonds between subunits in adjacent filaments.

The origin of the 160-Å spacing is more difficult to account for in terms of the arrangement of subunits, though there are indications that it may arise by displacement of every fourth subunit in each filament (this point has been discussed at some length elsewhere [9]). There

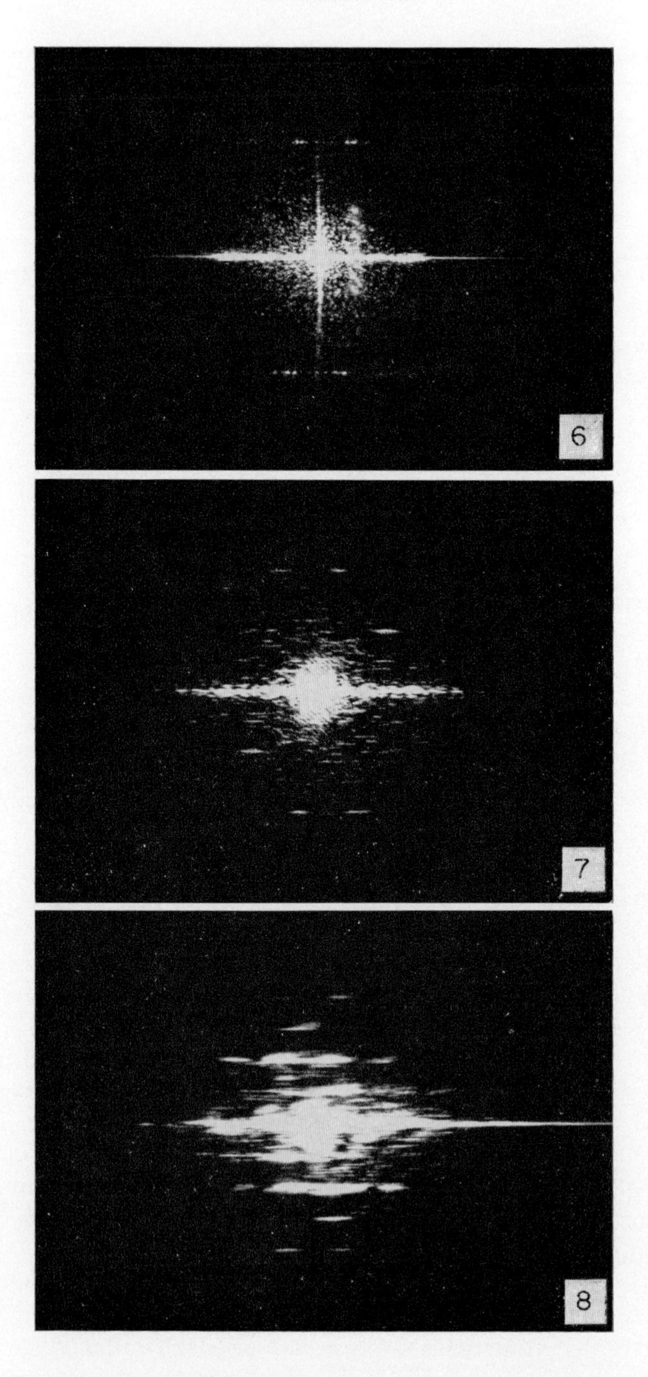
6
7
8

is no evidence to suggest that it is generated by additional material spaced at 160 Å.

The diffraction patterns obtained from intact fibers may be of formidable complexity, and while the 40-, 80- and 160-Å spacings are the most prominent features, it seems likely that in crystallographic terms these should all be viewed as orders of a still longer fundamental spacing of 480 Å. Spots of low intensity are present at this spacing in many of the diffraction patterns, as well as at 240 Å, and it seems likely that it will eventually prove possible to index all the spacings as orders of this fundamental repeating unit.

It will be obvious that at present a comprehensive description of the substructure of the outer fibers in terms of the arrangement of subunits (i.e., the distribution of matter) is not possible. Too little is known as yet about the characteristics of the outer fiber protein, and no X-ray data are as yet available. It is not unreasonable to predict, however, that the whole complex structure of the outer fibers will ultimately prove to be built up by interactions between one type of subunit. A parallel case is already available in some of the strains of tobacco mosaic virus described by Klug [10a].

The *central fibers* of flagella appear to be essentially similar in substructure to the outers, as judged by the diffraction patterns (Fig. 8). There are differences in the relative intensity of some of the spots, and the 80-Å spacing, in particular, is often more prominent than in the outer fibers. These differences, however, may well arise from the fact that the central fibers, if they are preserved at all in negatively stained preparations, usually survive intact. No examples of collapsed or fragmented central fibers have been found, an observation which suggests that if they begin to break down they disintegrate completely.

Some of the longer spacings demonstrated in flagellar fibers by means of diffraction analysis have been tentatively reported by previous work-

---

FIGS. 6–8. Optical diffraction patterns obtained from electron micrographs of negatively stained flagellar fibers. From [9].

FIG. 6. Pattern obtained from a collapsed outer fiber and showing only spots on the 40-Å layer line, and equatorial spots at 40–50 Å.

FIG. 7. Diffraction pattern obtained from a well-preserved outer fiber, showing prominent spots on the 80-Å layer line, in addition to less prominent ones at 40 and 160 Å.

FIG. 8. Diffraction pattern from a central fiber, with a complex array of spots at spacings corresponding to 40, 53, 80, and 160 Å. Note the prominence of the 80-Å spots. The multiplicity of spots on each layer line results from the fact that this is a pattern obtained from a "two-sided" image.

ers, studying both sectioned and negatively stained material (see [9]). It seems possible that they are not to be regarded simply as by-products of a complex arrangement of subunits, but may have a positive function in determining the spacing of other flagellar components. The 160-Å periodicity, for example, almost certainly corresponds to the spacing of the helically wound filaments that form the central sheath, encircling the two central fibers, and it could well determine that spacing. There may perhaps be similar relationships between periodicities in the outer fibers and the spacing of the arms [5] and the radial links, though more information is needed about these points. All that need be said here is that the substructure of the central and outer fibers appears to be sufficiently complex to serve as a basis on which the whole intricate morphology of the flagellar shaft might be constructed, and that it is not wholly implausible to envisage flagellar formation occurring by processes of sequential self-assembly, comparable to those that appear to be operative in bacteriophage formation.

## FORMATION OF SOME COMPLEX FIBRILLAR ORGANELLES

Centrioles and basal bodies in some organisms appear to have a special morphogenetic role, in that they seem to initiate the formation of other organelles. Many examples are available, apart from mitotic spindles, but only two need be considered here.

The ciliate *Nassula* possesses an elaborate organelle, the pharyngeal basket, which is used for ingesting strands of blue-green algae. It is a cylindrical structure, and a recent study of its fine structure by Tucker [15] has shown that it is largely made up of a series of longitudinal parallel rods, which form the wall of the basket, and each of which itself consists of a bundle of closely packed microtubules. Transverse sections of the basket show it to be almost perfectly radially symmetrical. The development of the basket has been studied by Tucker, who finds that the new rods develop from the proximal ends of certain basal bodies. These are not arranged in a circle, but in a more or less straight row, so that the initial disposition of the rods is quite different from that which they will ultimately assume. The rods detach from the basal bodies at some stage before the complete basket is formed. The detailed structure and morphogenesis of the basket are extremely complex but need not be considered here. The relevant point is that bundles of microtubules which will ultimately form the rods first manifest themselves as appendages of basal bodies, with which they maintain structural contact during the early stages of their growth.

A comparable situation can be found in many of the complex flagellates, of which *Trichomonas* is an example. In this organism at least four types of fibrillar organelle are present, apart from the mitotic spindle: flagella, axostyle, costa, and parabasal filament [1, 10]. Each of the three latter organelles has a distinct substructure and is clearly made of a different type of protein. The axostyle is a hollow cylinder of microtubules, while the costa and parabasal filament are solid, cross-striated structures, each with a characteristic band pattern and periodicity. Each of these organelles is connected, either directly or indirectly, to one particular member of a group of five basal bodies lying at the anterior end of the cell [7, 10]. At division the new organelles can be seen apparently growing out from these basal bodies (Fig. 9). In this respect *Trichomonas* is characteristic of flagellates in general, and it appears to be a rule that fibrillar organelles in these organisms first appear in connection with basal bodies.

The role of basal bodies in the production of these organelles seems most probably to be one of initation or "starting." There is no good evidence to suggest that basal bodies possess any synthetic abilities, and it seems more likely that they act in morphogenesis by providing specific sites on which protein molecules, or aggregates of such molecules, synthesized elsewhere in the cytoplasm, can begin to crystallize or assemble. To postulate such a role for basal bodies is one way of accounting for the precision with which the *number* of fibrillar organelles per cell is controlled: there is only one costa, one axostyle, etc. The hypothesis is that basal bodies provide special sites on which self-assembly is likely to occur with greater probability than elsewhere in the cell, and thereby act as unique foci of organelle formation.

It is, of course, obvious that such a process, if it occurs, is not in itself adequate to account for more than the first stage of formation of these organelles. The control of their subsequent growth and detailed shaping must depend on other factors, of which we are at present quite ignorant (see [6]).

This hypothesis would suggest that basal bodies must have highly differentiated surfaces, with specific and differing sites on which different types of protein molecule can begin to aggregate. There are indications of such complexity from other sources of evidence. Perhaps the best is the demonstration of complex and asymmetrical networks of fine fibers interconnecting some of the basal bodies in *Trichonympha* [4, 8].

## DISCUSSION

A common tactic in biological theorizing is to "explain" phenomena by building into the concepts that serve for explanation whatever

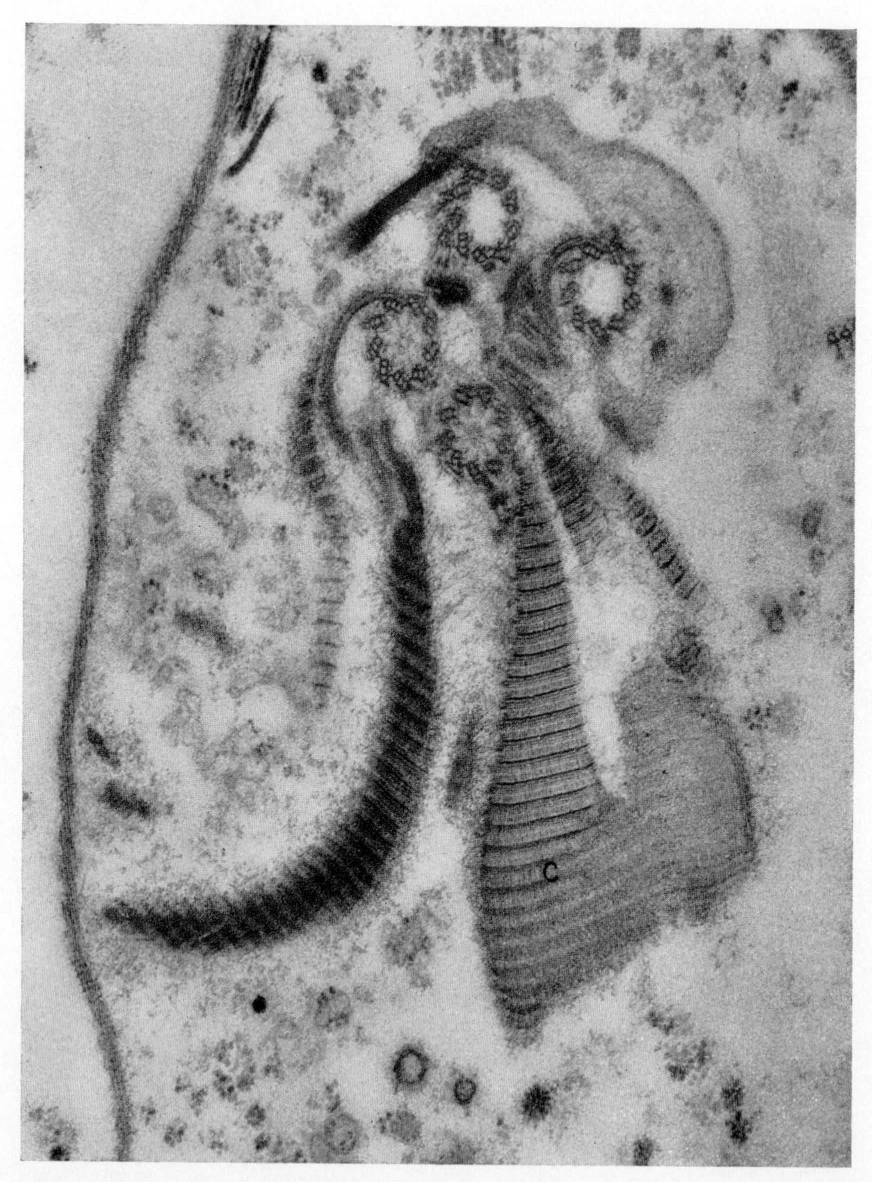


FIG. 9. Electron micrograph of the centriolar region of *Trichomonas termopsidis*, showing attachment of the costa (*c*) to a basal body. × 50,000.

properties are necessary to account for the phenomena in question. The supposed explanation is then, of course, no more than a tautology. Mainx [13] has drawn attention to this process and provides as one example the postulation of "organ-forming substances," held to be responsible for the development of particular regions of the egg into particular organs of the embryo. Many comparable examples can be found, especially in the fields of neurophysiology and psychology.

It could justifiably be argued that much of the current speculation about organelle formation, and in particular the tendency to "explain" morphogenesis in terms of self-assembly, displays the same fault. To ask how an organelle forms is to ask, "How did all these molecules come to be where they are?" And to answer, "They did it by self-assembly," is to say, essentially, "Because it is in the nature of these molecules to take up such mutual positions." By endowing the molecules with sufficiently complex shapes and arrays of binding sites it is not difficult to arrange for them to build almost any structure that one may be confronted with. At a more complex level it need hardly be pointed out that the role of basal bodies in initiating organelle formation, postulated above, is hardly more than a redescription of what is observed.

These criticisms are intended not to deny or disprove the importance of self-assembly in organelle formation—indeed, it is difficult to avoid the conclusion that it must play a fundamental role—but to point out the wide gap that exists at present between our knowledge of self-assembly in such simple cases as TMV (in which the process can be studied *in vitro,* with highly purified components, and in which the shapes and binding sites of the subunits are becoming known with increasing accuracy), and our present understanding of the enormously more complex organelles with which this article deals. For the latter the postulation of self-assembly must for the time being remain a working hypothesis (or an act of faith), rather than a proven mechanism.

## Acknowledgments

Figures 1, 2 and 4–8 are reproduced from *Journal of Cell Science* by permission of the Editors and Publisher.

## References

1. Anderson, E., and Beams, H. W., *J. Morphol.* **104,** 205 (1959).
2. André, J., and Thiéry, J. P., *J. Microscopie* **2,** 71 (1963).
3. Gibbons, I. R., *Arch. Biol.* (*Liege*) **76,** 317 (1965).

4. Gibbons, I. R., and Grimstone, A. V., *J. Biophys. Biochem. Cytol.* **11,** 697 (1960).
5. Gibbons, I. R., and Rowe, A. J., *Science* **149,** 424 (1965).
6. Grimstone, A. V., *Ann. Rev. Microbiol.* **20,** 131 (1966).
7. Grimstone, A. V., Unpublished observations, 1966.
8. Grimstone, A. V., and Gibbons, I. R., *Phil. Trans. Roy. Soc. London* **B250,** 215 (1966).
9. Grimstone, A. V., and Klug, A., *J. Cell Sci.* **1,** 351 (1966).
10. Joyon, L., *Ann. Fac. Sci. Univ. Clermont* **22,** 1 (1963).
10a. Klug, A., this symposium, p. 1.
11. Klug, A., and Berger, J. E., *J. Mol. Biol.* **10,** 565 (1964).
12. Ledbetter, M. C., and Porter, K. R., *Science* **144,** 872 (1964).
13. Mainx, F., *in* "International Encyclopedia of Unified Science" Vol. 1, No. 9. Univ. of Chicago Press, Chicago, Illinois, 1955.
14. Taylor, C. A., and Lipson, H., "Optical Transforms." Bell, London, 1964.
15. Tucker, J. B., Ph.D. Thesis, Cambridge Univ., Cambridge, England, 1966.

# IRRADIATION OF CELL ORGANELLES BY A LASER MICROBEAM: PROBLEMS AND TECHNIQUES

MARCEL BESSIS

*School of Medicine, University of Paris, Paris, France*

The destruction of cell organelles by radiation, in order to analyze their functions by the alterations thus produced, has been utilized since 1912, when Tchakhotine [11] described a method for the generation of a beam of ultraviolet radiation with a diameter of approximately 100 μ. A general discussion of this subject can be found in the reviews by Zirckle [13], Smith [9], and Bessis [3].

When lasers became available, it appeared that they could be used for the micropuncture of the cells. As will be shown, the use of the laser allows a greater versatility than the instruments previously used.

## PRINCIPLE OF LASER MICROIRRADIATION

The laser makes it possible to obtain more than 300 wavelengths of light from the ultraviolet to the infrared range. However, wavelengths of sufficient energy for cellular microirradiation are relatively few in number (Table I).

The characteristic feature of laser emission is the production of light that is coherent in time and space. The temporal coherence results in

TABLE I. *Some Wavelengths Used at Present for Biological Purposes*

| Type of laser | Wavelength (Å) | Range |
|---|---|---|
| Pulsed lasers | | |
| Neodymium | 10600 | Infrared |
| Ruby | 6943 | Red |
| Neodymium first harmonic | 5300 | Green |
| Neodymium second harmonic | 2650 | Ultraviolet |
| Continuous wave lasers | | |
| Ionized krypton | 6471 | Red |
| Ionized krypton | 5682 | Yellow |
| Ionized argon | 5145 | Green |
| Ionized argon | 4880 | Blue-green |
| Ionized argon | 4579 | Blue |

a high degree of monochromatism (for example, the ruby laser emits a wavelength of 6943 Å ± 0.01 Å). Spatial coherence is the quality of greater importance for microirradiation. It produces minimal divergence, a characteristic that makes it possible to focus all the emitted energy upon a very small space (of the order of 1 $\mu$ or less). This cannot be done if one employs a conventional light source. The pulse laser can produce very large quantities of power, as indicated in Table II. As a

TABLE II. *Energies, Emission Times, Powers Given by Ruby Laser Used for Microbeams*

| Operation | Energy (joules) | Emission (per second) | Power (watts) |
|---|---|---|---|
| Normal | 1 | $10^{-3}$ | $10^{3}$ |
| Quarterwave plate switched | 1 | $30 \times 10^{-9}$ | $3 \times 10^{7}$ |
| Pulsed, per pulse (100 pulses/second) | $50 \times 10^{-3}$ | $10^{-3}$ | 50 |

result, the absorbing portions of the cell are rapidly brought to a lethal temperature. Certain organelles may be vitally "infrastained" and thus selectively destroyed. This operation is very easily carried out and will probably be found to have significant value. Since enormous energy is delivered in a very short time (a few microseconds) with minimal heat dissipation, the dose of stain necessary for the destruction of an organelle is extremely minute (see below).

## IRRADIATION TECHNIQUES

The apparatus we have used was developed from one previously described for ultraviolet irradiation [4]. However, the specific characteristics introduced by the utilization of the laser had to be taken into account; in particular the very short duration of the laser emission, of the order of $10^{-3}$ second, sometimes makes it difficult to position the microspot. For this reason, the illustrated apparatus has been used [Fig. *1;* see also 6]. The microscope embodies a vertical illuminator. The object is placed on a dielectric mirror (*M*). The alignment is achieved by observing a second image of the object by autocollimation on the exit surface (*m*) of the laser (*L*), the brightness of which is controlled by two polarizers ($P_1$) and ($P_2$) and by a quarterwave plate (*Q*) that is withdrawn during operation. It is then possible to localize the spot for the alignment of the apparatus by means of a very weak image resulting from the reflection on the anterior face of the laser.

The laser rodes used in this work were cylinders 3 mm in diameter and 50 mm long. The posterior plane of the laser cylinder is silvered to the point of being almost opaque, allowing, however, the escape of a certain amount of light.[1] The anterior face of the laser is semitransparent with a transmission of 6%. The excitation is supplied by a spiral-shaped xenon flash bulb made of silica. This bulb is energized, for the ruby laser, by the discharge of a 160-microfarad condenser charged to 2500 V.

The described laser produces a spot 5 mm in diameter at the distance of 1 meter. This beam is therefore nearly parallel, with a divergence of the order of $5 \times 10^{-3}$. The fact that the diameter of this laser is smaller than the pupil of the objectives of the miscroscope can be used to advantage, and the laser is placed in front of the eyepiece. Under the circumstances, the beam is limited only by the eye ring, the diameter of which is $d_o = 500/6$ (NA). We have used an eyepiece of $6 \times$ magnification with a $100 \times$ objective; this combination diaphragms the laser by one-third. The calculated diameter of the spot under these circumstances is 2.5 $\mu$.

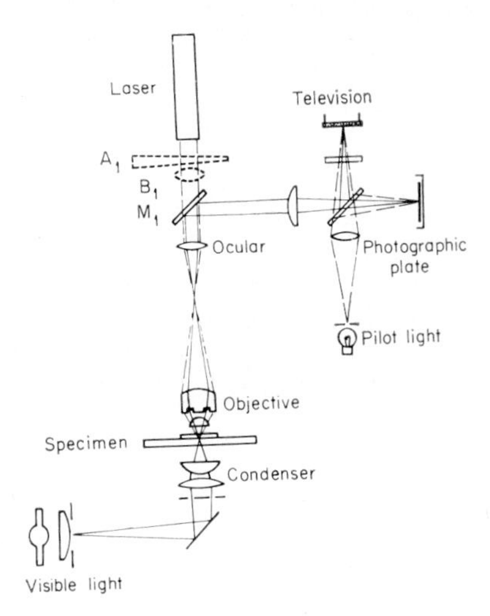


FIG. 5. Diagram of apparatus used for irradiation of cell organelles by a laser microbeam.

The position of the laser spot, before irradiation, is indicated by the image of a pilot light, which coincides with the "true" spot. The position of the "true" spot is determined either as described above or directly by its effect on an absorbing test object.

[1] The light escaping from the posterior face of the laser is measured by a photoelectric cell. This value is used in determining the total amount of light generated by the laser.

The preparation of living cells to be irradiated is carried out in the usual way, and they are illuminated by transmitted light by means of a phase contrast condenser.

In order to avoid possible damage to the eye subsequent to an unplanned firing of the laser, two precautions are taken: (1) the eye pieces are fitted with special screens that absorb the wavelength of the laser radiation; (2) the effect of micropuncture is observed by means of a closed circuit television system. The television vidicon tube must be protected by means of a screen absorbing the wavelength of the laser radiation.

### *Radiation Dosimetry*

The absolute quantitative determination of the energy flux of the laser beam impinging upon a very small surface is difficult. The calculation of the dose of radiation absorbed by the irradiated portion of the cell is even more complex. Various possible approaches to the solution of the problem have been considered. These include (1) the use of two photoelectric cells suitably placed to measure the energy flux before and after passage through the irradiated cell; (2) the utilization of a biological test object, for example the time of hemolysis of a human red blood cell under controlled conditions. These cells are very small in relation to the irradiation spot, and this method is already used in macroscopic tests. The inactivation of phages must be measured by using microdrops; this is a difficult technique, particularly if very small spots requiring immersion objectives are used. To date no satisfactory method of dosimetry has been developed.

## TECHNIQUES FOR EVALUATION

The results of micropuncture can be evaluated in three principal ways: study of the successive changes by time lapse cinemicrography; examination under an electron microscope after the cell has been carefully marked, embedded, and sectioned; and last, evaluation of the metabolic disturbances in the cell, as revealed by autoradiographic study of the incorporation of labeled material and by histochemistry.

### *Dynamic Studies of Sucessive Changes after Micropuncture*

These studies are done with a phase-contrast microscope and time-lapse cinematographic recording. It is difficult to observe a cell irradiated with a very small spot for longer than 3–6 hours, after which the cell dies spontaneously because the narrow confinement necessitated by the short focal length of the objective prevents proper cell development. On the other hand, with much larger spots it is relatively easy to observe the effects of irradiation of tissue cultures for several consecutive days. With this technique it is possible to examine changes of intracellular movements, alterations of mitosis, pathological morphology of cell structures, and the different forms of cell death.

### *Electron Microscope Studies*

Examination of an irradiated cell under an electron microscope is very difficult. The method of obtaining the cell, its fixation, embedding in various plastics, sectioning, and identification under the electron microscope have been described by Bloom [7] and Silvestre *et al.* [8]. A very reliable technique has more recently been described by Amy and Storb [1].

### *Evaluation of Metabolic Disturbances in the Cell*

These disturbances are evaluated principally by autoradiography. Labeled precursors (e.g., thymidine, uridine, amino acids) are introduced into the culture; neighboring nonirradiated cells serve as controls. The irradiated cell is examined for the absence of incorporation or an abnormal localization of a labeled substance.

## IRRADIATION AFTER VITAL STAINING

The concentration of dyes required for visible staining is much greater than that needed to obtain an effect after laser irradiation. It is therefore possible to use doses of vital stains so small as to be nontoxic. Because of such "infravisible" staining it is possible that hitherto unknown dyes will be discovered by laser irradiation.

The action of laser radiation enables certain dyes to be recognized in areas or cells where their presence was quite unsuspected. With an adequate, but still vital, concentration all dyes appear to fix themselves to all proteins. It is thus possible to sensitize the nucleolus as a whole, and even the hyaloplasm. Experiments with Janus green permit specific destruction of mitochondria without damaging the rest of the cytoplasmic organelles [2, 10, 12].

Table III shows the relationship between visible stain and sensitivity to the laser for several vital stains.

It is possible that specific stains will become available for organelles other than mitochondria (e.g., Golgi bodies, centrioles, nucleoli, lysosomes) that will bind in very low concentrations and permit the selective destruction of organelles.

TABLE III. *Limit of Sensitivity to the Laser after Various Vital Stains (irradiation of the granuloplasm)*

| Stains | Ratio toxic limit: visible limit[a] | Ratio toxic limit: laser limit[b] | Ratio visible limit: laser limit[c] |
|---|---|---|---|
| Gentian violet | 1.15 | 21.4 | 18.7 |
| Toluidine blue | 5.55 | 33.3 | 6.3 |
| Nile blue sulfate | 5.3 | 80 | 15 |
| Brilliant cresyl blue | 3.3 | 13.3 | 4 |
| Janus green B | 5.7 | 51.4 | 9 |
| New methylene blue | 3.3 | 18.3 | 5.6 |
| Methylene blue | 3.3 | 16.6 | 5.1 |
| Rhodamine B | 2 | 14 | 8.7 |
| Neutral red | 2 | 8 | 4 |
| Acridine orange NO | 2 | 2 | 1 |
| Auramine O | 0.67 | 1 | 1.5 |

[a] Ratio of the toxic stain concentration to the lowest concentration which gives a visible vital stain.

[b] Ratio of the toxic stain concentration to the lowest stain concentration which will give visible damage with the optical microscope after laser irradiation.

[c] Ratio of the weakest concentration which gives a visible stain to that which gives visible damage after laser irradiation.

## REFERENCES


1. Amy, R., and Storb, R., *Science* **150,** 756 (1965).
2. Amy, R., Storb, R., Fauconnier, B., and Wertz, R. K., *Exptl. Cell Res.* **45,** 361 (1967).
3. Bessis, M., *in* "Progress in Photobiology" (B. C. Christensen and B. Buchmann, eds.), p. 291. Blackwell, Oxford, 1965.
4. Bessis, M., and Nomarski, G., *J. Biophys. Biochem. Cytol.* **8,** 777 (1960).
5. Bessis, M., and Storb, R., *Nouvelle Rev. Franc. Hematol.* **5,** 459 (1965).
6. Bessis, M., Gires, F., Mayer, G., and Nomarski, G., *Compt. Rend.* **255,** 1010 (1962).
7. Bloom, W., *J. Biophys. Biochem. Cytol.* **7,** 191 (1960).
8. Silvestre, J., Burte, B., and Rousseau, R., *Bull. Microscop. Appl.* **11,** 109 (1961).
9. Smith, C. L., *Intern. Rev. Cytol.* **16,** 133 (1957).
10. Storb, R., Wertz, R. K., and Amy, R. L., *Exptl. Cell Res.* **45,** 374 (1967).
11. Tchakhotine, S., *Biol. Zentr.* **32,** 623 (1912).
12. Wertz, R. K., Storb, R., and Amy, R. L., *Exptl. Cell Res.* **45,** 61 (1966).
13. Zirckle, R. E., *Advan. Biol. Med. Phys.* **5,** 103 (1957).

# THE FORMATION, PHYSICAL STABILITY, AND PHYSIOLOGICAL CONTROL OF PAUCIMOLECULAR MEMBRANES

**J. F. DANIELLI**

*Unit for Theoretical Biology, School of Pharmacy, and Center for Theoretical Biology, State University of New York at Buffalo, Buffalo, New York*

## SUMMARY

It is shown that, as a result of assymmetry of orientation of the hydrocarbon chains in phospholipids composing a lipid membrane, there is an assymmetry of polarizability. This results in a free energy minimum at bimolecular thickness for a membrane composed of phospholipid bearing no net charge. The existence of this free energy minimum means that the lipid bilayer is a natural unit of structure in the same sense that the α-helix of protein and the double-stranded helix of DNA are natural units.

It is then shown that lipid in excess of that required to make a bilayer is unstable and will form either additional bilayer, or a lipid droplet in the interior of a bilayer. Lipid droplets with surface free energy of about 10 ergs/cm$^2$ will exist in equilibrium with bilayers with a much lower surface free energy. This accounts for the stability of Rudin-Mueller bilayer membranes.

It is shown that, if an oil droplet is to enter a bilayer, then the droplet must have a surface free energy for the oil-water surface of about 10 ergs/cm$^2$, or higher.

The functions of cholesterol and proteins in cell membranes is discussed, and a hypothesis for information storage on membranes is outlined.

Mechanisms for cellular control of membrane function are indicated which operate either by control of the bilayer $\rightleftharpoons$ micelle equilibrium or by causing a disparity between the surface pressures in the two monolayers of a bilayer.

## INTRODUCTION

As a result of studies made by electron microscopy over the last ten years, it has become evident that membranes of thickness between 50 and 100 Å are present not only in the plasma membrane, but as the principal method of subdivision of the cell and also as a structural component of cellular organelles. Such membranes then are primary structural units in cellular architecture. Consequently we must find an answer to the question: What physical principles render these membranes of such value in cell physiology that selection pressure, exerted over the whole course of terrestial evolution, has not produced alternative structures and mechanisms?

When I introduced the concept of the paucimolecular structure of the plasma membrane in 1934 [10, 11], it was defined as a thin layer of lipid, possibly bimolecular, with adsorbed layers of protein on either side. Consideration of the restrictions on packing in the lipid layer imposed by the strong interaction between lipid polar groups and water strongly favored a bimolecular structure as the fundamental unit for the lipid layer. Calculations of the permeability of bilayers showed that no advantage to the cell would ensue if the lipid layer were thicker than bimolecular. These and other arguments led to the conclusion that the lipid layer was bimolecular. Information concerning the protein layers was less decisive: it seemed possible that each protein layer consisted of a primary protein layer of "unrolled" protein (as in a protein monolayer at an air-water or oil-water interface) and adsorbed upon this was a layer of globular protein molecules. This work has been reviewed by Danielli and Harvey [12], Davson and Danielli [13], and Danielli [6, 7].

However, although the physical properties, including the packing properties, of phospholipids indicated that the basic lipid layer is bimolecular, the possibility remained that thicker membranes would occur, incorporating, between two oriented layers of phospholipid at the water-lipid interface, a more or less randomly oriented internal lipid layer. This internal layer would be composed of less polar lipids, such as cholesterol, cholesterol esters, triglycerides, carotenoids, hydrocarbons. It is unreasonable to expect electron microscopy to indicate whether such thicker membranes actually occur, unless they are much greater than bimolecular in thickness. Except in the case of the myelin sheath of nerve it is also difficult to use X-ray techniques to measure membrane thickness. However, we can now solve this problem by a theoretical approach, for it can now be shown that the free energy of formation of a bilayer is much less than of a membrane of any other thickness [8].

## THE FREE ENERGY OF FORMATION OF LIPID MEMBRANES

We are concerned here with membranes composed of both polar and nonpolar natural lipids. The polar lipids, such as the phospholipids, contain groupings so polar in nature that the only packings which will be stable will be those in which the polar groups are in contact with water, or in contact with one another: single phospholipid molecules cannot exist in isolation, as a molecular solution, in any nonpolar liquid except in extreme dilution. The relatively nonpolar lipids either have no hydrogen-bonding groups, or only weakly hydrogen-bonding groups, removal of which from water into hydrocarbon involves only a few thousand calories, so that they may exist randomly oriented in molecular solution in hydrocarbon.

Let us tabulate the surface free energies of the surfaces with which we shall be involved. The tabulated values are all for interfaces with water.

| | |
|---|---|
| Cell membranes | $<0.1$ erg/cm$^2$ |
| Phospholipid bilayer | $\approx 0$ |
| Hydrocarbon (decane) | $\approx 50$ |
| Cholesterol in hydrocarbon | 45 |
| Triglyceride in hydrocarbon | 20–30 |
| Phospholipid in hydrocarbon | 7–15 |

It will be observed that the phospholipids have the greatest surface activity of the lipids mentioned, but that when present at a hydrocarbon-water interface they will not reduce the interfacial free energy $\gamma_i$ below about 10 ergs/cm$^2$. When they are present as bilayers however, the free surface energy of the bilayer $\gamma_b$ is approximately zero, as is also true for cell membranes.

Consider Fig. 1, which illustrates the structure of a lipid-water inter-

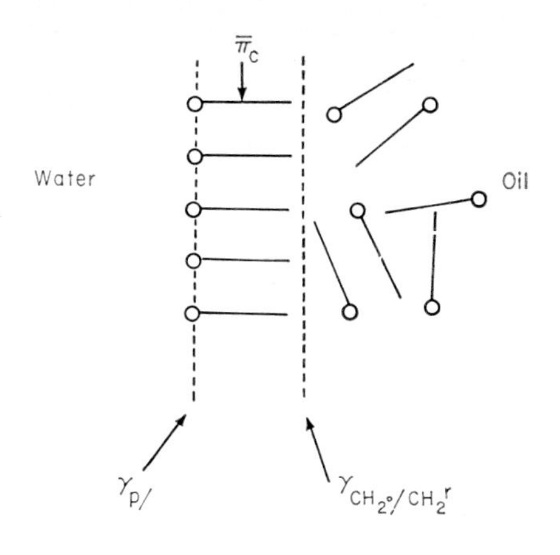


FIG. 1. Interface between a lipid droplet and water.

face. The surface layer of lipid is highly oriented, whereas the lipid in the interior of the droplet is oriented at random. Hence we have

$$\gamma_i = \gamma_{p/w} + \overline{\Pi}_c + \gamma_{CH_2^o/CH_2^r} \tag{1}$$

where $\gamma_{p/w}$ is the free energy of the polar group-water interface; $\gamma_{CH_2^o/CH_2^r}$ is the free energy of the interface between oriented paraffin chains and random paraffin chains; and $\overline{\Pi}_c$ is the excess cohesion pressure of the oriented chains.

Whereas $\gamma_{p/w}$ varies very widely according to the nature of the polar groups, $\gamma_{CH_2^o/CH_2^r}$ will be approximately constant for a given degree of packing of the chains. We can calculate the value of $\gamma_{CH_2^o/CH_2^r}$ in five different ways [9]:

1. from the contact angle observed when a droplet of liquid paraffin rests upon a suitable oriented hydrocarbon surface, e.g., a monolayer;
2. from the contact angles found with water droplets;
3. from the value of $\gamma$ at the inversion point for the transition from oil-water to water-in-oil emulsions;
4. from the difference in polarizability of hydrocarbon chains in the C—C direction compared with the direction perpendicular to the C—C bond;
5. from the critical surface free energy for penetration of an oil droplet into a bilayer membrane.

In all cases we find that $\gamma_{CH_2^o/CH_2^r}$ is of the order of 10 ergs/cm$^2$.

Now consider what happens when a thick uncharged lipid membrane, as in Fig. 2, becomes thinner. For the thick membrane, the surface free energy $\gamma_m$ is twice that of a single surface

$$\gamma_m = 2(\gamma_{p/w} + \overline{\Pi}_c + \gamma_{CH_2^o/CH_2^r}) \tag{2}$$

As the membrane is thinned, little happens until the oriented layers come within 10 Å of each other. At this distance the interaction between the two oriented hydrocarbon layers becomes significant, and it increases until the layer becomes bimolecular, at which point the term $\gamma_{CH_2^o/CH_2^r}$ vanishes and

$$\gamma_b = \gamma_m - 2\gamma_{CH_2^o/CH_2^r} \tag{3}$$

i.e., as a bilayer is formed from a thicker membrane the surface free energy falls by about 20 ergs/cm$^2$.

When the membrane is thinned to less than two molecules thick, no further decrease in free energy occurs; on the contrary, there is a

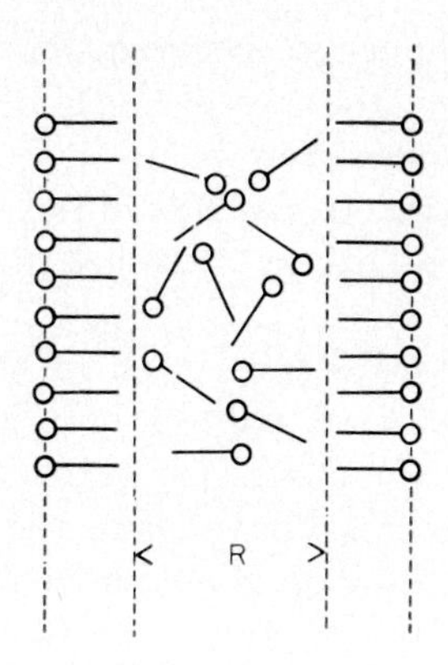


$$\gamma_m = 2\left(\gamma_{p/w} + \gamma_{CH_2^\circ/CH_2^r} + \bar{\pi}_c\right)$$

FIG. 2. Diagram of a membrane more than bimolecular in thickness.

sharp rise in the surface free energy. This results from the fact that in a bilayer all the polar groups are already in the water-lipid interface. Consequently as the membrane becomes less than two molecules in thickness the additional surface so formed can only be hydrocarbon-water, with a surface free energy of about 50 ergs/cm$^2$. The variation in $\Delta F^s$ is shown in Fig. 3. A very sharp minimum in $\Delta F^s$ is found at the bilayer thickness [8]. In this discussion we have ignored effects

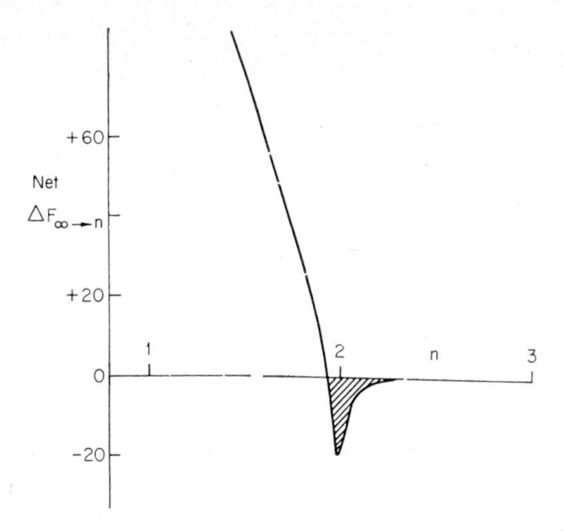


FIG. 3. Surface free energy $\Delta F^S$ of a membrane plotted against thickness expressed in number of molecules.

due to electrostatic interaction between the interfaces. The electrostatic interaction between the two surfaces of the membrane will not be negligible, but will not show sharp discontinuities such as have been found with the two terms just discussed. Hence the position of the minimum in free energy will not be much affected by the electrostatic interaction.

Thus we see that, for substances such as the natural phospholipids, the bilayer configuration is a natural unit of structure in the same sense that the $\alpha$-helix is for DNA. The physical shape of the molecules is conducive to packing in a bilayer, and the distribution of chemical groupings is such as to result in a minimum free energy at the bilayer thickness. Consequently, in the absence of other disturbing factors, such lipids will form bilayers spontaneously, whether in cells or otherwise. However, unlike the $\alpha$-helix and double helix, in which the directional effect of hydrogen-bonding confers a measure of specificity, the structure of the hydrocarbon moiety of a bilayer can have little specificity, because of the weakness of the physical forces involved and the lack of directional effect. For example the cohesion between the two lipid layers of a bilayer is only about one third of that between the polar groups of the bilayer and water. In fact bilayers arise not so much as a result of large forces between the component molecules, but because of the "squeezing out" effect of the high intermolecular forces between water, and other polar components. We can therefore make the generalization that generalized properties of cellular membranes may often arise from the hydrocarbon moiety of the phospholipids, but that the specific properties arise primarily from the macromolecules associated with the membranes.

## THE ORGANIZATION OF EXCESS LIPID AND OF NONPOLAR LIPID

If at a site, for example of synthesis, additional phospholipid is produced, the lipid will normally form additional bilayer membrane. However, with lipid which is relatively nonpolar, other possibilities exist. If the lipid has sufficient affinity for water, or if its energy of interaction with bilayers is high enough, the nonpolar lipid may form part of a bilayer, as is found with cholesterol. Otherwise the nonpolar lipid will accumulate between the monolayers of a bilayer forming a new surface between $CH_2^o$ and $CH_2^r$. But as this happens there is an increase in free energy (see Fig. 4). Consequently the nonpolar lipid does not disperse uniformly in the interior of the bilayer, producing a membrane more than two molecules thick, but accumulates as spherical droplets in the membrane, thus minimizing the surface free energy.

The consequence of this is that, in a system containing bilayers and excess lipid, that part of the lipid which is not present as bilayers will exist as spherical droplets. If such nonpolar lipid is produced at a number of sites on a bilayer membrane, an intramembrane droplet will form at each site. These droplets will exhibit Brownian movement in the membrane until two droplets meet, when they will immediately coalesce since the free energy of the resultant droplet is less than that of the two precursor droplets.

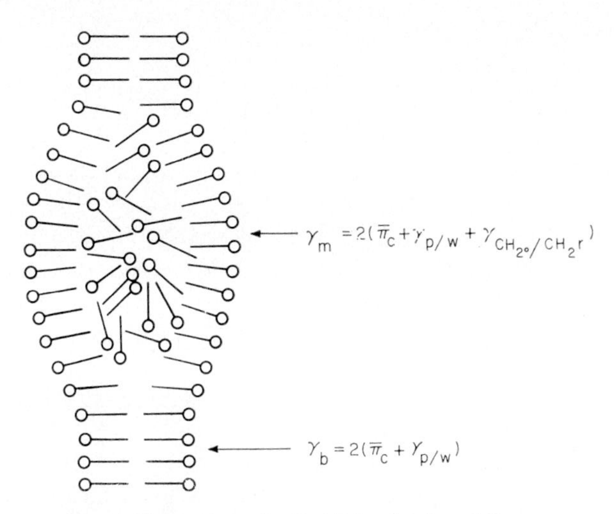


FIG. 4. Formation of a lipid droplet in a bilayer.

As an example of this we may take the Rudin-Mueller artificial bilayers, which are formed when a solution of, for example, phospholipid + tetradecane (hydrocarbon) in a solvent miscible with water is used to make a thick membrane across a hole in a piece of nonpolar plastic in water. Initially, as the solvent dissolves in the water, a thick tetradecane-phospholipid membrane is formed. But this is unstable for the reasons given above and breaks up into a mixture of bilayer and droplets. The droplets scurry around under Brownian movement until they coalesce as an annulus in contact with the plastic, as in Fig. 5. The surface free energy of the bilayer so formed is of the order of 0.5 erg/cm$^2$, and that of the droplet of the order of 10 ergs/cm$^2$. The tetradecane is necessary to form the $CH_2{}^r$ phase of the droplets. If it were not present all the phospholipid would form bilayer which would not be anchored to the plastic.

Recently I had the opportunity of discussing the problem of the formation of fat droplets in cells with Dr. S. Patton. Normally tri-

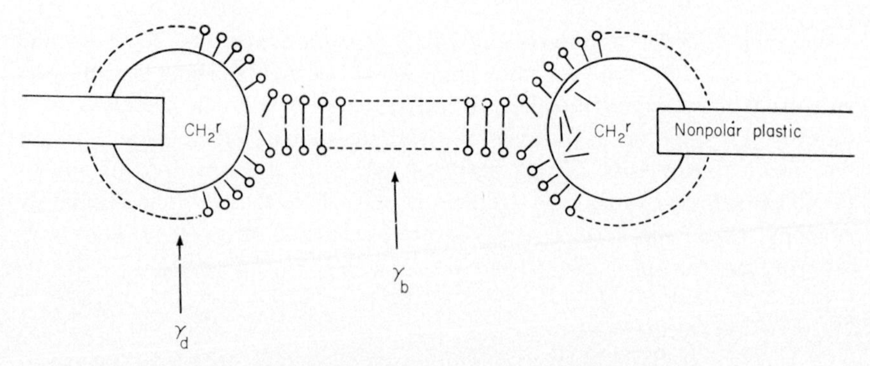


FIG. 5. Diagram of a Rudin-Mueller membrane. A bilayer ($\gamma_b \approx 0.5$ dyne) is in equilibrium with the annular droplet ($\gamma_i \approx 10$ dynes).

glyceride is formed from phospholipid, and phospholipids by acylation of glycerol derivatives. Thus the overall process must involve (a) enzymatic synthesis of phospholipid, probably by membrane-bound enzyme; (b) formation of additional bilayer by the phospholipid; (c) conversion of bilayer phospholipid to triglyceride by bilayer-bound enzyme; (d) formation of intramembrane fat droplet by triglyceride (Figs. 6 and 7).

Similar considerations enable us to understand the conditions under which oil droplets will pass into bilayer membranes. Consider Fig. 7a, in which a small oil droplet is presented to a bilayer membrane. The bilayer has $\gamma_b \simeq 0$, and the droplet has a surface free energy $\gamma_i$ which may lie anywhere between 0 and about 50 ergs/cm$^2$. If the droplet enters the bilayer, $\gamma_i$ will take up the value of $\gamma_{CH_2{}^o/CH_2{}^r}$, i.e.,

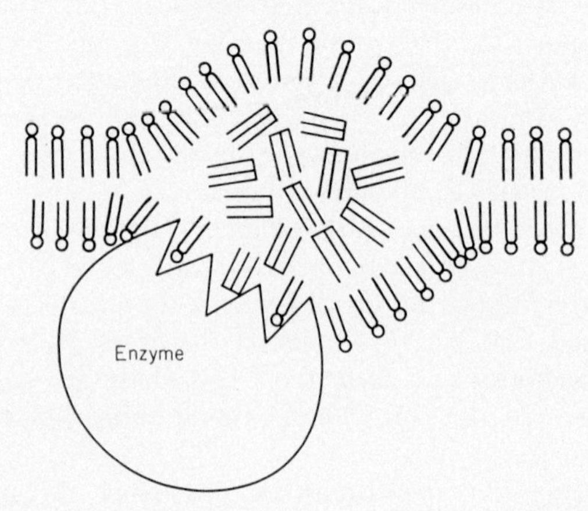


FIG. 6. Diagram of formation of a fat droplet from a phospholipid bilayer.

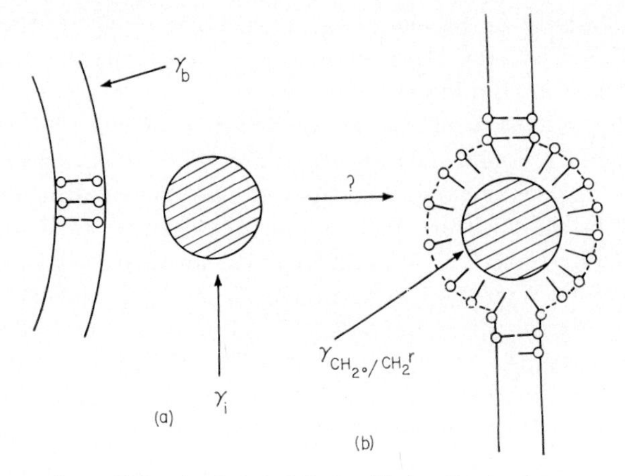


FIG. 7. Entry of an oil droplet into a bilayer. If the droplet has a $\gamma_i$ against water greater than $\gamma_{CH_2{}^o/CH_2{}^r}$, the droplet will pass into the interior of the bilayer.

about 10 ergs/cm$^2$. Consequently the droplet should enter if its initial $\gamma_i$ is $> 10$, but not if it is $< 10$. On searching the literature, I found, to my delight, that Chambers [2], and Kopac and Chambers [14], had conducted experiments of this type with echinoderm eggs, and had found that the threshold for penetration is 9.5 dynes/cm$^2$. Droplets with $\gamma_i < 9.5$ ergs/cm$^2$ do not penetrate.

## THE FUNCTION OF CHOLESTEROL IN BILAYER MEMBRANES

Cholesterol is not necessary for formation of bilayer membranes. Artificial membranes can be formed without cholesterol by phospholipid, but cholesterol can replace hydrocarbon as a stabilizing factor for Rudin-Mueller membranes. In cellular membranes the cholesterol content may vary from 0% in muscle mitochondria to 40% of the lipid molecules present in myelin membranes. Cholesterol is thus not a necessary constituent of bilayers, and so must have a particular rather than a general function.

Cell membranes are known to be liquid, and the natural phospholipids give rise to "expanded" monolayers (using N. K. Adam's nomenclature), in which the area per molecule is much larger than the minimum value of 20 Å per hydrocarbon chain. Biological membranes frequently must be capable of considerable variation in area without rupture. If the lipid bilayer constituent of such membranes were in the liquid expanded state it would be capable of sustaining such changes in area with relatively small changes in surface pressure. No other known type of coherent monolayer has such properties.

When cholesterol is added to an expanded monolayer, the area will decrease by 15 Å or more for each cholesterol molecule, if the monolayer is near the $L_1 \rightleftharpoons L_2$ liquid transition point. Thus it seems probable that, while the lipid bilayer of myelin is highly condensed, other bilayers such as those of mitochondria are expanded, with much looser molecular packing and higher entropy. An equal contraction of an expanded film can also be brought about by a fall in temperature of about 20°C, or by the addition of one $CH_2$ to each hydrocarbon chain. That is, lowering the temperature, increase in chainlength, or addition of cholesterol all produce about the same effect—smaller area per chain and a decrease in entropy. It seems reasonable to suggest that, where a membrane is required which shall be relatively compact, the required effect is produced by provision of an appropriate mole fraction of cholesterol.

But the mechanism whereby cholesterol acts is uncertain. Cholesterol itself has a low surface activity, being able to reduce the surface free energy of a hydrocarbon-water interface by only 5 ergs/cm$^2$, whereas lecithin reduces the free energy by 40 ergs [14]. This indicates that cholesterol by itself would not form stable bilayer, and if present in mixed bilayers would tend to associate as droplets rather than persist in the bilayer, whereas in fact cholesterol and phospholipids form stable mixed bilayers. Recently Chapman [3] has shown that addition of cholesterol to phospholipid suppresses the $CH_2$ band in the nuclear magnetic resonance spectrum of dispersions in water. Thus in some way cholesterol must suppress the free movement of the hydrocarbon chains of lipids [cf. Adam, 1] and procure the "condensation" of monolayer structure. Since the "condensing" effect of cholesterol is exhibited on monolayers of a large variety of molecular species, the basic action of cholesterol must depend upon hydrocarbon-hydrocarbon-interactions. Further, the effect of cholesterol upon a monolayer of a pure component is not marked unless the monolayer of the pure component is near to a phase transition such as the $L_1 \rightleftharpoons L_2$ transition. It is therefore probable that the "condensing" action of cholesterol is effected by intervention in the process of cooperative movement of clusters of hydrocarbon chains.

## PROTEINS AND BILAYERS

There is as yet no apparent reason for supposing that protein at the membrane-water interface is arranged in any way other than I originally suggested, i.e., as a primary layer in which adsorption is primarily due to adsorption of hydrocarbon side chains of protein, and a secondary layer in which the adsorbing forces are primarily polar [4, 5]. Unfortunately, precise information about the nature of membrane proteins is still scarce, and it is unsatisfactory even as regards such fundamental considerations as the prevalence of $\alpha$ and $\beta$ configurations, as is clearly shown by Maddy's recent review [15]. However, the fact that evidence is accumulating to show that membrane structural proteins contain a large proportion of nonpolar side chains adequately supports the view that the primary layer of protein is stabilized by nonpolar bonding to lipid.

Pollard has pointed out that frequently the bonding of protein to lipid will be sufficiently strong that the protein and lipid will no longer be independent kinetic units. This has an important consequence in determination of the surface free energy. The surface free energy of a surface of the membrane contains a term $\gamma_{p/w}$ which contains a surface pressure term $\pi_k$ due to thermal agitation. Insofar as lipid molecules are bound to a single protein unit, $\pi_k$ will be diminished in proportion to the reduction of the number of kinetically independent units. Furthermore, such binding will result in point-to-point variation in lipid composition, so that the distribution of phospholipids, ubiquinone, carotenoids, chlorophyll, etc. will no longer be random, but protein-determined (and therefore determined by genetics).

We can now consider the implications of membrane structure for transmembrane molecular interactions For example, there are at least three distinct modes whereby the presence of protein adsorbed upon one surface of a bilayer may significantly modify adsorption upon the opposite side of the membrane. First, if the presence of a protein organizes a population of lipid molecules in an area of one monolayer of a bilayer, this may in turn modulate the lipid population in the contiguous area of the second, opposing, bilayer. Second, if the adsorbed protein has a sequence of 10–20 amino acids with nonpolar, or mostly nonpolar, side chains, its polypeptide chain may extend right through the thickness of the membrane, possibly as an $\alpha$-helix. Third, the dielective constant of the lipid layer will be between 2 and 3, compared with 80 for water in bulk, consequently the electrostatic interactions between charges on proteins will be as great as though they were separated only by about 7 Å of water. Furthermore, since ions are so insoluble in hydrocarbon there will be little electrostatic shielding due to ionic atmospheres. Consequently electrostatic interaction across the thickness of the membrane will be much stronger than has commonly been supposed. Although the divergence of the electrostatic field of a macromolecule will cause the field to be significantly smoothed out at a distance of 50 Å, the residual patterning of the field must still be expected to require electrostatic complementarity in the macromolecule adsorbed on the far side of a bilayer.

This patterning of macromolecules introduces the possibility that information may be stored on membranes. For example, if we assume that 1 macromolecule = 1 bit, that there are $10^4$ alternative species of macromolecule for one site on a membrane, that 10% of the protein of a memory cell was utilized for information storage, and accept von Neuman's estimate of the number of bits of information taken in by a man in his life, then it can easily be shown that the whole memory bank of a man can be stored in about $10^2$ cells. This at first seems a surprisingly small number. But it appears less surprising when we recollect that the whole of the genetic information required for synthesizing a man is stored in the DNA of one cell—a space smaller than the minimum memory space given above by a factor of about $10^5$. We also gain insight into how it is possible for an insect such as a bee, which shows complicated behavior and extensive memory phenomena, to operate with a "brain" containing only about 20 cells. If information is stored in the postulated manner we can also see why cells in the central nervous system do not divide—each division would fragment part of a memory bank, so that if memories are to be retained over long periods it is probably more efficient to provide a high degree of redundancy, i.e., store the same information many times over, rather than attempt to replace dead cells by cell division.

## CELLULAR CONTROL OF MEMBRANE PHENOMENA

What has been outlined above constitutes an important part of the physics of lipoprotein membranes. But the picture so given is essentially static and does not, in the degree of development given so far, provide for the dynamic aspect of cellular membranes, nor for the degree of control by the cell which is a striking feature of cell physiology. But at this point we can readily see how chemical mechanisms can provide for control, through at least three mechanisms: (i) micelle formation, (ii) active flow of membrane, (iii) receptor systems.

Here I shall say little about receptor systems, because, as is the case with permeases, we know too little of the physics of the molecules concerned. But on micelle formation and membrane flow some interesting speculation seems permissible.

As I shall show elsewhere, a lipid bilayer must exist in equilibrium with micelles as indicated by the equilibria:

$$\text{bilayer} \rightleftharpoons \text{cylindrical micelle} \rightleftharpoons \text{spherical micelle} \rightleftharpoons \text{single molecules in solution}$$

With molecules such as the natural phospholipids, carrying no net charge, the bilayer is normally the stable form. The free energy change in forming cylindrical micelles from bilayers is about +6000 calories, and of formation of spherical micelles from cylindrical micelles also about +6000 calories—or about 12,000 calories for forming spherical micelles from bilayers. These $\Delta F$ (free energy changes) are unfavorable to formation of micelles and are sufficiently large, compared with $kT$ at physiological temperatures, for the bilayer structure to be the predominant form. However, the $\Delta F$ values are not so large that we can say that none of the membrane will be in micellar form. Nor are the $\Delta F$ values so large that extra constraints will not result in a preponderance of micelle formation. For example, if a charge is placed upon the lipid, or upon the associated protein, we shall see an expansion of the bilayer, a decrease in its thickness, and an increased proportion of micelles. A reverse constraint, antagonistic to micelle formation, will be imposed by any cross-binding agent, such as calcium. In consequence cellular control of the bilayer $\rightleftharpoons$ micelle equilibrium may be exercised by at least three procedures:

(1) varying lipid charge, as by interconversion of lecithin and phosphatidic acid; (2) phosphorylation and dephosphorylation of protein; (3) regulation of calcium level. It is well known that many active cellular membranes contain phosphatases, and it is becoming evident that phosphorylating systems are also present. So the enzymatic basis postulated certainly exists.

However, in addition to the relatively symmetrical changes postulated by micelle formation, there is also a requirement for assymmetry to provide for phenomena such as pinocytosis. Such assymmetry requires that the cell shall be able to act upon one side of a membrane preferentially. Consider Fig. 8a: each monolayer of the bilayer has its own surface pressure, $\pi^o$ for the outer monoplayer and $\pi^i$ for the inner. If a cellular process reduces $\pi^i$ below $\pi^o$, or raises $\pi^o$ above $\pi^i$, minimization of free energy requires that the area of the inner surface be reduced and that of the outer surface be increased. Hence curvature will be produced in the sense shown. This is conducive to invagination and will lead to membrane flow and formation of a pinocytosis channel as shown in Fig. 8b. If the membrane is liquid, it will be unstable if $\pi D$ becomes greater than $L$, and will break up into droplets free to move around in the cell. The necessary change in surface pressure can arise in several ways: e.g. (a) by conversion of phosphatidic acid on the inner monolayer to neutral phospholipid, or by removal of phosphatidylserine; (b) by the reverse of (a) on the outer monolayer; (c) by adsorption of a macromolecule on the outer surface protein,

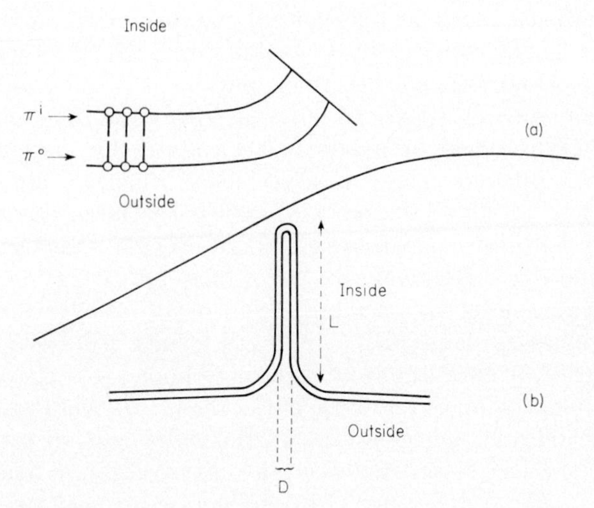


FIG. 8. Diagram to show how an assymmetry in membrane surface pressure leads to membrane flow and pinocytosis. Diagram (a) shows that if $\pi^o > \pi^i$, invagination follows; (b) shows a pinocytosis channel ready to break into droplets.

thereby releasing lipid from lipoprotein binding and raising $\pi^o$ by an increase in $\pi_k{}^o$; (d) by fusion of an assymmetrical vesicle with the membrane.

Direct action upon the cell exterior requires a special mechanism—e.g., that there shall be macromolecules passing through the thickness of the membrane, or that vesicles from the interior of the cell fuse with the membrane. Intracellular vesicles ordinarily have an internal enzymatic composition different from that of the general cytoplasm, so that the inner and outer surfaces may readily differ in charge density or in other ways. When such a vesicle fuses with the plasma membrane, the inner surface of the vesicle becomes part of the outer surface of the plasma membrane.

Thus we see that a variety of mechanisms are available which permit physiological control of cell membrane activity.

## REFERENCES

1. Adam, N. K., "Physics and Chemistry of Surfaces." Oxford Univ. Press, London and New York, 1941.
2. Chambers, R., *Biol. Bull.* **69,** 331 (1935).
3. Chapman, D., Advances in Chemistry Series, No. 63 American Chemical Society (1967).
4. Danielli, J. F., *Cold Spring Harbour Symp. Quant. Biol.* **6,** 190, (1938).
5. Danielli, J. F., *Proc. Roy. Soc.* (*London*) **B127,** 34 (1939).

6. Danielli, J. F., *in* "Surface Phenomena in Chemistry and Biology, p. 236. Macmillan (Pergamon), New York, 1958.
7. Danielli, J. F., *in* "Cytology and Cell Physiology" (G. Bourne, ed.). Academic Press, New York, 1964.
8. Danielli, J. F., *J. Theoret. Biol.* **12,** 439 (1966).
9. Danielli, J. F., *J. Theoret. Biol.* in press (1968).
10. Danielli, J. F., and Davson, H., *J. Cellular Comp. Physiol.* **5,** 495 (1935).
11. Danielli, J. F., and Harvey, E. N., *J. Cellular Comp. Physiol.* **5,** 483 (1935).
12. Danielli, J. F., and Harvey, E. N., *Biol. Rev. Cambridge Phil. Soc.* **13,** 319 (1938).
13. Davson, H., and Danielli, J. F., "The Permeability of Natural Membranes." Cambridge Univ. Press, London and New York, 1943.
14. Kopac, A., and Chambers, R., *J. Cellular Comp. Physiol.* **9,** 331, 345 (1937).
15. Maddy, A. H., *Intern. Rev. Cytol.* **20,** 1 (1966).

# THE ORGANIZATION OF PROTEIN IN THE PLASMA MEMBRANE

**A. H. MADDY**

*Chemical Biology Unit, Department of Zoology, University of Edinburgh, Edinburgh, Scotland*

## INTRODUCTION

The similarities between the properties of membranes and lipid films have resulted in the domination of theories of membrane structure by considerations of the arrangement of the lipids. As not much was known until quite recently about the supramolecular association of lipids, the possibilities for the membrane appeared to be relatively limited, and in any case the postulated bimolecular leaflet fulfilled the requirements for a membrane model [10]. During the last four or five years many different types of associations of lipid molecules have been discovered [3, 29, 30, 31, 62], and it is necessary to consider whether these new structures exist in the membrane, especially as they may explain some aspects of membrane behavior not accounted for by the classical theory. However, it is much easier to dwell on the inadequacies of any theory, to point out that such a new idea will resolve such an ambiguity and yet another idea illuminate some long-standing obscurity, than it is to construct an alternative theory taking into account all the facts even as well as the old. Nevertheless it is not possible to speak with certainty of the arrangement of lipids in the plasma membrane; indeed the membrane should not be considered as one unique structure, but rather as an entity with several alternative configurations, each configuration with its own energy level. The problem is therefore not whether the membrane is of this structure or that, but how much of this, and how much of that; not whether it is exclusively a bimolecular leaflet or exclusively a micellar organization, but to what extent a leaflet and to what extent a discontinuous micellar array.

It follows from the concept of the membrane as a molecular association of many interacting and reversible equilibria that any analytical procedure will tend to alter the state of equilibrium, and perhaps eliminate certain unstable states that are of exceptional metabolic importance, or induce a totally artificial equilibrium. This element of uncertainty arising from perturbations introduced by experimental operations may be unavoidable in the structural analysis of any cell organelle, but its importance will vary from case to case, and therefore some attempt should be made to assess its significance in any piece of work. For instance data obtained by the addition of detergents to plasma membrane preparations must be interpreted with great caution. Detergents have been successfully used to *dissociate* various molecular complexes, but complexes formed *by the addition* of detergent to membranes might well be artifacts. Thus the observation that molecular complexes are derived from membranes by the addition of detergents, especially when those complexes contain considerable quantities of the detergent (e.g., the membrane of PPLO [57]), cannot be simply interpreted. However, while it is desirable to be aware of the possible complexity of membrane transformations, one should not assume such a degree of complexity that further experimental analysis is impossible. The situation calls for cautious experiment rather than masterly inactivity.

While it is probable that membrane proteins influence the conformations of the lipid and perhaps control the metabolic activity of the membrane, very little is known of the mechanisms of the interaction or even the orientation of the protein relative to the lipid. The clearest prediction of the relationship between the lipid and the protein is found in the classical theory of membrane structure developed by Danielli, Davson, and Harvey in the 1930's [10, 20], where, as the lipid is postulated to be a continuous layer, the possibilities for the protein are limited to a layer on one or both sides of the lipid perhaps with some protein aqueous pores traversing the lipid [71]. The model therefore consists essentially of two protein layers separated by a layer of lipid which, as it is the permeability barrier of the cell, has the effect of separating the protein into two portions, one outside the permeability barrier of the cell and accessible from outside the cell, and the other inside the barrier and able to interact only with substances inside the cell or that pass through the lipid. When the model was first proposed it was generally held that proteins lost all secondary structure at interfaces and it was assumed that the protein adjacent to the lipid in the membrane "unrolled" into an extensive sheet of insoluble protein between the lipid and any globular proteins. This idea has become entrenched in the literature where the extended protein is frequently shown as a zigzag line, and it has recently been developed in relation to phase changes of the lipids [26]. As it is now known that secondary structures such as the $\alpha$-helix can persist at interfaces [39] this aspect of the original model must be reexamined (see the section on infrared spectroscopy).

Although several alternatives to the bimolecular lipid leaflet are recognized *in vitro,* it is not known whether any of these exist *in vivo,* and the relationships of proteins to such lipid arrays is quite obscure. It could occupy spaces between the purely lipid micelles [29] or be incorporated into a lipoprotein complex [17]. A probable consequence of either of these arrangements would be the presence of a regular array of components in the plane of the membrane, but such an appearance is not diagnostic, for a repeating pattern in surface views could equally well arise from an ordered packing of protein components outside a continuous lipid bilayer.

In addition to these discoveries, which have opened up many divergent lines of thought, there has been the quest for the "unit membrane" [58, 59, 61]. The "unit membrane" concept implies that all cell membranes are built on a common plan, in fact the Davson-Danielli structure. As a unifying concept this hypothesis has many attractions, but it is fraught with the danger of the unwarranted application of data obtained from one type of membrane to another. The evidence for the "unit membrane" as a generalized membrane structure is derived entirely from electron microscopy; the trilamellate pattern is concluded to represent the lipid and protein layers of the Davson-Danielli model, but this model was proposed before the advent of the electron microscope and it is based on evidence of an entirely different nature. The "tramlines" seen by electron microscopists *at cell surfaces* may confirm the Davson-Danielli theory of membrane structure, but the other data used by Davson, Danielli, and Harvey in their plasma membrane studies cannot be used to interpret "tramlines" in other cell membranes except where that data have been shown to apply. Similarly, work on the mitochondrion, where the evidence for a bilayer is weak and the balance possibly in favor of a micellar array, does not disprove the existence of a bilayer in the plasma membrane [14, 17].

It would be inappropriate to consider in further detail the current conflicting theories of membrane structure, but how can the structure be elucidated and the conflicts resolved? The procedure followed by organic chemists in the investigation of any complex substance may profitably be imitated. The chemist first collects as many data as possible from the complex substance itself; he then degrades it and characterizes the degradation products; and as final confirmation of its structure he synthesizes the original complex substance. These separate operations are rarely carried out in sequence; usually they proceed collaterally, information gleaned from one approach assisting the progress of another. A similar schedule is available for the analysis of membrane structure; the properties of the intact membrane, both in purified cell fractions and in the living cell can be studied by all available techniques; then the individual constituents may be separated and their behavior as simple molecules and as interacting elements observed; finally the intact, biologically active, membrane can be reconstituted from its component parts. Some aspects of this program are more feasible than others, some, especially the full realization of the last, perhaps do not belong to the foreseeable future, but greater or lesser progress has been made along all three avenues, and it is proposed to assess the current situation with particular reference to the proteins of the membrane.

## STUDIES ON THE INTACT MEMBRANE

The paramount significance of lipids in membrane structure was deduced from very early measurements of permeability and electrical resistance of intact cells, but this work is of limited relevance to the disposition of proteins in the membrane. Insight into the arrangement of the protein was obtained from studies of the birefringence of red cell ghosts [43, 63, 64] and work on model protein monolayers [1]. The observation that hemoglobin is essential for ghost birefringence has cast suspicion over the earlier results [54] and the conclusions derived from the model monolayers must be reappraised in the light of modern work on interfacial protein monolayers. Over a number of years electron microscopists have added to our knowledge of the structure of the membrane, and more recently infrared spectroscopy has been applied to the problem.

### *Electron Microscopy*

Electron microscopists have produced a body of evidence in favor of the Davson-Danielli model [35]. The trilamellate or "tramline" image has been extensively described [58, 59, 61, 66, 67, 68, 70], and, largely on the basis of electron microscopy of phospholipid–water and phospholipid–water–protein mixtures [72], it has been concluded that the clear central zone represents the nonpolar hydrocarbon chains of the lipids, and the electron dense lines represent their polar head groups and the proteins. However, the evidence applies to the fixed structures, and the conversion of all less stable states into the trilamellate structure during the processing of the tissue for electron microscopy cannot be excluded. As the Davson-Danielli model is a state where hydrophobic bonding between lipid and protein is probably at a minimum [21] and is favored by the maximum hydrophobic bonding between lipids (a condition satisfied by fully saturated lipids), the strong oxidative fixatives required for the demonstration of the unit membrane may convert other lipoprotein structures that depend on hydrophobic bonds and contain a relatively high percentage of unsaturated lipids into unit membranes. It might be significant that alternative lipid conformations have been displayed by highly polar, nonoxidative negative stains. For example, liver plasma membranes incubated at 37°C before negative staining in phosphotungstic acid appear as sheets of hexagonally packed facets with a center-to-center spacing of 90 Å. This micellar arrangement is absent if the preparations are not warmed before staining. A surface layer of globular units attached to the membrane by stalks and not affected by temperature is also apparent, but never in the same regions as the hexagonally packed complexes [4]. It is not known what changes the negative staining induces, but the knowledge that a phase transition occurs at 37° in phospholipid–water mixtures, the hexagonal phase being stable at higher temperatures [31], suggests that the observations of Benedetti and Emmelot are caused by similar temperature-dependent transitions in the conformation of the phospholipids in the membrane before the addition of the stain.

It has already been mentioned that a repeating pattern in the surface topography of a membrane might indicate, but not diagnose, a micellar organization. Such patterns are known in mitochondria [14, 18, 69], microsomes and cytomembranes [9, 69], chloroplasts [51, 53], and retinal rods [6, 49], but with the exception of some highly specialized structures, e.g., the synaptic vesicles of Mauthner cells [60] and the negatively stained liver membranes, a regular pattern is absent from the surface plasma membrane. The irregular plaque-like appearance of cell surfaces [8, 22] and microvilli [52] can hardly be used to argue one way or the other; the irregular particles could lie exterior to a lipid leaflet, or perhaps indicate an entirely different arrangement of a surface composed of a feltwork of looped fibers 300–500 Å in diameter, the top of each loop appearing in surface view as a plaque [16]. There is no information on the chemical nature of this structure except for the suggestion that the fibers consist of elenin [48].

### *Infrared Spectroscopy*

Measurements of birefringence and electron microscopy give some indication of the distribution of protein and lipid relative to each other. Infrared spectroscopy yields information relating to the states of the lipids [7] or proteins themselves, not to their relative positions. The spectrum is of interest to protein chemists as the position of a strong absorption band (the amide I band) in the spectra of proteins due to C=O stretching depends on the conformation of the protein, having a frequency of ca. 1660 $cm^{-1}$ in the $\alpha$-helix or random coil and ca. 1630 $cm^{-1}$ in the $\beta$ conformation. A second band of more complex origin and of less diagnostic reliability (the amide II band) usually shows a corresponding shift from 1540 $cm^{-1}$ to 1520 $cm^{-1}$. The presence of any $\beta$ conformation in the membrane will therefore be apparent from its spectrum.

Maddy and Malcolm [36] investigated the spectra of hemoglobin-free ox erythrocyte ghosts prepared as air-dried films on barium fluoride plates. There is no indication of any $\beta$ conformation in the membranes,

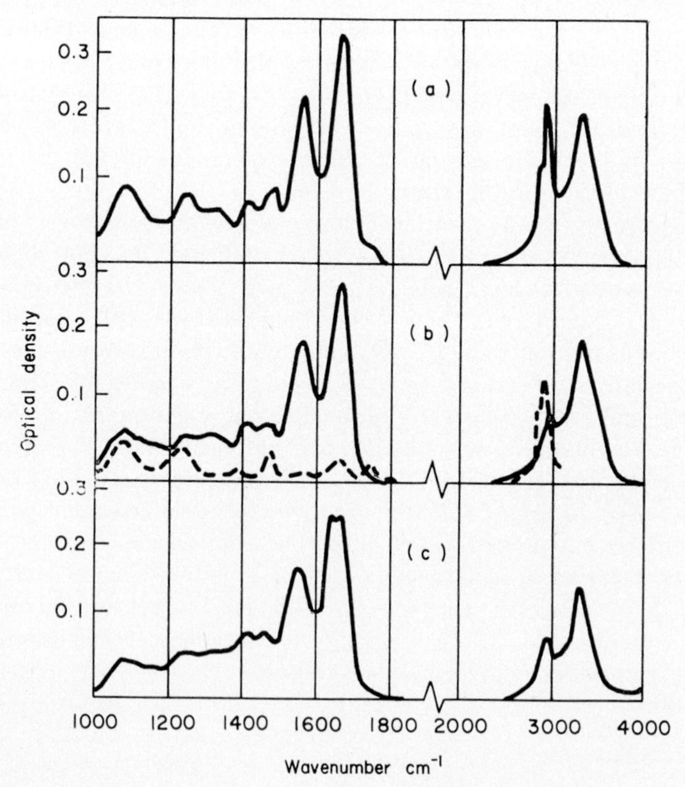


FIG. 1. Infrared spectra of air-dried specimens. (a) Red cell ghosts. (b) Solid line, extracted ghost protein; broken line, estimated relative contribution of lipid to spectrum of ghost. (c) Ghost protein denatured with ethanol, showing about 50% of $\beta$ conformation.

whose spectrum has a symmetrical amide I band at 1660 $cm^{-1}$ and an amide II band at 1540 $cm^{-1}$ (Fig. 1a). These bands are unchanged in dry films prepared from solutions of membrane protein, but the $\beta$ spectrum can be induced by heating the protein solution in 50% ethanol for 3 minutes at 70° (Fig. 1c). The high protein content of the membrane (60%) is reflected in its spectrum; the less conspicuous contribution of the lipid (40%) is also apparent (Fig. 1b). The two spectra summate to that of the intact ghost without any features indicative of lipid–protein interaction. It is improbable that the lipid is covered by $\beta$ protein but that this amount of $\beta$ sheet is too small to be detected in the infrared, for from a consideration of the relative areas of lipid and protein it may be calculated that 35% of the ghost protein must be in the $\beta$ state to form a continuous $\beta$-sheet over the lipid. This proportion of $\beta$ conformation would be clearly manifest in the infrared spectrum.

The suggestion that drying the membranes for spectroscopy destroys the $\beta$ components of the living membrane [37] is refuted by spectroscopy of suspensions of ghosts in deuterium oxide, which, unlike water, does not absorb in the relevant region of the spectrum. Spectra of fresh suspensions of the ghosts or solutions of the protein confirm the absence of the $\beta$ conformation. The amide I band of the deuterated material has a symmetrical peak at 1648 $cm^{-1}$ corresponding to the 1660 $cm^{-1}$ peak of the nondeuterated protein, while in the $\beta$ spectrum, which once again may be induced by thermal denaturation, this band is at 1632 $cm^{-1}$ [37].

Infrared spectroscopy can be supplemented by measurement of the optical rotatory dispersion (ORD) of aqueous solutions of membrane protein [36]. The ORD of the protein was measured between 588 $m\mu$ and 246 $m\mu$, and the results were fitted to the Moffitt-Yang equation with a $\lambda_o$ of 216 $m\mu$ [46]. From work on other proteins it is known that the value of the term $b_o$ is proportional to the helical content of the protein, so as the $b_o$ of the membrane protein is —90 as compared with —535 for a completely helical protein at $\lambda_o$ of 216 $m\mu$, it appears that the membrane protein has 17% $\alpha$-helix. This conclusion must be accepted with caution as the measurements may be affected by the 8% carbohydrate and 5% lipid in the preparation, and possibly by small amounts of $\beta$ protein not detected by infrared spectroscopy. It must also be stressed that these results apply to the conformation of the protein in solution, not to protein as it is associated with lipid in the membrane.

The infrared and optical rotatory dispersion results, while producing valuable information on the state of the protein in the membrane do not facilitate the choice of any particular membrane model. On the one hand, they do not contradict the essential features of the Davson-Danielli model, they only modify certain postulated details of the arrangement of the protein—i.e., that the lipid surface is covered by a distinct layer of "unrolled" and insoluble protein; on the other hand, the absence of a structure with as much intrinsic stability as a sheet of $\beta$ protein is consistent with the concept of the membrane as a delicately poised equilibrium between several alternative conformations.

## THE FRACTIONATION OF THE MEMBRANE

The technical problems of the analysis of the lipid constitution of membranes appear to be solved, but the characterization of the proteins has been dogged by the difficulty of dissociating them from the lipids in a native state. A subsidiary difficulty in the investigation of the proteins is the widely differing affinities they have for the physiological entity known as the membrane. This aspect has not hindered research but makes the distinction between true membrane components and contaminants of membrane preparations equivocal. The avidity of proteins for the lipid ranges from the one extreme of those proteins that no amount of aqueous washing can separate from the lipid, to those, for example the $\beta$-galactoside permease of *Escherichia coli* [28], where the binding is so weak that the protein is found in the cytoplasm. (It is quite conceivable that some metabolic constituents of the membrane, for instance this permease, are located in the membrane only during the short time required for the fulfillment of their catalytic role.) Many loosely bound proteins between these two extremes have been reported. The best-known case is that of hemoglobin in the red cell ghost, which adheres firmly to the ghost at pH 6.0, but can be removed at pH 8.0 [11]. More recently Green and his co-workers have reported an association between red cell ghosts and glycolytic enzymes indicated by the greater specific activity of the enzymes in partly washed ghosts than in the whole cell. However, such associations must not be accepted at face value, for while hexokinase is an intrinsic element of the plasma membrane of hepatoma cells, it is not a true part of the membrane of liver cells, but becomes associated with the membrane fraction during isolation. In contrast to the hepatoma cells, the enzyme of normal liver cells is washed off the membrane fraction by saline, and the association is due to a nonspecific interaction between the negatively charged membranes and positively charged enzyme [12]. Perhaps many of these loosely bound proteins are of more relevance to membrane activity than structure, the membrane functioning as a site for the localization of the metabolically active molecules.

The Davson-Danielli hypothesis implies that the membrane protein is separated into two layers by the lipid, and it should in principle be possible to distinguish between the two: in the first place, as all the sialic acid of the red cell is accessible to neuraminidase which does not pass through the membrane [13], any sialoprotein must be derived from the outer layer; and second, Maddy [32] has developed a fluorescent label for proteins outside the permeability barrier of the cell (Fig. 2). Therefore, once the bulk of the membrane protein has

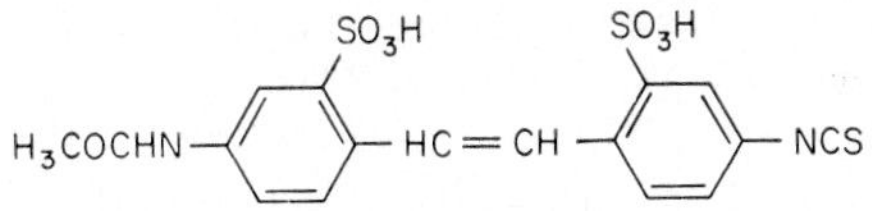

$$H_3COCHN-C_6H_3(SO_3H)-HC{=}CH-C_6H_3(SO_3H)-NCS$$

FIG. 2. 4-Acetamido-4′-isothiocyanostilbene-2,2′-disulfonic acid. This reagent may be used to label proteins outside the permeability barrier of cells. It reacts under physiological conditions without entering the cell and is easily detected by its fluorescence.

been solubilized, if the Davson-Danielli hypothesis is correct, it should be fairly easily separated into fractions with sialic acid and fluorescent label and without either. However, about 90% of the membrane protein after solubilization with butanol [33, 34] is a sialoprotein or a mixture of sialoproteins (Table I). This proportion of membrane protein passes straight through a carboxymethyl cellulose (CM) column at pH 6.0 but is irreversibly adsorbed onto diethylaminoethyl cellulose (DEAE) unless the sialic acid is previously removed by neuraminidase when up to 80% can be eluted at acid pH; in free-boundary electrophoresis

TABLE I. *The Chemical Composition of the Protein Obtained by Butanol Extraction of Erythrocyte Ghosts*

| Parameter | Protein | Ghost |
|---|---|---|
| Percentage lipid | 5 ± 1 | 43 ± 2 |
| Percentage nitrogen | 14.4 ± 0.4 | 9.1 ± 0.5 |
| Percentage phosphorus | 0.2 ± 0.02 | 1.3 ± 0.13 |
| N/P ratio | 340[a] | 15.5 |
| Sialic acid (μmoles/mg) | 0.067 ± 0.001 | — |
| Hexosamine (μmoles/mg) | 0.16 ± 0.01 | — |
| Hexose (μmoles/mg) | 0.28 ± 0.01 | — |

[a] Value after correction of protein phosphorus for phosphorus soluble in chloroform–methanol and inorganic phosphorus.

FIG. 3. Electrophoresis of butanol-solubilized ox erythrocyte ghost protein at pH 7.0 in phosphate buffer ($I = 0.1$), ascending limb, cathode to left of figure. The major component has a mobility of $-9 \times 10^{-5}$ cm$^2$ sec$^{-1}$ V$^{-1}$.

the protein moves as a rather broad band with a mobility at pH 7.0 of $-9 \times 10^{-5}$ cm$^2$ sec$^{-1}$ V$^{-1}$ (Fig. 3). The width of this peak may be due to a heterogeneity of size, for, as shown in Fig. 4, the protein separates in the ultracentrifuge into two major components (5 S and 10 S) with a certain amount of larger material. The rate at which the bands broaden as they pass down the cell suggests that the two major fractions are themselves made up of particles of slightly different sizes.

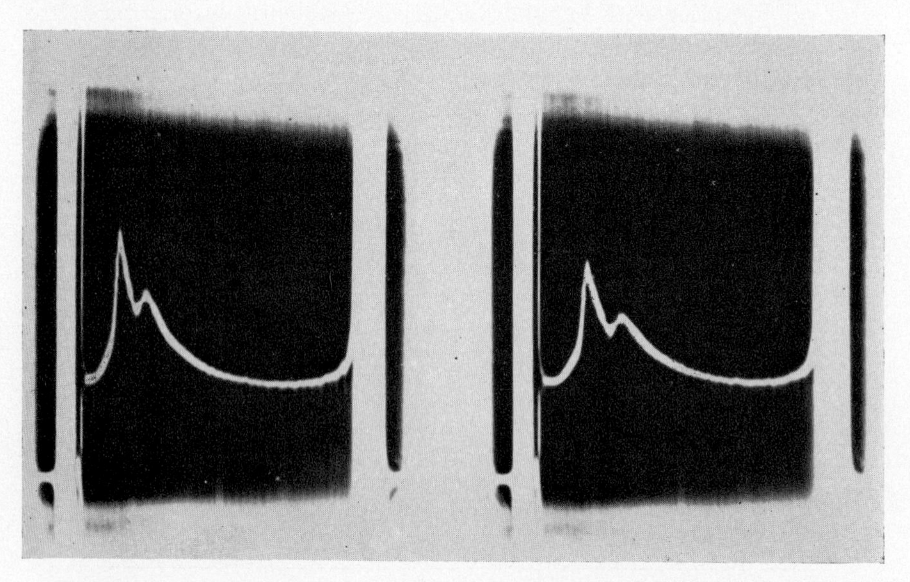

FIG. 4. Ultracentrifugation of butanol-solubilized ox erythrocyte ghost protein in Spinco Model E, 50760 rpm. Solvent: Phosphate buffer (pH 7.0, $I = 0.1$).

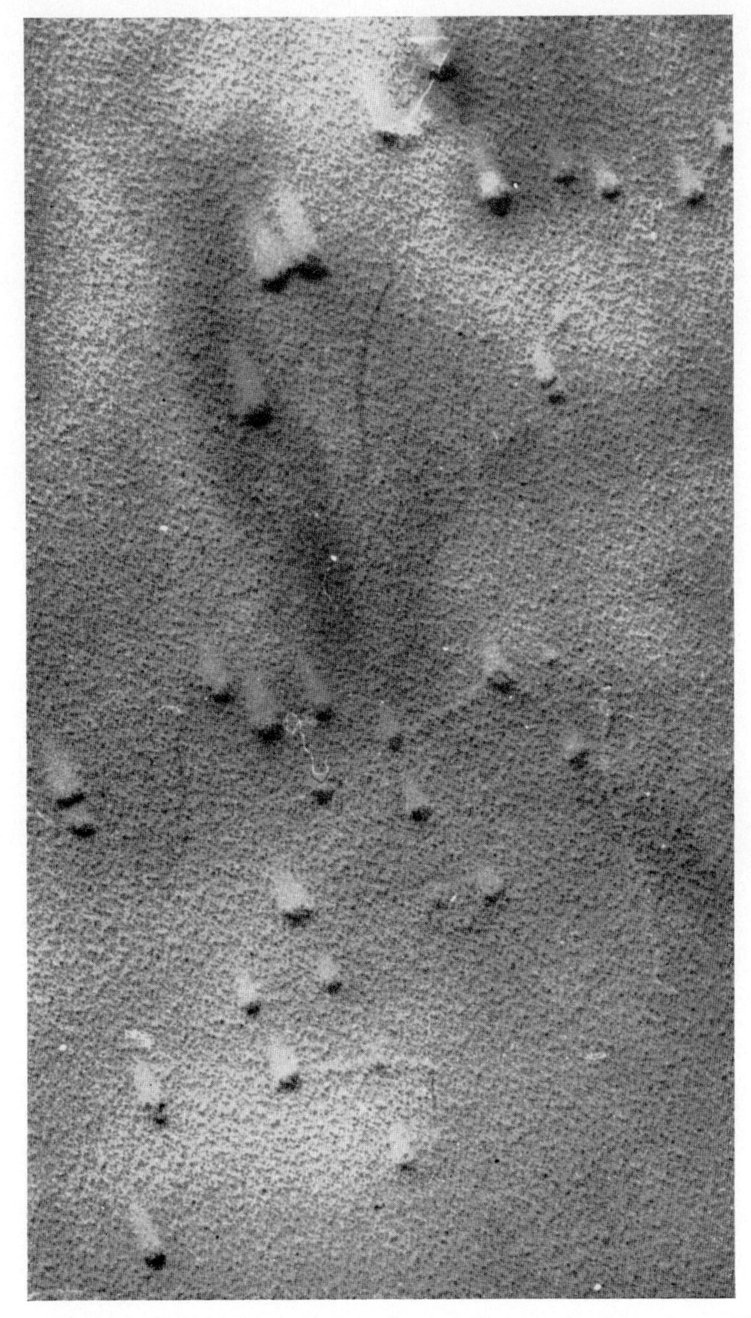

FIG. 5. Platinum-shadowed preparation of ox erythrocyte ghost protein. × 150,000. By courtesy Dr. G. H. Haggis.

This size distribution is confirmed by the preliminary findings of an electron microscope study of the preparation by Dr. Geoffrey Haggis. Shadowed preparations of the protein show two types of particles predominant, one with a diameter of about 90 Å, the other of about 180 Å, and a few still larger particles (Fig. 5). Sometimes the shadow shows that the 180 Å particles are made up of two smaller ones. Preparations with sodium phosphotungstate reveal a similar distribution of sizes, the frequent pairs of particles of slightly unequal size perhaps appearing as 180 Å particles in the shadowed preparations (Fig. 6). It should be stressed that these objects are proteins, virtually free of lipid, and not lipoprotein micelles. They do not relate to any

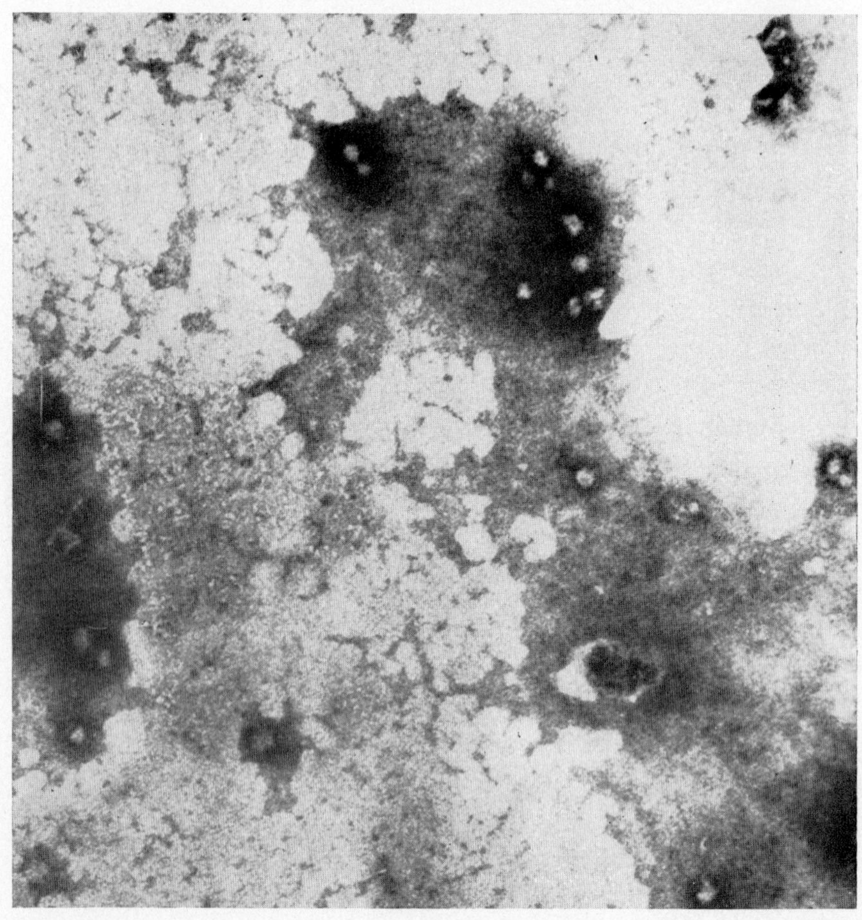

FIG. 6. Ox erythrocyte ghost protein negative stained in sodium phosphotungstate. × 184,000. By courtesy Dr. G. H. Haggis.

known structure of the intact red cell ghost although it is tempting to analogize them with the outer covering of 60 Å spheres described for liver plasma membranes by Benedetti and Emmelot [4]. As a lipid bilayer of about 60 Å thickness combined with one layer of 90 Å particles on only one side results in a membrane 150 Å thick, it may be argued that, on addition of protein to the other side, the thickness is incompatible with the value of about 75 Å for the fixed membrane and 90 Å for the wet unfixed membrane obtained by electron microscopy and confirmed by X-ray diffraction on both myelin [15] and the limiting membrane of cilia [65]. However, it seems more probable that the apparent similarity between the spacing of membranes and myelin is anomalous. The repeating unit of myelin, which is composed of 22% protein and 78% lipid [50], is interpreted as two closely apposed unit membranes; and the 180 Å spacing, as made up of two lipid bilayers of 50 Å each and two layers of protein. If the lipid is a constant frame of reference, it is difficult to see how the plasma membrane, which contains more protein (60%) than lipid (40%) can have the same overall thickness. More recent electron micrographs also suggest a greater thickness for the plasma membrane.

The observation that the bulk of the membrane protein bears sialic acid can be interpreted in a number of ways. (a) The ghost is largely devoid of protein on its inner surface. As the ghosts are exhaustively washed before butanol solubilization it is quite possible that protein organized on the inner surface *in vivo* (e.g., the glycolytic enzymes) has been washed away. The 10–15% of protein not accounted for as a sialoprotein may form a partial lining to the lipid and might be identical with the protein extractable from washed ghosts by 1.5 *M* saline. Mitchell *et al.* [40, 41] have extracted up to 30% by weight of the total protein of human ghosts by this method, but Maddy, who obtained a similar yield from human and sheep ghosts, finds that only about 12% of the protein of ox membrane can be extracted. The saline-extracted protein differs from that solubilized by butanol as it contains no detectable sialic acid (Maddy, unpublished results). (b) The two layers of protein aggregate during removal of lipid. Aggregation resulting from the butanol treatment cannot be excluded; the effect could be due to lipid binding sites in the protein binding with each other once the lipid is removed, and is not necessarily a deleterious consequence of butanol. (c) The protein is continuous throughout the thickness of the membrane. The Davson-Danielli model does not exclude some proteinaceous continuity through the lipid, although it would be vitiated by the demonstration of an extensive intussusception of the lipid by protein.

Fractionation of membrane proteins has been achieved by starch gel electrophoresis [2], but the protein must first be treated with urea and mercaptoethanol. The significance of this fractionation to membrane structure is not yet apparent, and it does not necessarily conflict with the observations on the butanol-solubilized protein.

Some workers have chosen to isolate lipoproteins from the membrane with a view to study lipid protein interactions in these less complex systems. Many years ago the lipid–protein–carbohydrate complex called elenin was isolated and described as a fibrous structure [48]. More recently lipoproteins have been separated from erythrocyte ghosts by high salt solutions [40] and by mild butanol treatment [47]. The high salt yields a lipoprotein fraction and a protein fraction; the butanol method also yields lipid-free protein and a lipoprotein complex containing 5% of a protein differing in its amino acid composition from the lipid-free protein. The investigation of the part these structures play in the structure of the whole membrane is awaited with interest.

## RECONSTITUTION OF THE MEMBRANE

It is premature to expect to be able to reconstruct a membrane with full metabolic activity, but as the protein of membranes can be separated from the lipid without apparent denaturation, their recombination could be attempted [38]. The preparations used were the membrane proteins as isolated by Maddy [34] and the lipid bilayers as prepared by Huang *et al.* [23, 24]. Interaction between the protein and lipid was observed when lipid films were spread in solutions of the protein (0.4 mg/ml) in the saline. The lipid layer did not thin in its usual characteristic fashion from the thick brightly colored mobile state to the black bimolecular state, but was "fixed" as a thick rigid disk in which the colored regions were static. At lower protein concentrations the lipid began to thin, but frequently the membrane shattered before the transition was complete. Even after a thousand-fold dilution of the protein there was a marked retardation of the thinning process (Table II). The prevention of thinning was a specific effect of the membrane protein; complete black films were formed in solutions of bovine serum albumin, $\gamma$-globulin, and cytochrome *c* at between 1 and 10 mg/ml, although, as Hanai *et al.* [19] have confirmed, these proteins do retard the process.

Because of the gross effects on the formation of the bimolecular film, in subsequent work the protein was added to preformed lipid bilayers by introducing equal volumes of protein solution to the saline on either

# HORMONE-MEMBRANE INTERACTION: THE ACTION OF INSULIN IN RED CELL SYSTEMS

T. L. DORMANDY

*Department of Chemical Pathology, Whittington Hospital, London, England*

## INTRODUCTION

Although the primary mechanism of hormone action remains obscure, it is generally conceived in metabolic terms, as a regulating device closely linked to enzymatic activity. This preconception explains why mature human red cells have long been regarded as the insulin-*insensitive* tissue par excellence. Since the earliest days of insulin research, the action of the hormone has been interpreted in relation to glycogen synthesis, protein synthesis, lipolysis, cellular growth, and respiration; and in red cells all these processes are rudimentary or nonexistent. The reason for choosing so unpromising a material as an experimental system was twofold. First, it seemed possible that hemoglobin might be used as an intracellular indicator dye. Second and more important, the possibility was envisaged that the familiar metabolic effects of insulin might be the consequences of an immediate physical change brought about by the hormone. In this context I would define "physical change" as an alteration in the physical properties of the cells under experimental conditions designed to reduce enzymatic activity to a minimum. For the demonstration of such a change the metabolic limitations of the system could be an advantage. A hypothetical example may illustrate this.

Let it be assumed that, by altering the electrical conductivity of the plasma membrane, insulin induced an immediate rise or fall in intracellular $H^+$-ion activity. Since most or all enzymatic equilibria would be affected by the pH change, such an action would entail extensive metabolic readjustments. By analogy with other biological regulating mechanisms one might expect that these readjustments would "backfire" on the site of the primary insulin-membrane interaction; and, again by analogy, they would tend to counteract the initial hormone effect. For instance, an initial swing toward a more alkaline intracellular milieu would be gradually neutralized by the resetting of metabolic pathways whose overall effect would be acidifying. In tissues with a wide metabolic reportory—e.g., adipose tissue, skeletal muscle—the primary change might become totally obscured. On the other hand, in structures incapable of such sweeping metabolic adaptation the immediate effects might be clearly demonstrable.

## PARADOXICAL IONIC FLUXES

In an exploratory series of experiments conducted by Dr. Zarday and me some years ago, our test systems were unwashed red cells suspended in various glucose-free saline buffers [5]. By "unwashed" I mean that after the initial centrifugation no attempt was made to eliminate all traces of trapped plasma from the suspending media. The experiments were performed at room temperature or at 4°C. Insulin was added to the cell suspension in catalytic concentrations; and a few minutes only were allowed for new equilibria to be established. The cell suspensions were then spun, and the supernatants and the lysed packed-cell deposits were analyzed. The differences between parallel insulin and control systems were transient but clear-cut and qualitatively reproducible. They were also paradoxical in the sense that they could not be interpreted in terms of the classical Gibbs-Donnan equilibrium. This is illustrated by the results shown in Table I.

Omitting a number of measurements and calculations, Table I shows the pH readings and the partition of inorganic ions between the hemolysed cell deposits and the extracellular fluids in two experiments. The suspending media were saline-bicarbonate buffers equilibrated with an

TABLE I. *The Effect of Insulin on the Extracellular pH and the Distribution of Inorganic Ions (in meq/l) between the Extracellular Fluids and the Separated Lysed Cell Deposits*[a]

| | Experiment 1 | | Experiment 2 | |
|---|---|---|---|---|
| | Insulin | Control | Insulin | Control |
| Extracellular | | | | |
| $Na^+$ | 165 | 165 | 164 | 163 |
| $Cl^-$ | *139* | *146* | *138* | *144* |
| $CO_2$ (cc) | 18 | 17 | 17 | 16 |
| pH | *7.48* | *7.40* | *7.51* | *7.43* |
| Intracellular | | | | |
| $Na^+$ | 14 | 15 | 13 | 14 |
| $K^+$ | 102 | 104 | 104 | 104 |
| $Cl^-$ | *80* | *76* | *81* | *77* |
| $CO_2$ (cc) | 12 | 11 | 12 | 11 |

[a] The insulin and control suspensions were treated in parallel. The insulin concentration was 5 μg/100 ml.

"infinite" gas space of 95% nitrogen and 5% $CO_2$. The first point which may be noted in the results is the more alkaline extracellular pH. Since the plasma membrane is virtually impermeable to cations but freely permeable to many anions, this change in ionization should entail an outward flux of $Cl^-$ and $HCO_3^-$. In the insulin systems the alkaline swing was associated with an ionic flux in the reverse direction. We explored this paradoxical response in some detail and eventually concluded that it could be explained only by accepting that insulin induced some form of immediate charge separation at the cell-extracellular interface. In more hypothetical detail we pictured the event as an insulin-induced outward flow of electrons across the plasma membrane (or as a transmembrane oxidation reduction) followed by the protonation of the interface from the outside.

In this tentative hypothesis there were two major gaps. First, the redox properties of cells are generally conceived as a function of oxidation-reduction-linked *enzymatic* processes: Yet under our experimental conditions active metabolism was virtually at a standstill. Second, a transmembrane oxidation-reduction implies a reversible redox system on *both* sides of the membrane; and in most of our experiments the extracellular component was apparently lacking.

Taking the second difficulty first, this could be resolved at a speculative level. Even in cells suspended in inorganic saline solutions one can conceive of a junctional (anatomically extracellular) zone with electrochemical properties transitional between those of the poised nonwatery intracellular compartment and those of the watery extracellular fluid. Such a zone might act as an "electron sink" to the cells: yet it could also exchange $H^+$ ions with the free watery extracellular fluid. It seemed possible, moreover, that the capacity of this region to mediate between the watery and the nonwatery phase—to generate a kind of hydrogen overvoltage—might be regulated by directional electron-transfer catalysts.

Turning to the first difficulty, this could be eliminated by assuming that intact cells maintain a definite reversible redox potential even at metabolic rest: in other words, that they behave as effectively poised systems. Since this is a subject of wider implications (and because red blood cells are particularly suited to its elucidation), I should like to digress and discuss the evidence in some detail. In particular, I should like to consider two properties of red cell systems which, I think, provide useful indirect parameters of the intracellular reducing potential. The first is the difference between the affinity of intracellular and of free hemoglobin for oxygen. The second is the apparent pH gradient between the cells and their extracellular milieu.

## HEMOGLOBIN AS AN INTRACELLULAR INDICATOR DYE

Whatever the metabolic limitations of red blood cells, they have one unique and valuable property: they possess a "built-in" redox-potential indicator dye. To appreciate this I must briefly review some of the molecular properties of hemoglobin.

It may be recalled that reversible oxygen binding by hemoglobin is accompanied by "molecular configurational changes" [8] which involve not only the four prosthetic heme groups but also the four polypeptide chains of globin: indeed, when heme is separated from globin it can only be oxidized, it cannot be oxygenated. These configurational changes or internal energy migrations affect the behavior of the pigment in relation to its milieu. The best-known example of this is the Bohr effect: a molecule of oxygenated hemoglobin is a stronger acid and therefore binds more base than a molecule of deoxygenated hemoglobin. In more recent years it has been shown that oxygenation also entails—or *can* entail—the unmasking of reversible reducing groups [1, 2, 8, 9]. These are commonly but somewhat misleadingly referred to as —SH groups. What the unmasking probably means is that reversible electron-donor sites become accessible to complementary electron-acceptor sites in the milieu. It may be worth emphasizing the difference between the change in pK or acid-base function and this change in $E_h$ or redox function. In a hemoglobin solution or in a hemolyzate one can normally assume the presence of water; and, as the hemoglobin becomes oxygenated or deoxygenated, the watery solvent will accommodate a greater or lesser number of $H^+$ ions. In other words, the pH of the milieu can change and therefore the pK of the hemoglobin can change. This is not necessarily true of reducing groups. Only if the milieu can accept electrons—i.e., if it has reversible redox properties—can the electron-donor groups of hemoglobin contribute to the stabilization of the molecule.

In describing these changes we sometimes represent the oxygenation of the hemes as the central event; at a fixed $pO_2$ the hemes combine with a certain amount of oxygen and the combination determines the acid and the reducing functions of the molecule as a whole. In fact, of course, the three functions are reciprocally related. In a simple watery buffer at a fixed oxygen tension hemoglobin behaves like any pH indicator dye: provided that it is neither fully oxygenated nor fully deoxygenated, it changes color and combines with more or less oxygen depending on the pH. If a suitable reducing or oxidizing agent is added to the system, the titration curve changes: the pigment will then behave as a redox-indicator dye, responding both to the pH and to the $E_h$ of the milieu. If the partial pressure of oxygen is raised or lowered, both titration curves will shift. In short, the molecule tends to maintain its internal stability as best it can by reacting in three different ways to three environmental variables—the $pO_2$, the pH, and the $E_h$. None of these reactions is linear; and they cannot at present be expressed in the form of an extended mathematical equation. Nevertheless, useful qualitative deductions can be made from their interplay.

### *Hemoglobin as a Structural Cell Constituent*

The interpretation of the behavior of hemoglobin as an intracellular indicator is based on a number of facts and concepts [3,6].

First, on a weight-for-weight basis hemoglobin accounts for approximately 40% of the red cell mass. It is not simply contained in the cells, but it is anchored more or less firmly to the protein or lipoprotein stroma. From direct evidence virtually nothing is known about these bonds. From indirect evidence I would suggest that they can be conceived as nonwatery complexes between electron-donor and electron-acceptor sites. By way of these "electron-pressure points" redox potential changes in the stromal component are transmitted to the hemoglobin component and vice versa.

Second, a great deal of indirect evidence suggests that metabolic redox-potential changes in organized biological systems entail continuous poising by the supporting macromolecular structure in the same way as metabolic acid production entails continuous buffering. In bacterial cultures both functions can be monitored by electrodes in contact with the culture medium. In red cells suspended in saline buffers the position is different. Here the buffering is shared more or less equally between the intracellular and the extracellular compartments. The poising, on the other hand, is confined to the intracellular compartment and perhaps to the cell-extracellular interface.

Third, even when cells are not obviously breaking down glucose or some other high-energy substrate (i.e., under conditions which I have loosely described as representing a state of metabolic rest), the sheer maintenance of their shape and integrity involves a steady expenditure of energy. This means that within their fabric electrons are continuously transferred from weaker to stronger acceptors, the only type of chemical reaction that can be tapped for biological work.

Fourth, although hemoglobin is distributed fairly evenly inside red cells, when the pigment is used as an intracellular indicator dye the term "intracellular" must be more precisely defined. Intracellular then refers to the compartment which has hemoglobin "wired into" its structure by means of stromal-hemoglobin electron-transfer bonds: in other words, it is used as an equivalent for "hemoglobinated."

### *The Effects of Structural Incorporation*

The effects of intracellular structural incorporation on the molecular properties of hemoglobin can be studied only "in reverse"—i.e., by observing the changes that accompany cell lysis. One of these changes is the release of oxygen at a constant external $pO_2$. In the series of experiments summarized in Fig. 1, cell suspensions were equilibrated with

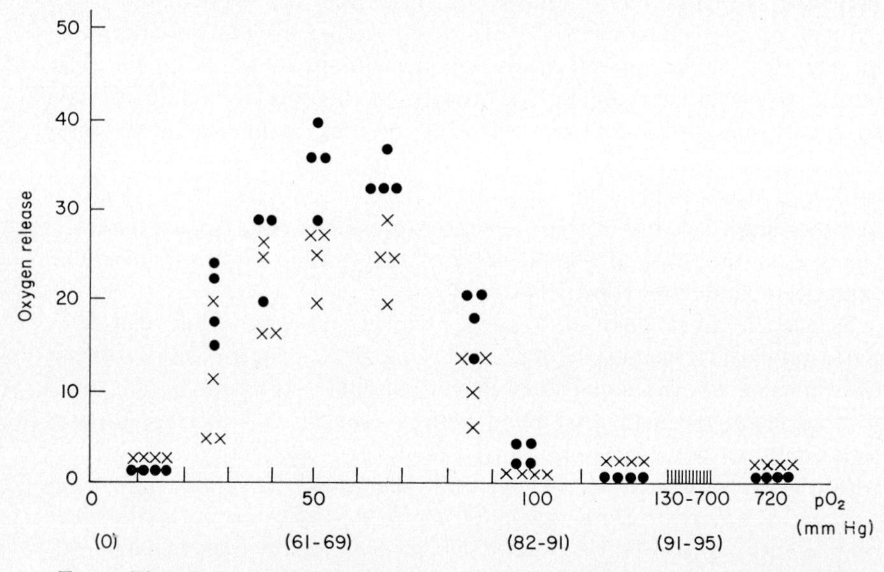


FIG. 1. The effect of red cell lysis on the binding of oxygen by hemoglobin. Two series of cell suspensions in phosphate buffers were equilibrated with 10 gas mixtures representing various oxygen partial pressures ($pO_2$). (The corresponding oxygen saturations as measured by reflection oximetry in the intact cells are indicated in parentheses.) The volumes of oxygen evolved on lysis are plotted in microliters per cubic centimeter of packed cells.

known gas mixtures in closed Warburg-type vessels. The cells were then lysed and the volumes of oxygen evolved were measured. Perhaps surprisingly the gas evolution was greatest (in absolute as well as in relative terms) not at full oxygen saturation—i.e., by cells equilibrated with high oxygen partial pressures—but in the intermediate $pO_2$ and oxygen-saturation range. My interpretation stems from the assumption that the pigment molecule can maintain its internal stability in various ways under different conditions. The stabilizing configurational changes which depend on the molecule combining with more or less oxygen cannot come into play in an oxygen-free atmosphere; nor, it seems, is the molecule capable of discharging "excess" oxygen when the outside oxygen pressure is high. This particular form of internal molecular adaptation has real scope only within a limited $pO_2$ range, approximately the range which corresponds to the steep part of the hemoglobin-oxygen dissociation curve.

## *The Effects of Metabolic Activity*

The curves shown in Fig. 2 represent the changing oxygen saturation of cells incubated at 37°C in a saline medium at a constant $pO_2$. The increase in oxygen saturation (as reflected by the color change) which occurs in cell suspensions at less than 100% initial saturation after the addition of glucose cannot be attributed to any of the factors that are known to influence oxygen binding in free hemoglobin solutions or in hemolysates: in particular, the drift toward a more acid pH which accompanies glucolysis would tend to diminish rather than to increase

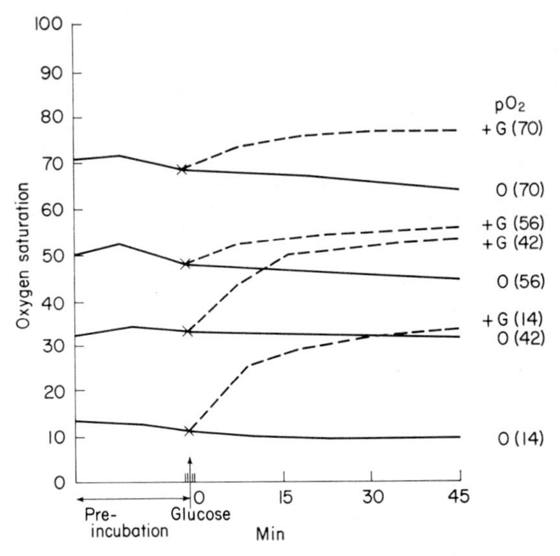


FIG. 2. The effect of glucose on the oxygen saturation of incubated cell suspensions. Duplicate samples of bicarbonate-buffered cell suspensions were incubated. The $pO_2$ (mm Hg) in the equilibrating gas mixtures is indicated on the right. After a 30–60 minute preincubation period glucose (to 200 mg/100 ml) was added to one of the duplicate samples (interrupted line). Oxygen saturations were read at 10–15-minute intervals thereafter.

the affinity of the pigment for oxygen. My interpretation of the extra oxygen bound by intracellular, structural hemoglobin and of the further increase in oxygen binding during active glucolysis depends on the concept of nonwatery cellular electron donor–acceptor complexes. It is as though these complexes provided the pigment molecule inside the cell with an extra stabilizing foothold for maintaining its internal balance. This foothold is lost in a simple watery solution, i.e., after cell disruption by lysis. The support these structural bonds provide depends on the $E_h$ at which the cell structure is poised; and this, in turn, depends on the "energy level" (i.e., the rH or the combined pH and $E_h$) at which its metabolic foci operate [3, 4].

## THE CELL-EXTRACELLULAR pH GRADIENT

Before illustrating the effect of insulin on cellular $E_h$ as reflected by hemoglobin oxygenation, I want to introduce my second indirect parameter, the apparent cell-extracellular pH gradient.

It has been known since the early studies of Henderson, Van Slyke, and others that so-called intracellular fluid—that is, the lysed red cell mass—is much more acid than the extracellular fluid with which it is (or was) in equilibrium [7, 10]. I have called this pH difference "apparent" because I do not think that it necessarily exists before the cells are lysed—that is, between the extracellular pH and the pH of true cell water. An explanation commonly given for the low intracellular pH is that under physiological conditions cell-bound proteins are negatively charged and that the partition of the freely diffusing $OH^-$ anions merely conforms to the postulates of the Gibbs-Donnan equilibrium. There may be some truth in this, but it is certainly not the whole truth. If it were the whole truth then the $OH^-$ ion concentration ratio should be at least of the same order as the intracellular-extracellular $Cl^-$ and $HCO_3^-$ ion ratios. In fact, the $OH^-$ concentration or pH difference is many times greater. More significantly, not only the pH of the cell suspensions as a whole, but also the apparent cell-extracellular pH difference, is strongly influenced by such variables as the state of oxygenation of hemoglobin, the metabolic activity of the cell, and the presence of hormones such as insulin in the extracellular milieu. None of these factors should affect equilibria governed solely by the Gibbs-Donnan law.

If both the affinity of intracellular hemoglobin for oxygen and the apparent cell-extracellular pH gradient are influenced by the $E_h$ of the cell, the two parameters themselves should be related. That this is so is illustrated in Tables II and III. In both tables the apparent cell-

TABLE II. *The Effect of Hemoglobin Oxygenation on the Apparent Cell-Extracellular pH Gradient*

| Expt. No. | Parameter | Gas phase | | |
|---|---|---|---|---|
| | | 95% $O_2$, 5% $CO_2$, 0% N | 10% $O_2$, 5% $CO_2$, 85% N | 0% $O_2$, 5% $CO_2$, 95% N |
| 1 | Extracellular pH | 7.03 | 7.10 | 7.12 |
| | Apparent pH gradient | 0.58 | 0.28 | 0.46 |
| 2 | Extracellular pH | 7.63 | 7.68 | 7.71 |
| | Apparent pH gradient | 0.32 | 0.13 | 0.18 |
| 3 | Extracellular pH | 7.50 | 7.52 | 7.58 |
| | Apparent pH gradient | 0.42 | 0.25 | 0.32 |
| 4 | Extracellular pH | 7.00 | 7.01 | 7.04 |
| | Apparent pH gradient | 0.77 | 0.34 | 0.49 |

[a] Samples of the same red cell suspension were equilibrated with the gas mixtures indicated and the extracellular pH's and the pH's of the lysed packed-cell deposits were measured.

TABLE III. *The Effect of Insulin and of Oxygenation on the Apparent Cell-Extracellular pH Gradient*

| Expt. No. | Gas phase | | | | | |
|---|---|---|---|---|---|---|
| | 95% $O_2$, 5% $CO_2$, 0% N | | 10% $O_2$, 5% $CO_2$, 85% N | | 0% $O_2$, 5% $CO_2$, 95% N | |
| | Insulin | Control | Insulin | Control | Insulin | Control |
| 1 | 0.72 | 0.65 | 0.41 | 0.38 | 0.53 | 0.42 |
| 2 | 0.74 | 0.67 | 0.38 | 0.34 | 0.55 | 0.41 |

extracellular pH difference is smallest when the cells are partly oxygenated: this is the oxygenation state which allows hemoglobin the widest scope for molecular adaptation by oxygen release.

I have suggested earlier that cell lysis involves the disruption of nonwatery intermolecular (hemoglobin-stromal) electron-transfer complexes. I would also suggest that this entails the generation of both acidic and basic groups. One can reasonably assume that in their new watery milieu the complementary electron-donor and electron-acceptor sites will behave as proton donors (acids) and proton acceptors (bases), respectively; but there is no reason why the two functions should be equivalent. Experimental evidence suggests that the more reducing the intracellular $E_h$ the more acid the watery cell lysate: i.e., a more negative (reducing) intracellular $E_h$ is "translated" into a more acid pH. This concept is supported by findings set out in Table IV.

## THE ACTION OF INSULIN

As mentioned earlier, our preliminary experiments suggested that in suitably treated red cell suspensions insulin has an immediate

TABLE IV. *The Changes in Extracellular pH and the Shifts in the Apparent Cell-Extracellular pH Gradient in Red Cell Suspensions in the Course of Incubation*[a]

| Conditions | Time (min) 0 | | 60 | | 120 |
|---|---|---|---|---|---|
| No glucose, no insulin | 7.28<br>0.65 | → | 7.26<br>0.60 | → | 7.21<br>0.59 |
| Glucose +, no insulin | 7.37<br>0.63 | → | 7.34<br>0.83 | → | 7.20<br>0.96 |
| No glucose, insulin + | 7.32<br>0.69 | → | 7.30<br>0.65 | → | 7.27<br>0.63 |
| Glucose +, insulin + | 7.38<br>0.65 | → | 7.34<br>1.00 | → | 7.31<br>1.13 |

[a] Incubation was at 37° with and without glucose and with and without insulin. The as phase in all systems was oxygen 20%, nitrogen 75%, $CO_2$ 5%.

effect on the cell-extracellular interface; and the findings could be interpreted in terms of a flow of electrons from the cell interior to a transitional junctional zone on the outer face of the plasma membrane. Such an electron flow would inevitably lead to a change in intracellular $E_h$; and this should be reflected by both the indirect $E_h$ parameters discussed.

### *The Effect of Insulin on Oxygen Bindng*

Figure 3 shows the oxygen binding and oxygen release in a single experiment in parallel insulin and control cell suspensions. Both cell suspensions were incubated in Warburg-type flasks at 24°C in the

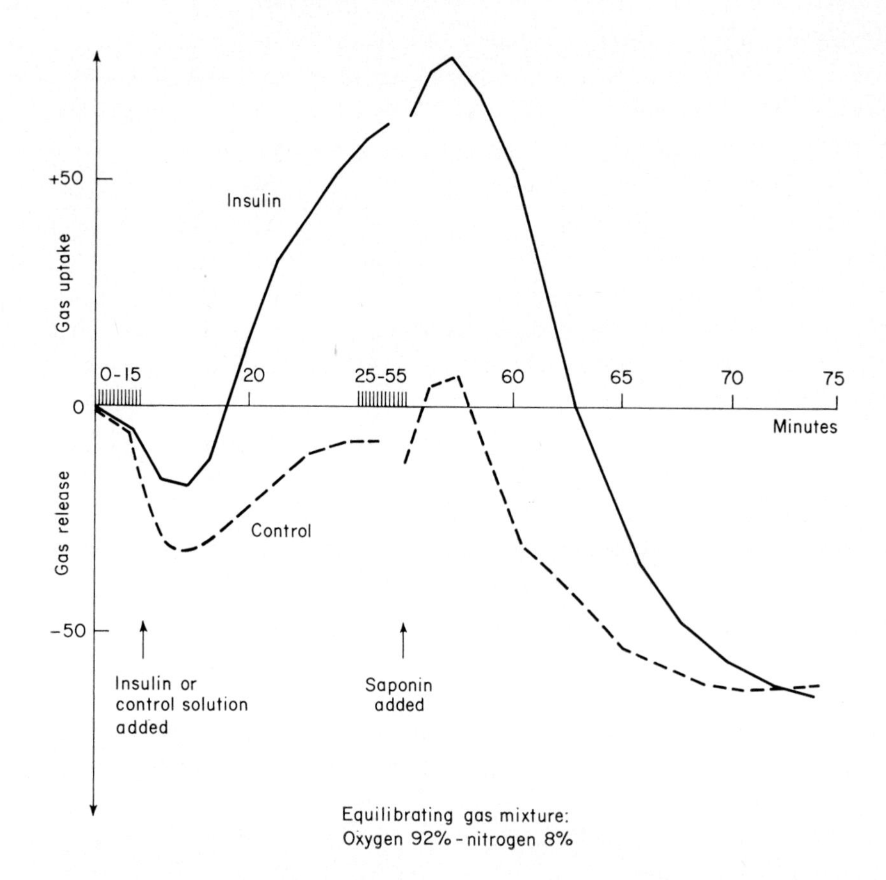


FIG. 3. The effect of insulin and lysis on oxygen binding. Samples of phosphate-buffered saline suspensions of red cells were incubated at 24°C in Warburg flasks with two sidearms. Insulin (or control) and saponin solutions were added as indicated. Continuous line = manometric readings in the insulin system. Dotted line = manometric reading in the control system. Insulin concentrations 2 $\mu$g/100 ml. Both gas uptake and gas output are indicated in microliters per milliliter of cell suspension.

absence of glucose. The left side of the graph shows the uptake of oxygen when insulin was added from the first sidearm: it was virtually immediate and I do not believe that it could have been in any way related to metabolic or enzymatic changes. What I think the oxygen uptake reflects is an immediate change in intracellular $E_h$ due to the "leakage" of electrons across the plasma membrane. The right side of the graph shows the release of oxygen when a lysing agent was added to the cell suspensions from a second sidearm. It should be noted that the two curves meet at the same level—i.e., insulin had no effect on

TABLE V. *Oxygen Uptake (in μl/cc Packed Cells) on Adding Insulin to Red Cell Suspensions Equilibrated with Three Different Oxygen-Nitrogen Gas Mixtures*[a]

| Expt. No. | Gas phase | | |
|---|---|---|---|
| | 4% $O_2$ | 8% $O_2$ | 16% $O_2$ |
| 1 | 9 | 33 | 1 |
| 2 | 0 | 38 | 0 |
| 3 | 7 | 27 | 0 |
| 4 | 17 | 46 | 8 |
| 5 | 10 | 28 | 0 |
| 6 | 18 | 42 | 3 |
| 7 | 10 | 39 | 1 |
| 8 | 0 | 36 | 0 |
| 9 | 0 | 29 | 0 |
| 10 | 0 | 32 | 0 |

[a] Different hematocrit values (46–68%) and different buffering (pH 7.14–7.56) account for the differences in the 10 experiments. The insulin added gave final concentrations of less than 2 μg/100 ml in all cell suspensions.

oxygen binding by free hemoglobin; it affected the affinity of the pigment for oxygen only so long as the pigment molecules were part of the cell structure. Table V shows the extra oxygen taken up by a series of red cell suspensions on addition of insulin. The effect was greatest in cell suspensions equilibrated with oxygen partial pressures at which hemoglobin was partially oxygenated. Table VI shows the volume of oxygen released by a series of parallel insulin and control cell suspensions when the cells were lysed in closed Warburg vessels.

TABLE VI. *The Effect of Insulin on the Volume of Oxygen Released on Cell Lysis*[a]

| Expt. No. | Gas phase | | | | | |
|---|---|---|---|---|---|---|
| | 2% $O_2$ | | 4% $O_2$ | | 16% $O_2$ | |
| | Insulin | Control | Insulin | Control | Insulin | Control |
| 1 | 9 | 0 | 46 | 33 | 8 | 2 |
| 2 | 4 | 0 | 49 | 27 | 2 | 3 |
| 3 | 6 | 2 | 43 | 29 | 0 | 0 |
| 4 | 3 | 1 | 43 | 22 | 0 | 1 |
| 5 | 8 | 6 | 51 | 32 | 0 | 2 |

[a] Parallel insulin and control cell suspensions were equilibrated with the same gas mixtures and then lysed. The figures express the oxygen evolved in microliters per cubic centimeter of packed cells. All the experiments illustrated in Tables V and VI were performed at 4°C; the insulin effects were immediate.

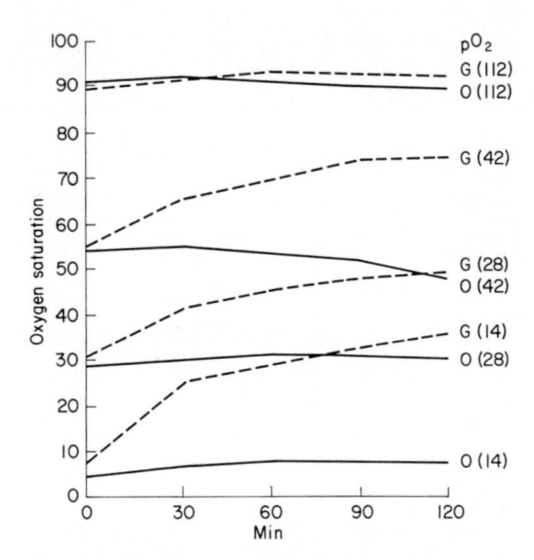


FIG. 4. The effect of insulin on the changing oxygen saturation in incubated cell suspensions. Two cell suspensions in bicarbonate-saline buffer with glucose (*G*) or without (*O*) were incubated. The $pO_2$ (mm Hg) in the equilibrating gas mixtures is indicated on the right. Parallel cell suspensions were incubated with insulin (added to 2 $\mu$g/100 ml) (interrupted line).

Figure 4 shows the effect of insulin on the increasing oxygenation of incubated, metabolically active cells.

### *The Effect of Insulin on the Cell-Extracellular pH Gradient*

Table V illustrates the effect of insulin on the apparent cell-extracellular pH gradient. If this gradient does, in fact, depend on the reducing potential at which the cells are poised, the results are at least consistent with the hypothesis of a transmembrane electron flow and an uptake of protons from the extracellular watery milieu.

## CONCLUSIONS AND SUMMARY

From the experimental findings I have outlined it is impossible to formulate a detailed molecular concept of hormone-membrane interaction; but, more or less tentatively, one can draw a number of general conclusions.

First, little doubt remains that insulin in physiological concentrations does have an immediate physical effect on suitable red cell preparations; and this effect can be studied independently of its metabolic sequelae. I have assumed that the actual site of the cell-hormone interaction is the cell-extracellular interface. The experiments I have described provide no conclusive proof of this; but circumstantial evidence in favor of such an interpretation is strong. In particular, it would be difficult to reconcile the immediacy of the insulin effects with the idea of this relatively large molecule passing across the cell membrane. It must also be emphasized that the immediate physical insulin effects can be explained only on the assumption that each insulin molecule interacts with a unit very much larger than another molecule. Theoretically this unit could be the cell as a whole, behaving as a cross-linked single particle; but it is easier to visualize the interaction with a membrane structure constructed of aligned lipoprotein complexes. Of course the distinction is somewhat artificial since the molecular architecture and electrochemical properties of the membrane must be governed by the cell as a whole.

Second, both the experimental results summarized in the present paper and current work that will be reported elsewhere suggest that the essential requirement for the physical action of the hormone is an interface between a watery and a nonwatery system; and that the primary insulin effect involves a change in the electron-pressure (or proton-pressure) gradient between the two phases. Although the phenomenon of hydrogen overvoltage, characteristic of nonwatery electrodes in contact with watery solutions, is itself ill understood, it may represent a physical analogy. It may be recalled that it accounts for the fact that a greater electromotive force is necessary to induce a current to flow across such junctional regions than would be expected from calculations based on standard electrochemical potentials. One mechanism that may be envisaged is the formation and stabilization of hydrogen atoms as a result of trace impurities at the interface interfering with either proton release or hydrogen-gas evolution. The physical action of insulin could be pictured as a similar mechanism altering the capacity of the cell-extracellular interface to develop and maintain such an overvoltage.

Third, the experiments described show that enzymatic metabolic activity is unnecessary for insulin action: in other words, insulin action is a "biological" phenomenon only in the sense that its target, the cell-extracellular interface, is a product of biosynthesis. This raises the question of how far even metabolically resting cells are essential. Experimental work now in progress suggests that simple protein aggregates in a two-phase solvent can also provide a suitable target system.

Fourth, I would cast doubt on the concept of insulin-sensitive and insulin-insensitive *cells*. "Insensitive" red cells can be made sensitive by manipulating their extracellular milieu. Conversely, I doubt that such "sensitive" cells as skeletal-muscle cells or the cells of epididymal fat pads would respond to insulin if they could be thoroughly cleansed of interstitial fluid and resuspended in saline. What we are dealing with—or, rather, what the hormone is acting on—is an interface or a critical potential *difference* between two compartments. Experiments with radioactive-labeled insulin suggest that the relationship may be reciprocal: if the hormone is one of the factors which "sets" the potential gradient prevailing across the cell-extracellular interface, it is also possible that the initial reducing-potential gradient determines the affinity between the hormone and the interface. In other words, one can conceive of a self-regulating mechanism designed to maintain not a reducing potential in one or other biological compartment, but a relatively fixed potential *gradient* between two compartments.

## References


1. Antonini, E., Wyman, J., Moretti, R., and Rossi-Fanelli, A., *J. Biol. Chem.* **237,** 2773 (1962).
2. Benesch, R. E., and Benesch, R., *Biochemistry* **1,** 735 (1962).
3. Dormandy, T. L., *J. Physiol.* (*London*) **183,** 378 (1966).
4. Dormandy, T. L., *Lancet* **i,** 755 (1966).
5. Dormandy, T. L., and Zarday, Y., *J. Physiol.* (*London*) **180,** 708 (1965).
6. Editorial, *Lancet* **ii,** 427 (1966).
7. Henderson, L. J., "Blood." Yale Univ. Press, New Haven, Connecticut, 1928.
8. Manwell, C., *in* "Oxygen in the Animal Organism" (F. Dickens and E. Neil, eds.). Macmillan (Pergamon), New York, 1964.
9. Riggs, A. F., and Wolbach, R. A. *J. Gen. Physiol.* **39,** 585 (1956).
10. Van Slyke, D. D., and Neil, J. M., *J. Biol. Chem.* **61,** 523 (1924).

# THE DEGREE OF ORGANIZATION IN THE BACTERIAL CELL

**ERNEST C. POLLARD**

*Biophysics Department,*[1] *The Pennsylvania State University, University Park, Pennsylvania*

## INTRODUCTION

In a conference devoted to cell organelles one may wonder why we discuss the bacterium. Leaving aside the possibility that a mitochondrion is the overfed remains of a symbiotic bacterium, there is good reason to examine the bacterium for order. If molecular order is required in so small a cell, then the same kind of order is to be expected in larger and more representative cells.

## BACTERIAL CELL: "GEMISCH" OR ORDER?

Does the selective membrane of the bacterial cell contain a wholly mixed, unordered and nonstructured synthetic and metabolic apparatus? Under these circumstances the regulated shape and behavior would have to result from specificity of selection working against chaos, using the speed of molecular diffusion and the firmness of covalent bonding to give vast choice and, yet, exact decision for each "right" reaction to occur. Figure 1 shows an imaginary thought picture of the "gemisch" cell. Alternatively, is the cell a completely ordered entity with a linear code script prescribing the location and amino acid order of each enzyme in the cell? This is suggested by the presence of codes for neighboring molecules on one operon. For example, the $\beta$-galactosidase operon contains the code for the enzyme $\beta$-galactosidase, located just below the cell membrane, and the permease for the substrate just above. On the histidine operon the codes for the 10 or more enzymes involved in the metabolic pathway for histidine are transcribed together and, therefore, possibly "laid down" together in the cell.

Some early previous studies [3, 4] suggested that the cell could be unstructured and still function. Later considerations have tended to modify this idea [5]. This paper presents evidence that the DNA in

[1] The background material for this paper was developed under a grant NsG-324 from NASA.

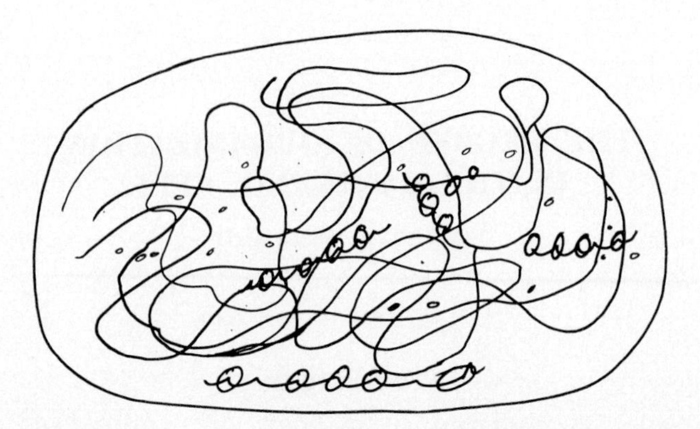

FIG. 1. Imaginary picture of a bacterial cell as a "gemisch." Only random collision in a complete mixture occurs. Specificity of bonding alone maintains the form of the living cell.

the cell must be so organized as to be able to hold a triphosphate pool in place. If this is so, there is a measure of compartmentation even in the bacterial cell. Two points—theoretical and experimental—regarding internal pools will be made which suggest the probability of an internal DNA pool. Then, a brief discussion of what is known about the order of the DNA in the bacterial cell will follow.

The philosophy which arises from the structure of the bacterial cell suggested by these studies does not necessarily require the concept of detailed order in a cell. Rather it suggests that the bacterial cell (and other cells in consequence) started out relatively unordered, but gradually evolutionary mechanisms have given their components order and position to aid growth and create a more efficient cell. The cell can probably perform its functions without the order suggested in the bacteria, but the rate of functioning would be lowered. Thus, the suggestion that there are localized pools for the DNA inside cells does not really mean that all cells require an exact order to function.

The same reasoning which seems to give a rather plausible approach to the bacterial cell suggests that a mammalian cell, a plant cell or any cell which has a diameter exceeding 5 $\mu$ will require separation into organelles. Thus, the necessity for organelles can be seen in terms of the efficiency of synthesis which is given by a small size. This point will be discussed briefly later.

## THE RATE OF DNA SYNTHESIS

New bacterial DNA is synthesized usually at one site (polymerase) per bacterial chromosome. These synthetic points may proliferate to two or even four, but conditions in which there is only one polymerase per nucleus are very easy to secure and most normal. The bacterial nucleus has $3 \times 10^6$ bases, and under optimal conditions these can replicate in a little over a thousand seconds, giving a rate of synthesis of 3000 bases per second. Phage DNA can be shown to be made at about the same rate per polymerase—or possibly even faster.

From a consideration of collision kinetics one can derive the molecular concentration necessary to provide collisions at a rate fast enough to cause the reaction to go at the observed speed [3, 4]. Here, it will be looked at slightly differently, but equivalently, to derive similar figures.

Figure 2 shows a schematic picture of the motion of a particle under-

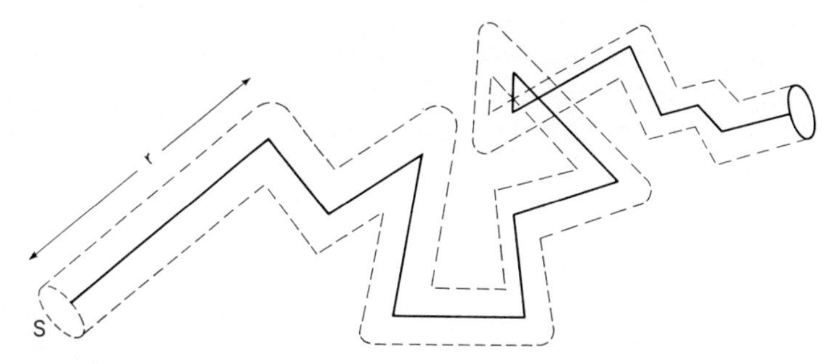


FIG. 2. Collision volume swept out by a molecule undergoing Brownian movement.

going Brownian movement. It is thought of as having an average square displacement $\bar{r}^2$ whose square root is called $\bar{r}$. In the picture $\bar{r}$ is developed as resulting from a series of displacements of widely different size. However, as one looks at this, it is clear that the particle sweeps out a certain volume as it moves and that the ratio of this volume to the volume in which it is confined is very important as a measure of the efficiency of collisions.

We can make a rough estimate of this ratio as follows. Let $V$ be the whole reactive volume. This would be the whole cell in "gemisch" or a part of the cell in a compartmented cell. Now, if $N$ is the number of "jumps" made, the length of the path traversed is roughly $N\bar{r}$ and $v$, the volume swept out, is $N\bar{r}S$, where $S$ is the cross section for collision of the molecule. Therefore,

$$v = N\bar{r}S.$$

But we also have the relation between $D$, the diffusion coefficient, $\bar{r}^2$ and the time $t$, namely:

$$Dt = \frac{1}{6} N\bar{r}^2$$

$$N\bar{r} = \frac{6\,Dt}{\bar{r}}$$

$$v = N\bar{r}S = \frac{6\,DtS}{\bar{r}}, \text{ so } \frac{v}{V} = \frac{6\,DtS}{\bar{r}\,V}$$

In each of these estimates of collision rate there is a parameter which can be estimated only with difficulty; in this case it is $\bar{r}$. We can briefly discuss the basis on which reasonable limits for $\bar{r}$ can be chosen. If one considers the actual rapid motion of a particle which is being buffeted by many molecules, then the meaning of the value of $\bar{r}$ is that an excess of impacts will swing it in one direction and then an excess will swing it in a different direction. The approximation used in this kind of derivation, an approximation due originally to Einstein, is to say that there are a series of short displacements that carry the particle along in this tortuous and random path. Clearly, if we think of the small segments of the random path as being *extremely* small, then the contribution to the swept-out volume will not be great because there will be a good deal of back and forth motion without covering of new volume. On the other hand, if we consider these distances as rather large, then it is quite possible that for a large particle the chance of such a considerable displacement may be a good deal less. In the aggregate, as shown by Einstein, there is no great difficulty about deriving the diffusion constant, but there is some difficulty in assigning a particular value of $\bar{r}$ as has to be done in this case. Thus, the weakness of this method of thinking is that it gives an uncertainty which can be as high as a factor of 10. However, for our discussion purposes we are concerned with relatively small molecules, the molecules of the triphosphates which make up the DNA, and the need for considering extremely small displacements does not arise. Since we are concerned mostly with the concept, we choose as a rough estimate 1 Å for the value of $\bar{r}$.

## THEORETICAL ESTIMATION OF POOL SIZE

Bacterial "cytoplasm" is very viscous if measurements of disrupted cell viscosity have any meaning. Such measurements have been made [2], and the viscosity value estimated is around 100 poise. Almost certainly a great deal of this high viscosity can be attributed to the disrupted cell wall and the partially sheared DNA. Because of this, the actual diffusion of molecules may well be quite fast. Measurements by Lehman and the author [2] yielded the values in Table I.

TABLE I. *A Comparison of Diffusion Constants of Sucrose, Dextran, and β-Galactosidase in Water and E. coli Cell Extrudate*

| Component | Diffusion constants ($cm^2/sec$) | |
|---|---|---|
| | Water | *E. coli* cell extrudate |
| Sucrose (M. W. = 342) | $2.4 \times 10^{-6}$ | $1.0 \times 10^{-6}$ |
| Dextran (M. W. = 16,000) | $1.0 \times 10^{-6}$ | $3.0 \times 10^{-7}$ |
| β-Galactosidase (M. W. = 350,000) | $3.5 \times 10^{-7}$ | $2.5 \times 10^{-8}$ |

It can be seen that only as the size of the diffusing molecule becomes quite large does the diffusion constant drop toward the slow value expected for high viscosity medium. This reinforces the idea that in the interstitial zones of the cell the movement of molecules is relatively free. Thus, we can use values such as those above to estimate $v/V$.

To describe $S$ we must remember that it involves the size of the molecule and the extent of bonding. Very probably the function of hydrogen bonding is to act over a rather large range to stop a molecule in collision. In any event an assumption of 10 Å² cannot be very far out. Thus, in a mammalian cell of dimensions 10 μ on a side for a molecule like sucrose

$$\frac{v}{V} = \frac{6\times10^{-6}\times10^{-15}t}{10^{-8}\times10^{-9}} = 6\times10^{-4}t$$

For a molecule to "scavenge" the whole cell, $v/V$ should be about 1 and this should require about 1600 seconds. It is quite clear that this is too long and that the cell has adapted to it. Some time ago it was suggested by the author [7] that the purpose of the endoplasmic reticulum is to constrain diffusion from 3 dimensions to 2. It can be seen, in terms of the above analysis, to reduce $V$ to the volume of the region either within the membrane layer or between membranes.

This reasoning points up the value to the cell of keeping in small confines those regions in which rapid operation must be performed. Thus, it makes sense in terms of diffusion kinetics to have chloroplasts and mitochondria relatively delimited. Both of these are regions of intense chemical action with very high rates of turnover. Such high rates are more readily secured if diffusing molecules do not have a large space over which to roam.

In the bacterium $V = 2 \times 10^{-12}$cc and $v/V = 3 \times 10^{-1}t$. Thus, it takes one-third of a second to scavenge the cell. Nevertheless, to function at $3 \times 10^{3}$ bases assembled per second we must have a pool of the order of $10^{4}$ per cell to operate. The alternative approach of collision kinetics gives the figure of $4 \times 10^{4}$ per cell.

Note that for larger molecules, although $D$ is less, $S$ is larger. Even if we suppose that $R$ is one-quarter the molecular radius making $R$ larger, we find $v/V$ is not unfavorably effective. Thus, if for $\beta$-galactosidase we put $\bar{r} = 5$ Å then for $D = 2.5 \times 10^{-8}$ and $S = 2500$ Å$^2$ we get roughly $v/V = 0.3t$ as before. Thus, the process of self-assembly is really only hindered by the approach to the diffusion constant of the macroscopic viscosity, not the interstitial viscosity.

This is really rather an important conclusion when applied to the assembly of such elements as microtubules. If an analysis is made in terms of an equivalent system in the bacterial cell, the formation of the *pili,* then the actual figures for the rate of formation of the *pili* and the known amounts of the protein subunits which must be present in the cell actually suggest that collision kinetics could make the assembly go faster than is observed. Thus, there is absolutely nothing about the complication of collision to interfere with the self-assembly of a simple organelle, so long as the parts from which it is assembled are not spread over too large a cell volume. As long as the assembly is made in a region of the order of 2 $\mu$ on each side, assembly can be relatively fast. Note also that it is necessary that a loose complex form to permit rotational motion to find the binding site.

To summarize: theory requires $4 \times 10^{4}$ thymidine triphosphate molecules per cell if the cell is a "gemisch" and the whole cell volume is used.

## EXPERIMENTAL DETERMINATION OF POOL SIZES

The method employed in our laboratory has been to isolate the triphosphate pool on a DEAE column using a marker of deoxythymidine triphosphate. The marker was put on in sufficient amount to give an easily observable absorption at 260 m$\mu$, and this absorption was used to indicate where the pool was to be found in the separation procedure of the column.

Cells of *Escherichia coli* $T^{-}L^{-}$, which require both thymine and leucine, were grown on thymine until an equilibrium condition was established. They then were subjected to osmotic shock by putting them into distilled water at 100°C. This has been shown by many workers to release all the pools in the cell. The solution which remains when the cells are spun off in the centrifuge is then put on the column, and the column is eluted with sodium chloride covering the range 0.1–0.2 *M*. A typical elution pattern is shown in Fig. 3. The marker of the thymi-

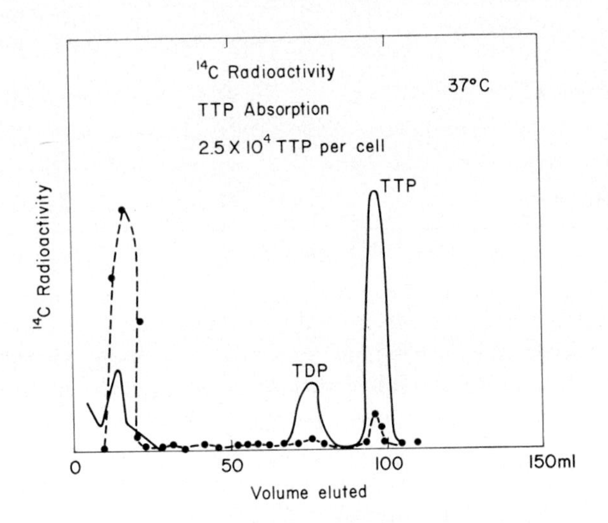


FIG. 3. An elution curve for thymidine triphosphate (TTP) labeled with $^{14}C$.

dine triphosphate shows quite clearly together with a little contamination of diphosphate which we find hard to avoid. The radioactivity corresponding to the thymidine triphosphate is easily visible at that point. By carrying through a calculation in terms of the known concentration of thymine it is not difficult to calculate the size of the pool per cell. Experiments were conducted many times to measure this quantity, and the results of this and somewhat similar work on the phosphorus pools are shown in Table II. In these experiments conditions were varied in such a way as to change the rate of DNA synthesis. It is interesting that as the rate of DNA synthesis goes up, the size of the pool goes down. Thus, when high temperatures or previous irradiation are used to diminish the rate of DNA synthesis the size of the pool rises. It is also of interest that the phosphate pools, which are not separated with regard to whether they are deoxy or ribo, are not relatively large. They are, indeed, probably higher. But ATP is used at an average of 4 molecules per synthetic molecule per unit synthesized throughout the cell, and this pool size is not large enough to permit the ATP in the cell to be uniformly distributed. Therefore, both

TABLE II. *Size of Some Nucleic Acid Pools in the Cell under Various Conditions*[a]

| Component | Environmental conditions | Molecules per cell |
|---|---|---|
| Thymidine triphosphate | 37°C | $2.5 \times 10^4$ |
| | 45°C | $6.4 \times 10^4$ |
| | 25°C | $2.2 \times 10^4$ |
| | ↓ $\gamma$-ray, 37°C | $1.2 \times 10^5$ |
| Triphosphate | 37°C | $3.0 \times 10^5$ |
| | 45°C | $5.5 \times 10^5$ |
| | ↓ 48°C | $8.9 \times 10^5$ |
| Thymine | 37°C | $1.0 \times 10^5$ |

[a] ↓ indicates diminishing rate of DNA synthesis.

the pool of triphosphates for assembly of DNA and the pools of triphosphate which are needed in chemical synthesis throughout the cell are inadequate to permit collision kinetics to function smoothly. This gives rise to the alternative suggestion that somehow the cell is able to separate out both the pool necessary for the synthesis of DNA and the various subpools which are necessary for the synthesis of protein and cell wall and membrane.

## POOL SIZE AND RECOGNITION

The problem of recognition is illustrated in Fig. 4. The scheme for synthesis of DNA as suggested by Kornberg is illustrated. This is effec-

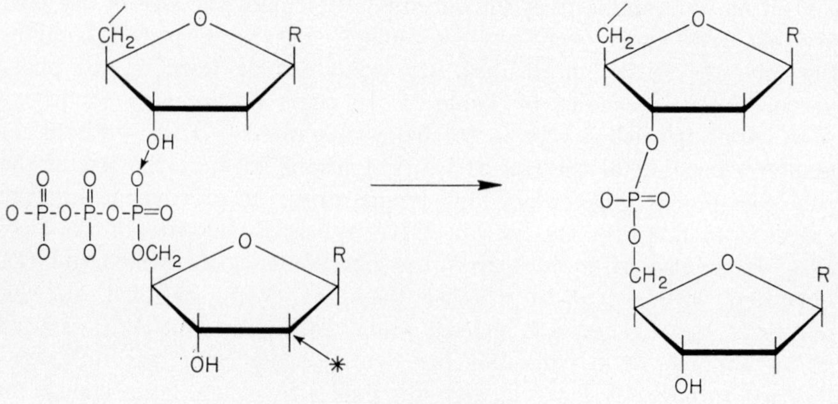

FIG. 4. Reaction scheme for DNA synthesis.

tively taken from Davidson [1]. If we simplify the equation very greatly then it looks as though

$$\underset{(a)}{R'O{-}\overset{\overset{O}{\|}}{\underset{\underset{O}{|}}{P}}{-}P_2O_7} + \underset{(b)}{ROH} \longrightarrow R'O{-}\overset{\overset{O}{\|}}{\underset{\underset{O}{|}}{P}}{-}OR + HP_2O_7$$

The actual zone of action requires that (a) and (b) approach within 1.5 Å of one another. While the energetics of this process are not readily available, it can be assumed that when the combination occurs there is a release of some 5000 calories per mole. This suggests that if (a) approaches (b) within 1.5 Å the double phosphate $R'$-$PO^4$-R occurs and the synthesis advances. Now, if this occurs for *any* phosphate this is a hazard to the cell, for there are about 60 times as many "wrong" triphosphates available, all of which can bond instantly. The polymerase mechanism (or the cell, depending upon one's point of view) solves this problem somehow. A brief attempt will be made to examine the mechanism by which this is accomplished.

First, it would seem that the ribo triphosphates must be rejected from DNA synthesis other than by compartmentation. A uracil ribo triphosphate is very close in structure to thymidine triphosphate; indeed, 5 BUdR is readily accepted. Thus, even with any but *perfect* compartmentation the error rate would be far too high. Suppose we draw a surface grid on which we can imagine the triphosphates to be assembled (Fig. 5). Four points, as shown, are absolutely essential

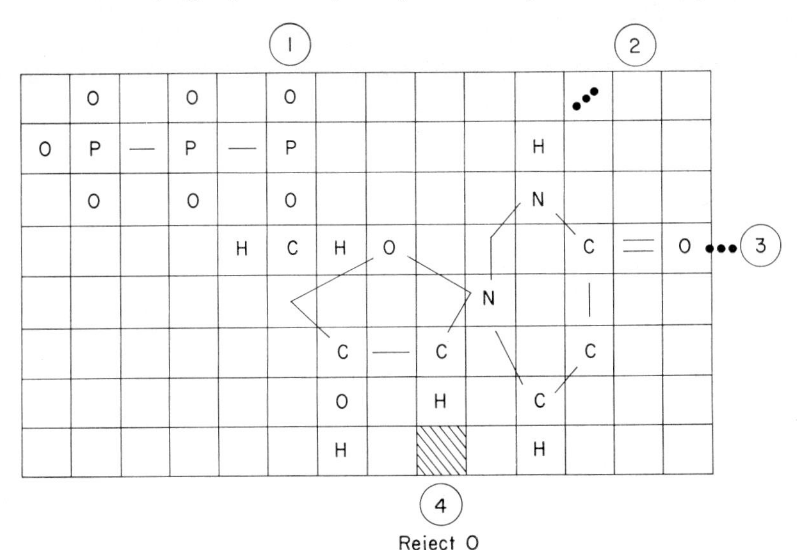


FIG. 5. Thought diagram for recognition in DNA synthesis.

to the recognition scheme. There are many secondary points, also. We can look at this diagram from the point of view of information theory. We have to put four exact selections (a minimum value) into these. Each one takes $H$ bits of information where $2^H$ is equal to 128 and so $H$ is roughly 7 bits. Thus, the total recognition problem requires about 28 bits. Now the relation between information theory and entropy [6] states that entropy is information/0.73; so we have a requirement for 28/0.73 or about 35 entropy units per molecule. At 300°K temperature this is about 9000 calories per mole, about what we have available in the reaction. While not very rigorous, this estimate is a warning that very elaborate recognition processes may prove to be excessively costly in entropy.

If the enzymatic surface is exposed to bombardment by all the triphosphates, the collision rate, $\phi$, is given by the Smoluchowski formula as

$$\phi = 2\pi DRC_\infty$$

where $D$ is diffusion constant, $R$ is collision radius, and $C_\infty$ is concentration. If we put in the concentration of all possible triphosphates, we find that the collision rate is about 65,000 per second. Thus, if the synthetic surface is to be exposed to all this, the correct decision to reject must be made at the rate of something like 10 $\mu$sec for each collision. We may estimate whether this is possible in terms of rotational Brownian movement as follows. The formula for rotational diffusion states that

$$\Theta^2 = \frac{2kTt}{8\pi\eta R^3}$$

and

$$D = \frac{2kT}{6\pi\eta R}$$

so that

$$\Theta^2 = \frac{3Dt}{4R^2}$$

where $k$ is Boltzmann's constant, $T$ is °K, $\eta$ is viscosity, and $R$ is radius.

If we make the reasonable assumption that the radius of collision is 10 Å and if we use the modified formula with the diffusion constant in it as shown, we can see—approximately—that in the time available the molecule will move through 40 radians. This is an ample exposure of the surface of the molecule to the critical surface and should mean the recognition problem is easily solved. On the other hand, suppose there is no order to the metabolic process; in other words, the cell is really just a complete mixture of everything. Then there is a rain of all kinds of metabolites and the enzymes have to reject all the wrong ones and operate only the right kind. This is once again "random collision and specific selection."

Now, there are $1.5 \times 10^8$ metabolic units per cell [6]. This will then give a concentration of around $10^{20}$ per milliliter and this means a collision rate of something like 30,000,000 per second. If we apply the equation to the problem of making a recognition in one hundred-millionth of a second, which is what we are asking in this case, then we find an altogether different situation. The angle turned through by each colliding metabolite is less than one radian and so the recognition time is not really adequate. Therefore, we can see the importance in a cell of the question of the rejection time if the cell is only operating as a complete "gemisch."

## SUGGESTION FOR THE STRUCTURE OF THE DNA INSIDE THE BACTERIAL CELL

From the point of view of the measured pools and from the point of view of recognition time, the suggestion arises that the DNA is able to act as its own method of compartmentation. It is suggested that inside the bacterial cell the DNA forms a coil and that this coil contains within it all the mechanisms for the synthesis of the deoxy triphosphate into DNA and none of the mechanisms for the normal metabolism of the cell nor the mechanisms for the formation of RNA. Those mechanisms are on the outside of the DNA. They form part of a much tighter and more highly concentrated unit which is actually outside the "nucleus." How the DNA is actually wound in this way is not easy to say. A very simple suggestion is that it is simply wound in one helix per "nucleus" and that the new "nuclei" form new helices.

If this idea is right it is just possible that some differential release of the pools from a bacterial cell should be possible. We have tried this by exposing the cells to lower temperatures and intermediate salt concentrations in the process of pool release. Some differences are seen as evidenced by Fig. 6 where it can be noticed that the triphosphate pool, which contains TTP, is more readily released at high temperatures. One cannot conclude from this experiment alone that there is a nuclear compartmentation but, at least, it suggests that a study of the differential release of pools is worthwhile.

## INFORMATION ON DNA ORGANIZATION

An attempt at assembling all the present information on DNA organization in bacterial cells indicates some criteria for DNA organization.

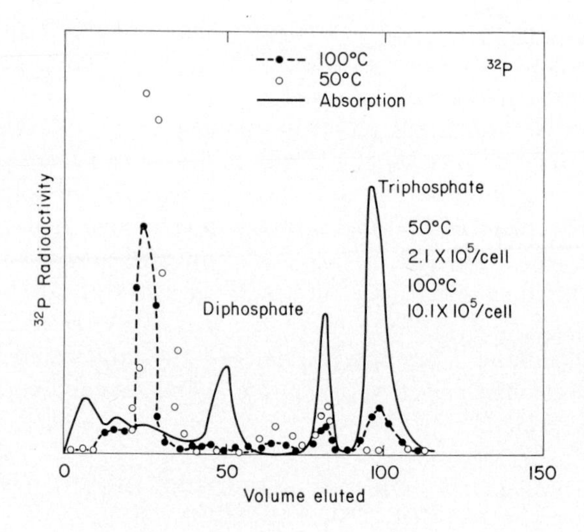


FIG. 6. Effect of temperature on release of thymidine triphosphate (TTP) pool.

1. The DNA must be 980 $\mu$ long, in one piece with an initiator and polymerase.
2. It must go into a space about 1 $\mu$ long and 0.25 $\mu$ radius.
3. It must show no directional light absorption.
4. It must be capable of replication of 3000 bases per second.
5. All DNA must be available for transcription essentially at once.
6. No more than one error in $10^8$ bases can occur.
7. Under centrifugal forces of 50,000 $g$ no change in rate of synthesis is observed.
8. It must be capable of repression and depression with no apparent selectivity.
9. The pool of DTTP, etc., is low, very marginal for collision rates.

Based on current knowledge, all these criteria would seem to be fulfilled by a single helical coil of DNA fastened to the membrane by the polymerase with the pools confined to the inside, and the synthetic process occurring from the inside.

## RELATIONSHIP OF THIS ANALYSIS TO ORGANELLES IN CELLS

The purpose of this conference is to discuss the origin and fate of organelles. This consideration of the bacterial cell shows that even in a cell far smaller than the cells that we are normally thinking of there has been an evolutionary advantage to what might be called an economy of arrangement. Thus, while the cell can probably function slowly with no internal order at all, the indication seems to be that the DNA has formed into an organized part of the cell and has acted to separate functional units from one another. If this process can occur in a cell as small as a bacterial cell, then there can be no doubt that a much greater advantage will accrue to any larger cell which must function at similar synthetic rates. On this very straightforward basis it is easy to see that simply to produce rapid chemistry the cell must have organelles within which separate specific processes can occur at high speeds. It is likely, following the bacterial analogy, that even these organelles have an evolved special internal arrangement. Thus, it would be very likely that in a mitochondrion the organization is not simply that which is sufficient for the purpose of respiration, but that there may also be an organization which is effective in producing the structural protein necessary for the accretion of the enzymatic protein.

The result of the deliberation on the cells in this small way is to suggest

1. That the synthesis by self-assembly of an organelle in itself could occur at a reasonable rate by collision if the parts were suitably ready, but

2. That kinetics require that there be organelles, for other synthetic reactions, and

3. That even within organelles it will be in place to look for functional order.

### Acknowledgment

The author wishes to thank Mrs. Pat Weller for very valuable technical assistance.

### References

1. Davidson, J. N., "The Biochemistry of the Nucleic Acids," p. **216**. Wiley, New York, **1965**.
2. Lehman, R. C., and Pollard, E. C., *Biophys. J.* **5**, 109 (1965).
3. Pollard, E. C., *J. Theoret. Biol.* **1**, 328 (1961).
4. Pollard, E. C., *J. Theoret. Biol.* **4**, 98 (1963).
5. Pollard, E. C., *Am. Scientist* **53**, 437 (1965).
6. Quastler, H., ed., "Essays on the Use of Information Theory in Biology." Univ. of Illinois Press, Urbana, Illinois, **1953**.
7. Setlow, R. B., and Pollard, E. C., "Molecular Biophysics," p. **503**. Addison-Wesley, Reading, Massachusetts, **1962**.

# INHERITANCE OF CYTOPLASMIC ORGANELLES

AHARON GIBOR

*Rockefeller University, New York, New York*[1]

## INTRODUCTION

The last several years witnessed a greatly increased interest in the role of the cytoplasm in transmission of information from one cell generation to the next. Of special interest are the inheritance of organelles such as plastids, mitochondria, and centrioles. Interest in this field is due to obvious implications from these studies to the problems of normal and pathological tissue development and differentiation. For complete discussion of various aspects of cytoplasmic inheritance the reader is referred to several recent monographs and reviews [3a, 9, 13, 18, 27a, 39].

In this paper I will consider the following:

1. The type of evidence which led to the hypothesis that organelles such as plastids and mitochondria are endowed with a degree of genetic autonomy.
2. Some evidence which suggests that the molecular bases of these hereditary properties are similar to those established for nuclear genes.
3. Some hypothesis on the evolutionary origin of these organelles and the significance of the genetic compartmentalization.

Because of the close analogy between plastids and mitochondria [9] evidence obtained from studies on either organelle is considered here as relevant to the other as well.

## GENETIC AUTONOMY OF THE ORGANELLES

Before considering the hypothesis that the organelles are endowed with genetic autonomy, it is important to emphasize that many nuclear genes were mapped which affect the development of plastids [38]. These nuclear genes might control the availability of specific metabolites required by the organelles or the synthesis of specific enzymes which are subsequently incorporated into the organelles [14]. There are, how-

[1] Present address: Department of Biological Sciences, University of California, Santa Barbara, California.

ever, other properties of the organelles that are not inherited according to the classic Mendelian laws; some of these are thought to be controlled by factors that reside in the organelles. It is therefore a genetic semiautonomy of the organelles that we are considering.

Several categories of observations led to the hypothesis that genetic factors reside in the organelles themselves. I will briefly consider some of these categories.

### *Genetic Studies on Higher Plants*

Studies of inheritance of variegation in higher plants indicated the unequal contribution of the male and female parent to the progeny. It is known that most of the cytoplasm and organelles of the zygote are contributed by the egg cell. The maternal inheritance of variegation was attributed therefore to the fact that most of the plastids of the zygote are derived from the egg cell. If all the plastids of the egg cells were defective, i.e., white, the developed zygote would be a white plant. Introduction of a small number of normal plastids by the pollen into the zygote would result in green patches in a generally white plant. The pattern of somatic segregation of the green and white cells conforms to a predicted probability of segregation if a relatively small number of reproducing particles are involved. The required number to give such patterns of segregation is similar to the number of plastids per cell. This suggests that it is the plastids themselves that determine their own state of development rather than the cytoplasm or the nucleus.

Mixed cells which contained two types of plastids in the border regions between green and white regions of a variegated leaf were discovered. These mixed cells are the most convincing evidence that the differences between the plastids are due to intrinsic differences in the organelles rather than to differences in the cytoplasm or nucleus of the cell [12].

It is important to establish that the differences between the organelles of a single cell are perpetuated in subsequent divisions of the plastids, i.e., the differences are genotypic and not phenotypic.

Differences in the population of mitochondria of a single yeast cell were demonstrated by Avers *et al.* [1] and in rat heart tissue cells by Ogawa and Barrnett [27]. Reduction of tetrazolium salts and accumulation of formazan crystals or products of the diamine reaction over some mitochondria and not over others indicated such differences. Such differences, however, are not necessarily due to genetic differences between the organelles. We recently discovered [6] biochemical differences between the plastids of a single *Nitella* cell. The cells were exposed to $^{14}CO_2$ for 30 minutes under light. By radioautography of the fixed chloroplasts we found that some perfectly healthy looking chloroplasts did not fix an appreciable amount of carbon whereas most others did. The distribution of the non-carbon-fixing chloroplasts among the rows of normal carbon-fixing plastids indicates that the difference is not inherited but may represent a quiescent stage in the life cycle of the chloroplasts.

Convincing evidence that different plastids of the same cell perpetuate their difference is provided by the observations of Van Wisselingh [37] on *Spirogyra*. He found cells that contained two different plastids, one normal with pyrenoids, the other lacking pyrenoids. Subsequent cell divisions produced a chain of such cells, each containing two types of plastids. The perpetuated difference between the plastids must be attributed to factors within the plastids rather than to factors in the cytoplasm or nucleus.

For a detailed review of the evidence of plastid inheritance from genetic studies on higher plants the reader is referred to the monograph by Hagemann [13].

### *Physical Continuity of the Organelles*

Plastids and mitochondria appear to arise by growth and division of preexisting organelles, and not *de novo*. This type of reproduction is easily demonstrated among lower algae which possess few organelles per cell. In *Micromonas,* an organism which possesses a single plastid and a single mitochondrion per cell, these organelles were shown by Manton [23] to divide with each nuclear division and separate into the two daughter cells. By a cinematographic study, Green [11] demonstrated the regular replication of the chloroplasts of expanding cells of *Nitella,* to form long rows of genealogically related chloroplasts.

Mitochondria also arise by growth and division of existing mitochondria. Luck [22] demonstrated this by prelabeling the mitochondria of *Neurospora* cells. In subsequent multiplications of these cells in a standard medium the population of mitochondria was found to be always homogeneous with respect to the original label. All the mitochondria were equally labeled rather than having two populations of mitochondria as would be expected if *de novo* origin took place. This indicates that the mitochondria do not arise *de novo,* but by growth and division of the originally labeled organelles.

### *Microsurgical Demonstration of Genetic Properties*

Direct demonstration of hereditary properties that are associated with mitochondria were reported recently by Diacumakos *et al.* [4]. They demonstrated that the injection of mitochondria, which were obtained from a slow-growing *Neurospora* strain and were deficient in cytochromes, into a hypha of a normal strain could cause the transformation of the injected hypha into a slow-growing cytochrome-deficient clone. The details of the interaction of the introduced mitochondria with the organelles which were originally present in the cells are not yet known. The fact that defective mitochondria that are introduced into a cell inhibit the native normal organelles suggests the possibility of an intracellular genetic interaction between the organelles.

### *Mutability of Plastids of Euglena*

Studies on the mutability of plastids of *Euglena* also indicated that a nonnuclear genetic system controls the development of the plastids [8]. A slight dose of ultraviolet light, if irradiating the cytoplasm, caused a permanent loss of the ability of the plastids to green. Irradiating the nucleus alone did not cause such mutations.

Differences between strains derived from ultraviolet irradiated *Euglena* cells suggest that the genetic system of the plastids is complex and multigenic [9].

## THE MOLECULAR BASIS OF THE GENETICS OF THE ORGANELLES

### *Presence of Nucleic Acids*

The presence of nucleic acids in plastids and mitochondria was demonstrated by cytological staining techniques, by incorporation of labeled precursors, by electron microscopic techniques, and by biochemical isolation procedures. A detailed review of this evidence was presented elsewhere [10, 17]. Briefly the evidence indicates the presence in the organelles of DNA and RNA and enzymes for RNA and protein synthesis. The DNA of the plastids or mitochondria was found in most cases to be qualitatively distinguishable from the DNA of the nucleus by physical properties such as buoyant density and melting temperature. These physical properties reflect differences in ratios of the bases composing the DNA [24, 32]. An increase in the ratio of guanine to adenine is directly correlated with an increase in density and elevation in melting temperature.

Chloroplasts and mitochondrial DNA were reported to be easily renatured after heat denaturation. This fact suggests that the DNA of the organelles is more uniform than the nuclear DNA and is perhaps circular [35, 36].

### *Synthesis of DNA of the Organelles*

Is the DNA of the organelles synthesized independently or is it derived from the nucleus?

Indirect evidence that the DNA of the plastids is not derived from the nucleus was deduced from the experiments on ultraviolet bleaching of *Euglena* [8]. When the nucleus was shielded from irradiation and only the cytoplasm was irradiated, bleaching occurred. Conversely, if the nucleus only was irradiated, no bleaching resulted. Because a healthy nucleus did not cure the irradiated cytoplasm and an irradiated nucleus did not bleach the plastids, it was presumed that the target DNA was not derived from the nucleus.

Direct evidence that chloroplast DNA can be synthesized in the absence of the nucleus has been obtained in experiments on the alga *Acetabularia* [7]. Bacteria-free cultures were enucleated and exposed to $^{14}CO_2$ for 7 days. From chloroplasts which were isolated from these enucleated cells it was possible to isolate radioactive DNA. After enzymatic hydrolysis of this DNA radioactive deoxynucleosides were identified. Radioactivity was also found in the free bases derived from these nucleosides. It is apparent that the enzymes for the synthesis of deoxynucleosides, their phosphorylation and polymerization are present in the cytoplasm of *Acetabularia*.

Other evidence for an independent synthesis of chloroplast DNA are the studies of Chiang *et al.* [3]. They studied the replication of $^{15}N$-labeled DNA of *Chlamydomonas*. Both nuclear and chloroplast DNA were reported to replicate in a semiconservative fashion. However the chloroplast DNA did not replicate in synchrony with the replication of nuclear DNA.

That mitochondrial DNA replicates autonomously is indicated from the studies of Reich and Luck [28]. They reported that different pools of precursers are used for the synthesis of nuclear and mitochondrial DNA in *Neurospora*. This was indicated from experiments in which $^{15}N$-labeled cells were returned to an $^{14}N$ culture medium. The DNA of the mitochondria which was subsequently isolated was found to continue to incorporate $^{15}N$ in the following three cell divisions, while the

These studies do, however, indicate that the loss of ability of the plastids to develop into normal chloroplast is apparently associated with some modification in the DNA of these organelles.

### *"Metabolically Active" DNA*

There are several reports which suggested to some that perhaps the DNA of the organelles has a "metabolic" rather then a "genetic" function [17]. These reports stress the rapid incorporation of labeled precursors, and subsequently a rapid loss of specific activity of the DNA of the organelles in the absence of nuclear DNA synthesis. The result of such experiments could also be interpreted as due to active proliferation and death of the organelles rather than a metabolic turnover of the DNA itself. The number of plastids in nondividing cells of expanding leaves increases severalfold. An extreme case is *Acetabularia,* in which the zygote contains a single nucleus and two chloroplasts. Before further nuclear divisions take place the number of plastids increases up to about $10^7$. In natural habitats portions of the cell of *Acetabularia* die at the end of the first growth season and active plastid proliferation starts again in the following spring. Multiplication and death of organelles in this case is obviously independent of nuclear divisions. The available data do not justify at present the claim for the presence of DNA that takes part in an active turnover.

## THE EVOLUTIONARY ORIGIN AND SIGNIFICANCE OF ORGANELLES

The question of the evolutionary significance of the compartmentalization of the hereditary material of the cell is most intriguing. I wish to consider two hypotheses on the origins of this phenomenon.

### *Origin from Endosymbionts*

An old and often repeated hypothesis [2, 29] considers the organelles to be well-integrated symbionts that invaded a primitive cell early in evolution. The appeal of this hypothesis is the fact that such a process of integration of a symbiont into the life of a cell is well known to occur. Different degrees of such relationships are known. Examples of such symbionts are the cytoplasmic particles of *Paramecium* [34] and the algal symbionts of many invertebrates [2]. The fact that the DNA of the plastids and mitochondria differs qualitatively from the nuclear DNA is also cited as support to a foreign origin of these organelles. Another argument used in support of this hypothesis is the fact that the plastids and mitochondria are separated by a double membrane from the rest of the cell, and remain so during the entire life of the cell [33]. Also the organization of the DNA fibers of the organelles, as revealed by electron microscope studies, resembles the DNA organization in a bacterial nucleoid rather than the chromosomes of the nucleus of higher organisms [29].

However, the fact that the organelles of different species differ in their DNA would require that different symbionts enter different cells and all proceed to evolve in an identical pattern. Another difficulty is the finding that some of the enzymes of the organelles are coded for by nuclear genes and are probably synthesized in the cytoplasm and then integrated into the organelles [14]. Furthermore some of the RNA of the mitochondria appears to complement nuclear DNA, suggesting that it was synthesized in the nucleus and transferred into the mitochondria [16]. The double membrane is thus only an apparent isolation rather than a true separation from the cytoplasm.

### *Origin from a Specialized Region of the Plasma Membrane*

An interesting new hypothesis was presented recently by Haldar *et al.* [14]. They note that a specialized region of the bacterial membrane, the mesosome, is rich in cytochromes. It was also suggested that the mesosome functions in separating the DNA molecules at the time of cell division. This region thus appears to combine the roles of mitochondria and centrioles, the DNA containing cytoplasmic organelles of the higher cells [10]. Haldar *et al.* suggest therefore that the mesosome area of bacteria is similar to the area of the membrane of a primitive cell from which the cytoplasmic organelles evolved after separating from the outer membranes of the cell. A segment of the DNA molecule presumably containing the code for a DNA replicase was incorporated into the separated vesicle. These vesicles then evolved into the familiar cytoplasmic organelles.

### *Evolutionary Significance of Compartmentalization of the DNA*

Several reports indicate a possible advantage to the cell in the proliferation of only parts of its genome.

Douthit and Halvorson [5] studied the DNA of the spore-forming bacteria, *Bacillus cereus.* DNA which was isolated from spores at an early stage of germination contained a satellite DNA which composed up to 18% of the total DNA of the cells. However this satellite DNA could not be detected in DNA isolated from cells in their log phase of growth, about 1 hour after germination.

It was previously suggested [30] that the ability of these bacteria to form spores is correlated with the presence of an episome-type of element in the cells because treatment with acridines, which supposably eliminates episomes, caused the loss of ability of the treated cells to form spores. The proliferation of this episome-like DNA appears therefore to be necessary for the production of spores. It is remarkable that this proliferation of part of the genome is a coordinated and controlled process in the life cycle of the cells.

Higuchi *et al.* [15] studied the induction of bacteriochlorophyll synthesis in *Rhodopseudomonas spheroides* when the bacteria were transferred from dark aerobic to dark anaerobic conditions. Although no cell proliferation occurs, DNA synthesis can be demonstrated during the adaptation period. Inhibition of DNA synthesis by antibiotics, such as mitomycin, phleomycin, and by acriflavins blocked also the synthesis of chlorophyll. DNA synthesis was not required for the simultaneous induction of ALA synthetase and catalase synthesis. These findings suggest the possibility that in this case too an episome-like segment of the DNA codes for the organization of the photosynthetic lamella and it proliferates separately in response to an environmental stimulation.

There are several examples of the proliferation of a small part of the DNA genome among higher organisms. Lima De Faria and Moses [21] studied an extrachromosomal DNA body in the oocytes of the fly *Tipula*. From cytological studies they concluded that this DNA body represents a many-fold multiplication of a small segment of the genome, the nucleolar-organizer region.

Miller [25] found that amphibian oocytes contain a high number of nucleoli in the periphery of their nucleus. Each of these was found to contain a DNA segment. Miller suggested that these nucleoli represent the products of the repeated multiplication of a small region of the genome. The segment is presumably required in excessive quantities for the synthesis of the proteins of the large egg cell.

It is implied in the above interpretations that the rate of synthesis of a specific protein might be limited by the rate of transcription from the DNA template.

A multiplicity of specific templates will provide a cell with two obvious advantages. Ontogenetically the cell is provided with an efficient and rapid mechanism for the synthesis of essential proteins; phylogenetically the multiplicity of templates can function to stabilize very important enzyme systems from the hazards of deleterious mutations.

## CONCLUSION

The genetic properties that are ascribed to plastids and mitochondria may now be attributed to the DNA molecules in these organelles. No genetic analysis of the genes which this DNA represents has been accomplished as yet.

We are yet to discover an appropriate biological system in which this intracellular genetics can be conveniently studied.

### References

1. Avers, C. J., Rancourt, M. W., and Kin, F. H., *Proc. Natl. Acad. Sci. U.S.* **54,** 527 (1965).
2. Buchner, P., "Endosymbiosis of Animals with Plant Microorganisms." Wiley (Interscience), New York, **1965.**
3. Chiang, K. S., Kates, J. R., and Sueoka, N., *Genetics* **52,** 434 (1965). Abstr.
3a. Cytoplasmic Units of Inheritance, Symp., *Am. Naturalist* **99,** 193 (1965).
4. Diacumakos, E. G., Garnjobst, L., and Tatum, E. L., *J. Cell Biol.* **26,** 427 (1965).
5. Douthit, H. A., and Halvorson, H. O., *Science* **153,** 182 (1966).
6. Gibor, A., *Science* **155,** 327 (1967).
7. Gibor, A., *in* "Biochemistry of Chloroplasts" (T. W. Goodwin, ed.), Vol. II. Academic Press, New York, **1966.**
8. Gibor, A., and Granick, S., *J. Cell Biol.* **15,** 599 (1962).
9. Gibor, A., and Granick, S., *Science* **145,** 890 (1964).
10. Granick, S., and Gibor, A., *Progr. Nucleic Acid Res.* **6,** 143 (1967).
11. Green, P., *Am. J. Botany* **51,** 334 (1964).
12. Hagemann, R., *in* "Genetics Today," Proc. 11th Intern. Congr. Genet. (S. J. Geerts, ed.), Pp. 613–625. Macmillan (Pergamon), New York, **1963.**
13. Hagemann, R., "Plasmatische Vererbung." Fischer, Jena, **1964.**
14. Haldar, D., Freeman, K., and Work, T. S., *Nature* **211,** 9 (1966).
15. Higuchi, M., Goto, K., Fujimoto, M., Namiloi, O., and Kikuchi, G., *Biochim. Biophys. Acta* **95,** 94 (1965).
16. Humm, D. G., and Humm, J. H., *Proc. Natl. Acad. Sci. U.S.* **55, 114** (1966).
17. Iwamura, T., *Progr. Nucleic Acid Res.* **5,** 133 (1966).
18. Jinks, J. L., "Extrachromosomal Inheritance." Prentice-Hall, Englewood Cliffs, New Jersey, **1964.**
19. Lang, N., *J. Protozoology* **10,** 333 (1963).
20. Leff, J., Mandel, M., Epstein, H. T., and Schiff, J. A., *Biochem. Biophys. Res. Commun.* **13,** 126 (1963).
21. Lima De Faria, A., and Moses, M. J., *J. Cell Biol.* **30, 177** (1966).
22. Luck, D. J. L., *Am. Naturalist* **99,** 241 (1965).
23. Manton, I., *J. Marine Biol. Assoc. U.K.* **38,** 319 (1959).
24. Marmur, J., and Doty, P., *J. Mol. Biol.* **5,** 109 (1962).
25. Miller, O. L., *J. Cell Biol.* **23,** 60A (1964).
26. Mounolou, J., Jakob, H., and Slonimski, P., Personal communication, **1966.**

27. Ogawa, K., and Barrnett, R. J., *Nature* **203,** 724 (1964).
27a. Plasmatic Inheritance, Symp., *in* "Genetics Today," Proc. 11th Intern. Congr. Genet. Vol. I (S. J. Geerts, ed.). Macmillan (Pergamon), New York, 1963.
28. Reich, E., and Luck, D. J. L., *Proc. Natl. Acad. Sci. U.S.* **55,** 1600 (1966).
29. Ris, H., and Plaut, W., *J. Cell Biol.* **13,** 383 (1962).
30. Rogolsky, M., and Slepecky, R. A., *Biochem. Biophys. Res. Commun.* **16,** 204 (1964).
31. Schiff, J. A., and Epstein, H. T., *in* "Reproduction: Molecular, Subcellular, and Cellular," 24th Symp. Soc. Study Develop. Biol. (M. Locke, ed.). Academic Press, New York, 1965.
32. Schildkraut, C. L., Marmur, J., and Doty, P., *J. Mol. Biol.* **4,** 430 (1962).
33. Schnepf, E., *in* "Probleme der biologischen Reduplikation" (P. Sitte, ed.). Springer, Berlin, 1966.
34. Sonneborn, T. M., *Advan. Virus Res.* **6,** 229 (1959).
35. Tewari, K. K., and Wildman, S. G., *Science* **153,** 1269 (1966).
36. Tewari, K. K., Votsch, W., Mahler, H. R., and Mackler, B., *J. Mol. Biol.* **20,** 453 (1966).
37. Van Wisselingh, C., *Z. Induktive Abstammungs-Vererbungslehre* **22,** 65 (1920).
38. Von Wettstein, D., and Erikson, G., *in* "Genetics Today," Proc. 11th Intern. Congr. Genet. (S. J. Geerts, ed.), p. 591. Macmillan (Pergamon), New York, 1964.
39. Wilkie, D., "The Cytoplasm in Heredity." Wiley, New York, 1964.
40. Woodward, D. W., and Munkres, K. D., *Proc. Natl. Acad. Sci. U.S.* **55,** 872 (1966).

# CYTOPLASMIC GENES AND ORGANELLE FORMATION

**RUTH SAGER**

*Department of Biological Sciences, Hunter College of the City University of New York, New York, New York*

## INTRODUCTION

The role of genes in the organization of subcellular structures has been a matter of strong dispute and sparse data for many decades. By 1900, the uniqueness of subcellular organelles, such as chloroplasts and mitochondria, was well established by cytologists; and the question of whether these organelles arose *de novo* or were formed from preexisting organelles of the same kind was already under debate. Cytologists such as Meves [16] postulated that mitochondria contained hereditary factors controlling their reproduction, and this point of view was strengthened by Correns' demonstration, published in 1908 [3], of the existence of non-Mendelian genes affecting chloroplast formation. Despite the early recognition and rather clear formulation of this problem, the methodology for its solution is only now becoming available.

We approach the problem now from the vantage point of powerful new understandings and experimental tools from the realm of molecular genetics. We recognize two classes of transcribed RNA's: the ribosomal and transfer RNA's which function in the machinery of protein synthesis, and the messenger RNA's which determine the specific sequence of amino acids in polypeptide synthesis. We also know of at least two modes of metabolic regulation operating at the DNA level: the control of each round of DNA replication which serves to maintain the constancy of DNA per cell; and the control of the rate of protein synthesis operating at the level of mRNA transcription.

The problem of organelle formation embodies these mechanisms and adds a further level of complexity, the assembly of macromolecules into structures. At the present time, we know of no template processes by which preexisting two-dimensional heterogeneous arrays of macromolecules determine the formation of copies of themselves. Approaches to this problem in terms of self-assembly are discussed in this Symposium by Klug and by Glauert and Lucy, and have been recently considered also in other symposia [11, 34, 36].

I shall take a different tack and discuss evidence for the role of cytoplasmic genetic systems in organelle formation. Ultimately we shall have to integrate these diverse approaches. Indeed, hints of the direction of this integration are already beginning to appear. In a recent report a cytoplasmic gene in *Neurospora* has been shown to code for a structural protein component of cytoplasmic membranes [37, 38]. In *Paramecium,* the arrangement of kineties (rows of cilia) in the cortex has been shown to be under control of a genetic system located not in the micronucleus, macronucleus, or free-flowing cytoplasm, but rather in the cortical region itself [1]. Subsequently a careful search revealed the presence of DNA in the basal bodies of the cilia [31]. What role, if any, this DNA may have in cortical organization is not known but can now be investigated.

In this context, mention should be made of the phage T4 system in which about 75 genes have been identified and mapped, and in which the genetic control of phage assembly is under investigation [5, 7]. By the brilliant use of conditionally lethal mutations, electron microscopic analysis of phage lysates and *in vitro* assembly of infectious virus from mixtures of lysates of cells infected with different mutants, it has been possible to distinguish two classes of genes. Some have been shown to code for structural components like the head and sheath proteins. Others, referred to as morphogenetic genes, do not seem to contribute a recognizable component, but each is essential for a particular step in phage assembly. Perhaps these genes code for proteins which act as joining elements in assembling the basic structure, but are present in very small amount and therefore difficult to detect experimentally. Perhaps their role is even a subtler one. Whatever the solution, the identification of the specific molecular role of morphogenetic genes in phage T4 may be expected to shed considerable light on the general problem of how genes control the assembly of structures.

## FORMAL CONSIDERATIONS

Reports of non-Mendelian genes are scattered through the literature of classical genetics [2, 4, 19]. In the 1920's and 1930's higher plants provided most of the evidence, and the phenotypes involved included not only chloroplast abnormalities, but also pollen sterility and impaired growth. Since 1940, numerous instances of cytoplasmic heredity have been reported in the higher microorganisms, i.e., the fungi, algae, and yeasts, and many sorts of phenotypes have been implicated. Cytoplasmic genes have been identified principally by non-Mendelian behavior in crosses and secondarily by the demonstration of somatic segregation of the parental cytoplasmic genes during vegetative growth of the progeny [13, 35].

Thus the existence of cytoplasmic genes was inferred initially from formal genetic analysis. To those familiar with the material, the hypothesis of a cytoplasmic genetic system seemed entirely reasonable. As with Mendelian genes in the early 1900's, the evidence needed corroboration from other lines of experimentation, such as cytochemistry. In the absence of independent confirmation, some critics proposed alternative explanations of the data.

In the 1950's the principal alternative hypotheses were the following: (A) non-Mendelian factors represent particulate genes composed of nucleic acids with structural and functional properties analogous to, if not identical with, those of chromosomal genes; and (B) non-Mendelian inheritance is based upon cytoplasmic metabolic systems, so-called steady state systems, consisting of interlocking series of enzymatic reactions stable under particular environmental conditions. (Such a system might be indefinitely self-perpetuating until a sufficient change of environmental conditions shifted it to a different stable phenotype.) The data newly reported in the last ten years support hypothesis A and provide no evidence for hypothesis B. However, while experimental findings can negate hypothesis B in each specific instance, they cannot do so in general. There is a strong theoretical objection to the steady state hypothesis, and it seems worthwhile to review it briefly here.

Let us recall the classical separation in biological systems between the genotype and the phenotype, or more accurately, between the replication of genetic determinants and the ongoing metabolism of the cell in its totality of reponses to a changing environment. The genotype represents the sum total of genetic determinants transmitted with great precision from generation to generation. The phenotype results from developmental interactions between the genes as they are expressed in their biological context and in a particular environment. Thus the phenotype shows great flexibility, the genotype essentially none. To maintain evolutionary stability, it would seem to be essential to separate the genes from the gene products so that the phenotype may vary while the genotype remains relatively fixed.

A steady state system of cytoplasmic inheritance, based not on genes, but on gene products, would be a system in which no separation has occurred between the genetic instructions and their translation. Indeed, in the steady state system the phenotype and genotype are identical. This feature alone casts great doubt on the evolutionary selective advantage and survival potential of such a system.

## POTENTIALITIES OF ORGANELLE GENETICS

Current studies of cytoplasmic genetic systems are rooted in two generalizations, long anticipated but very recently established. (a)

Organelles (chloroplasts and mitochondria) contain double-stranded DNA of high molecular weight as well as the machinery for protein synthesis. (b) Organelle biogenesis is controlled by interactions between chromosomal and cytoplasmic genes.

These two generalizations, while deriving from very different sorts of experiments, both stem from a similar conceptual frame of reference: recognition of cytoplasmic genetic systems and attempts to elucidate their role in organelle biogenesis and function. Relevant experimental approaches have included not only genetic analysis, but at the biochemical level physical studies of DNA, *in vitro* studies of RNA and protein synthesis in isolated cell fractions, hybridization of cytoplasmic DNA's and RNA's, and structural studies of individual organelle proteins. These lines of investigation are complementary: the biochemical approach describes the machinery while the genetic approach provides specificity by identifying and describing the function of individual genetic components of the system.

In addition, nucleic acid hybridization provides a means to identify genes that serve as templates for the transcription of RNA's present in large quantity, like the ribosomal and transfer RNA's. Hybridization is particularly interesting because it represents the first method for the identification of unmutated wild-type genes. Ultimately it may become possible to identify all genes by their gene products without recourse to the use of mutants.

As yet, however, the methods of genetic analysis—identification of genes by comparison of mutant and wild-type forms, and analysis of their structural organization by mapping—provide the only experimental approach to the recognition of individual genes and investigation of their functions.

At present genetics is also the analytical tool with the greatest sensitivity for the detection of DNA and for the identification of sites of cytoplasmic DNA beyond the resolving power of other methods. The identification of organelle DNA has been based on the technically fortunate circumstance that in many organisms minor DNA components have buoyant densities in cesium chloride considerably different from that of nuclear DNA. The lower limit of detection of minor components present in a DNA extract by banding in a CsCl gradient is about 1–5%, depending on the actual density differences. Randall and Disbrey [18] detected DNA in the basal bodies of the cilia of *Tetrahymena* by acridine orange fluorescence, estimating the amount as $2 \times 10^{-16}$ gm per basal body. Their qualitative results were confirmed in *Paramecium* by the same method as well as by autoradiography [31]. In this Symposium Dr. Stubblefield mentioned preliminary evidence of the presence of DNA in centrioles and Dr. Bessis suggested the use of laser techniques for the detection of stained or pigmented macromolecules present in very low amounts.

With genetic methods, however, single genes can be recognized, provided only that one has a suitable genetic system for the induction and detection of mutations. A single nucleotide change in a single gene can be recognized provided the altered gene product, for example, an altered protein, causes a detectable phenotypic change in the organism.

In principle, genetic analysis can provide the following kinds of information:

1. With the use of suitable mutagenic and selective systems it should be possible to identify every cytoplasmic gene.
2. By means of recombination analysis, it should be possible to elucidate the physical organization of these cytoplasmic genes. Are they organized into linear structures like chromosomes? Are there one or many linkage groups, etc?
3. With appropriate biochemical methods, it should be possible to locate the position of the mutant block and thereby to specify the function of each cytoplasmic gene in cellular metabolism.

Thus, in general terms, the use of genetic techniques in organelle biogenesis does not differ from the use of genetic analysis in the study of whole organisms.

If this appraisal of the problem is correct and geneticists have been aware of the existence of non-Mendelian genes since 1908, why has there been so much controversy and so little progress? There have been two classes of technical impediments, both rather formidable. First, it has been very difficult to induce mutations of cytoplasmic genes. Most cytoplasmic mutants so far studied have occurred spontaneously; a few have been induced with acridine dyes [6, 13, 32] and with nitrosoguanidine [8]. The only mutagenic agent as yet reported to induce a wide spectrum of cytoplasmic mutations is the antibiotic streptomycin acting on the single-cell green alga, *Chlamydomonas* [21]. The mode of action of streptomycin in this system is as yet unclear, although there is some evidence that it may act as an intercalating agent binding to DNA and interfering with replication, somewhat in the manner of the acridine dyes [15]. Why the drug should specifically act on the cytoplasmic system and not as a mutagen of chromosomal genes is as mysterious as the peculiar pattern of specificity of different acridine derivatives. (Acridine orange blocks replication of episomes, but not of the bacterial chromosomes; euflavin produces petite mutants in yeast, but proflavin is relatively inactive in that system; proflavin is a potent mutagen in phage-infected bacterial systems, but many other acridine dyes are not.)

Beyond the puzzling features of mutagen specificity, however, cytoplasmic genes as a class appear to be insensitive to irradiation and to conventional chemical mutagens. Whether this appearance reflects chemical differences in the molecules themselves or in their protection by other cell components such as DNA repair systems or more subtle features of their organization remains an open question. Obviously it is a problem of great importance both for general understanding and for technical availability of cytoplasmic genetic systems to investigation in any organism of choice.

The second serious impediment to cytoplasmic genetic analysis has been the maternal pattern of transmission of most cytoplasmic genes. In the higher plants and in the *Ascomycetes* most of the cytoplasm is contributed by the female parent in crosses, and the fact that cytoplasmic genes in these systems show maternal transmission has been interpreted as evidence of their cytoplasmic location. So long as transmission is purely maternal, there is no opportunity for recombination analysis or for mapping. Indeed, the genetic analysis of maternal inheritance is exceedingly dull and uninformative.

In *Chlamydomonas* the gametes are isogamous, both contributing equal amounts of cytoplasm in the cell fusion process leading to zygote formation. Nevertheless, cytoplasmic genes in *Chlamydomonas* exhibit maternal inheritance. Cytoplasmic genes from the female parent ($mt^+$) are transmitted to all the progeny in crosses, whereas usually those from the male parent ($mt^-$) disappear in the zygote and never reappear among the progeny or their descendants.

A few years ago, working with *Chlamydomonas* we found exceptions to the pattern of maternal inheritance [25, 26]. In these exceptional zygotes, which occurred with a frequency of 0.01–0.1%, nonchromosomal genes were transmitted from both parents to all the progeny and recombination analysis was possible. This finding marked the inception of recombination analysis of cytoplasmic genes. Recently we have developed a method to increase the yield of exceptional zygotes to almost 100% [27]. The studies which led to this finding have also thrown some new light on the mechanism of uniparental transmission of nonchromosomal genes in *Chlamydomonas* and may have general implications for other systems.

The results reveal the presence of a DNA-based system which controls maternal inheritance by blocking the transmission of male cytoplasmic genes to the progeny. As will be described below, the blocking mechanism can be abolished by a low dose of UV irradiation to the female parent before mating. The UV effect can in turn be reversed by photoreactivation with visible light, demonstrating that nucleic acids, probably DNA's, are the effective UV targets.

The presence of such a system indicates an evolutionary importance to the organism in maintaining uniparental transmission of nonchromosomal genes. In higher organisms, the same effect is achieved by decreasing the cytoplasmic contribution of the male parent. However, it is not clear how complete has been the elimination of cytoplasmic components. In most higher plants which have been examined, cytoplasmic genes show strict maternal inheritance. In a few, however, appreciable amounts of male transmission have been detected. In sperm the mitochondrial mid-piece enters the egg at the time of fertilization, at least in some species. It should be possible to discover whether mitochondrial DNA from the male parent enters the egg and what is its fate. It would be of interest to discover whether higher organisms retain an enzymatic mechanism to ensure the maintenance of a pattern of uniparental transmission of cytoplasmic genes. If so, blocking that mechanism might facilitate cytoplasmic genetic studies.

According to present day understandings of the evolutionary process, mutation and recombination provide the genetic variability for natural selection. Both the difficulty of inducing mutations and the impediment to recombination provided by maternal inheritance indicate that cytoplasmic genetic systems have been designed to minimize evolutionary variability [23]. Unfortunately for the geneticist, these blocks to variability also represent blocks to conventional genetic analysis.

At the present time, *Chlamydomonas* represents the only experimental system in which these blocks have been overcome: with the use of streptomycin as a mutagen for cytoplasmic genes and with the development of the UV method for induction of biparental inheritance. The rest of this paper will summarize the present state of our knowledge concerning the organization of cytoplasmic genes in *Chlamydomonas*.

## ON THE MECHANISM OF MATERNAL INHERITANCE

Cytoplasmic genes in *Chlamydomonas* were originally detected by their maternal pattern of inheritance [20]. New mutations can readily be classified as Mendelian (i.e., chromosomal) or non-Mendelian (i.e., cytoplasmic) by their pattern of transmission in crosses with the wild type as shown in Fig. 1. We employ the distinction between 2 : 2 and 4 : 0 segregation for routine preliminary classification of mutants. In biparental zygotes in which nonchromosomal genes are transmitted from both parents to the progeny, a further criterion of non-Mendelian be-

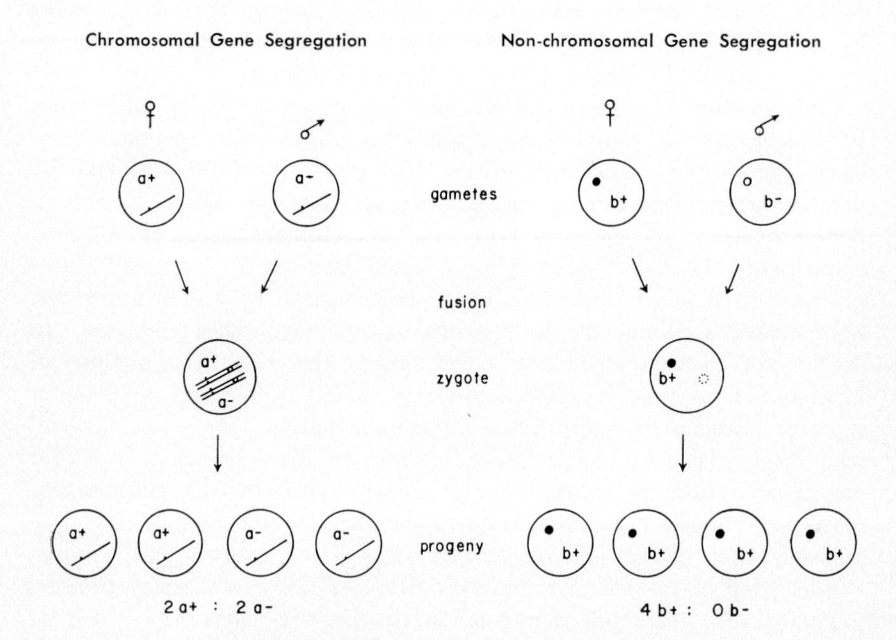


FIG. 1. Chromosomal and cytoplasmic gene segregation in a haploid organism. The segregation of chromosomal gene *a*, present as the $a^+$ allele in the female parent and as the $a^-$ allele in the male parent, depends on the mechanics of chromosome duplication and segregation at meiosis. Each chromosome duplicates, and the homologous chromosomes from the two parents pair, giving rise to four strands for each chromosome of the haploid set. In meiosis the strands are distributed equally to the four progeny cells. In the same cells, the cytoplasmic genes, represented as $b^+$ in the female and $b^-$ in the male, enter the zygote, but the $b^-$ complement is generally not transmitted to the progeny. (After Sager [22].)

havior is found in the pattern of segregation and recombination of cytoplasmic genes which occurs during the mitotic multiplication of haploid zoospore clones. An example of this behavior is shown in Fig. 2.

In biparental zygotes, whether spontaneous or induced by UV irradiation of the female parent before mating, cytoplasmic genes from both parents are transmitted to the four products of meiosis. These cells, called zoospores, are haploid for their chromosomal complement, but contain cytoplasmic genes from both parents. Thus these zoospores contain twice the customary content of cytoplasmic genes, and, perhaps for this reason, segregation occurs during the first several mitotic doublings, leading eventually to the appearance of progeny carrying only

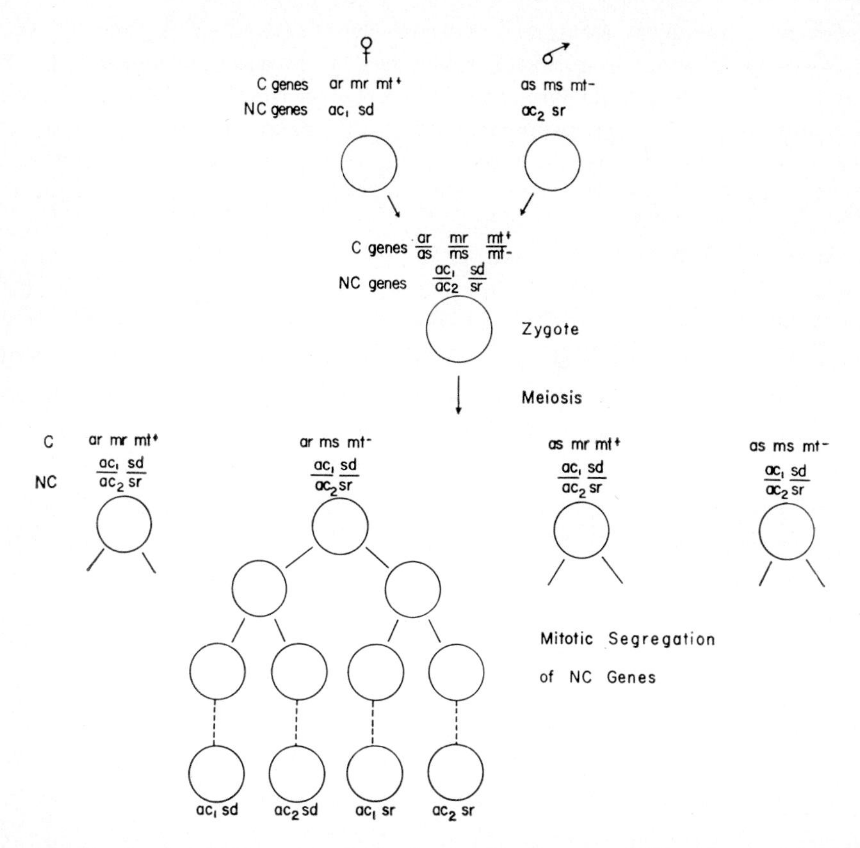


**Cross showing C and NC gene segregation in exceptional zygotes of Chlamydomonas (haploid)**

FIG. 2. Cross showing chromosomal and cytoplasmic gene segregation in exceptional zygotes of *Chlamydomonas*. In this cross the female parent differs from the male by three pairs of unlinked chromosomal (C) genes and two pairs of cytoplasmic (NC) genes. The zygote is diploid, containing all the genes from both parents. In meiosis the C genes segregate as determined by chromosome behavior, giving rise to four genetically different products. The NC genes do not segregate in meiosis, and in these exceptional zygotes the haploid progeny initially still contain NC genes from both parents. NC gene segregation occurs in the mitotic divisions of each vegetative clone after meiosis. (After Sager [22].)

one or the other of each parental cytoplasmic gene. The segregation process will be discussed below.

In the past, our studies of segregation were hampered by the low frequency of occurrence of biparental zygotes. For this and other reasons, we instigated a search for ways to enrich this fraction. We obtained about a 10-fold enrichment by placing freshly formed zygotes at temperatures of 37–40°C and found that the effective period for the high-temperature treatment occurred very soon after mating. This information was valuable, but the increased yield was inadequate. The results with UV irradiation were far more satisfactory [27].

Irradiation of the female parent before mating produced a dramatic rise in the percentage of the biparental zygotes, whereas similar treatment of the male parent had no effect. The effect of UV on survival of female gametes and on the conversion from maternal to exceptional zygotes is shown in Figs. 3 and 4. In both systems there is a strong

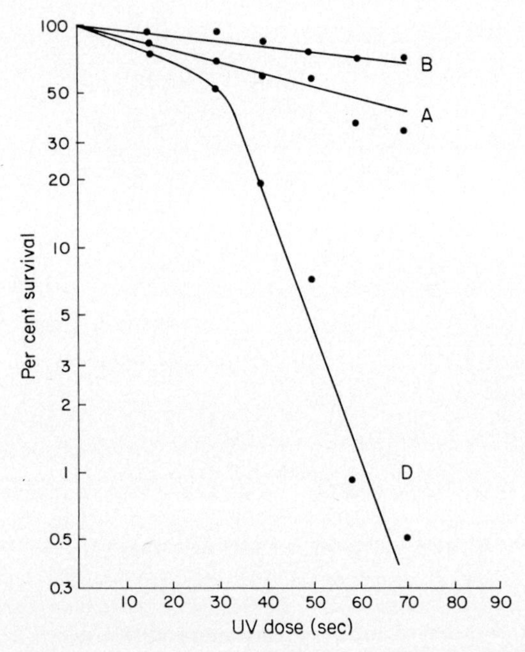


FIG. 3. Effect of UV on survival of ♀ gametes. Gametes were irradiated, plated at dilution, and incubated in dark (curve *D*), in light immediately after plating (curve *B*), or in light after 2 hours in dark (curve A) from Sager and Ramanis [27].

sensitivity to photoreactivation by visible light, and the effectiveness of photoreactivation is shown in the dose reduction curves in Fig. 5. The dark survival curve for gametes is comparable to that obtained with other strains of *Chlamydomonas*, and the slope of the dose reduction curve indicates that the effect of photoreactivation is in line with that recorded for other organisms [17, 30]. The dose reduction curve for the conversion of maternal to exceptional zygotes has a similar slope, suggesting that the same process, presumably the photoreversal of pyrimidine dimers, is involved.

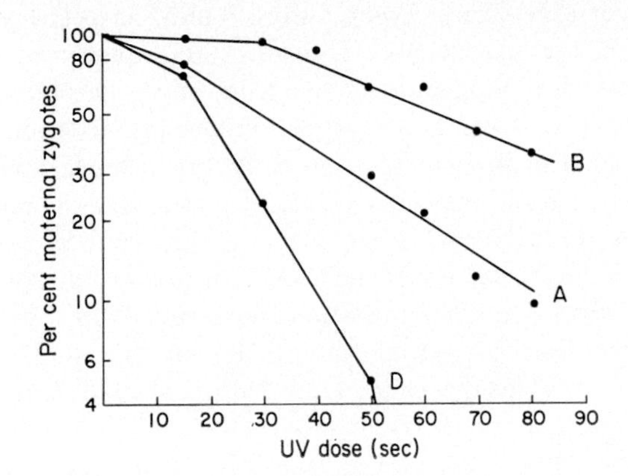


FIG. 4. Effect of UV on conversion of maternal to exceptional zygotes. Female gametes were irradiated, mated with unirradiated males, and kept in dark until zygote formation was completed. Zygotes were plated at dilution, incubated in dark (*D*), or in light (*A*). In curve *B*, mating gametes were exposed to light and all subsequent plating was done in light. After 24 hours in light, *A* and *B* series were placed in dark with *D* series for 1 week, then all plates were exposed to light for germination of zygotes from Sager and Ramanis [27].

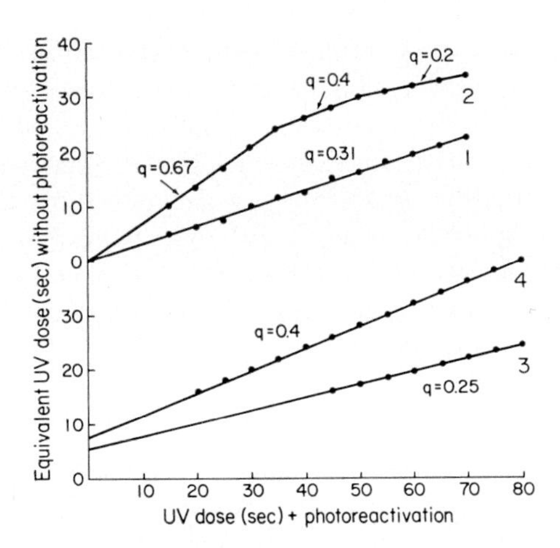


FIG. 5. Dose reduction curves for gametes and zygotes, based on data of Figs. 3 and 4. Curve *1:* Gametes were photoreactivated immediately. Curve *2:* Gametes were photoreactivated after 2 hours in dark. Curve *3:* Exceptional zygotes recovered when $mt^+$ gametes were photoreactivated before mating. Curve *4:* Exceptional zygotes recovered when zygotes were photoreactivated after mating.

*Abscissa:* UV dose + photoreactivation producing a given yield of survivors (gametes) in curves *1* and *2;* exceptional zygotes in curves *3* and *4. Ordinate:* UV dose without photoreactivation producing the same yield of survivors at each corresponding point from Sager and Ramanis [27].

Two classes of exceptional zygotes were found. At relatively low doses of UV most of the exceptional zygotes were biparental, that is, the zoospores received cytoplasmic genes from both parents. As the UV dose was increased, there appeared an increasing fraction of paternal zygotes in which the progeny carried only the cytoplasmic genes from the male parent. With photoreactivation, biparental zygotes could be converted back to maternal ones, and the paternal zygotes could be converted back both to biparental and to maternal ones. Thus we recognize two effects of UV irradiation, both reversible by visible light: (a) the preservation of cytoplasmic genes of the male parent, and (b) the loss of cytoplasmic genes from the irradiated female parent at higher UV doses.

We have postulated that at the time of mating the female gametes normally form a gene product, possibly an enzyme, required for the loss of the male cytoplasmic gene. UV presumably blocks the synthesis of this gene product or inactivates it. The process might be analogous to the UV induction of phage which results from a block in the synthesis of the phage repressor substance. At higher doses, UV also interferes with the replication of the female cytoplasmic genes. *Chlamydomonas* is particularly suitable for studies of this sort because the replication of chloroplast DNA occurs immediately after mating and several days before the replication of the chromosomal DNA which occurs at the time of meiosis [33]. We have not detected any effect of this irradiation upon the chromosomal genetic complement.

The response of cytoplasmic genes to UV irradiation, and particularly to photoreactivation, provides strong evidence for the nucleic acid nature of the cytoplasmic genes which we have been investigating. These results are consistent with the evidence concerning segregation and recombination, which previously led us to postulate the nucleic acid nature of these determinants [26].

## CLASSES OF CYTOPLASMIC MUTANTS

Some fifty different cytoplasmic mutants have been identified by our standard crossing procedure following mutagenic treatment with streptomycin. Mutant types selected for include: (a) acetate-requiring mutants, a heterogeneous class unable to carry out photosynthesis and therefore requiring acetate as a carbon source for heterotrophic growth; (b) mutants responding to streptomycin and related antibiotics—the wild-type is sensitive and mutants exhibit a number of levels of resistance and dependence; (c) slow-growing strains which do not respond to supplements in the medium; and (d) temperature-sensitive strains that grow at 25°C but not at 35°C.

Temperature-sensitive mutants are particularly interesting because with them it should be possible in principle to detect every gene that codes for a protein. In known proteins, single amino acid substitutions at many locations can induce thermolability such that the protein shows an altered behavior only at elevated temperatures. With mutations of this type, genes can be identified regardless of their essential function. In phage T4, for example, the gene that codes for DNA polymerase has been identified by means of temperature-sensitive mutants, so-called conditional lethals [7]. Studies in *Chlamydomonas* are recent, but already they have revealed the presence of numerous temperature-sensitive cytoplasmic mutants.

It seems likely that the acetate-requiring mutants are located in chloroplast DNA, but critical evidence in support of this view is not yet available. The location of the cytoplasmic genes affecting antibiotic resistance and dependence is also not established, but they too may reside in chloroplast DNA. Chloroplast ribosomes are 70 S and resemble bacterial ribosomes in some properties [24]. In bacteria, genetic changes in ribosomal structure have been correlated with changes in response to streptomycin in *in vitro* polypeptide synthesis [10]. In *Chlamydomonas* we found that the 80 S cytoplasmic ribosomes from streptomycin sensitive, resistant, and dependent strains are insensitive to streptomycin in an *in vitro* system examining polypeptide synthesis and miscoding [29]. A similar study of chloroplast ribosomes has just become technically possible. If chloroplast ribosomes do respond to streptomycin, this class of cytoplasmic genes that influence streptomycin resistance and dependence might indeed be located in the chloroplast.

As will be noted below, the acetate and streptomycin markers appear to be unlinked, but the data do not exclude the possibility that they are widely separated on the same linkage group. It should also be pointed out that chromosomal genes are known with phenotypes similar to the cytoplasmic ones. Thus it seems that in *Chlamydomonas* as in other organisms, nuclear and cytoplasmic genetic systems interact in the control of organelle formation.

## SEGREGATION AND RECOMBINATION OF CYTOPLASMIC GENES

Detailed studies of segregation and recombination have been focused primarily upon two genetic regions. One of these contains the cluster of mutations that alter the cellular response to streptomycin and related antibiotics, and the other the cluster of mutations leading to acetate requirement. We have previously reported that the two regions segregate independently [25, 26]. We have also noted that contrasting alleles from the two parents (for example, $ac_1/ac_2$ and *sr/sd* in cross 1, Fig. 2) segregate on the average 1 : 1 among the progeny.

Intragenic recombination has been found to occur between pairs of different acetate-requiring mutants [26]. Hybrid zoospores from crosses of $ac_1 \times ac_2$ give rise to $ac^+$ and $ac_1 - ac_2$ recombinants during clonal multiplication. Mutants $ac_1$ and $ac_2$ are considered alleles because they complement in hybrid cells, giving better growth in the absence of acetate than either mutant alone, while the double mutant is an extreme acetate-requiring type.

Recombinants between $ac_1$ and $ac_2$ arise at a constant rate of about 1–5% per doubling of hybrid cells. Other acetate mutants have been crossed with $ac_1$ and give rise to recombinants at lower frequencies, each characteristic of a particular pair of alleles. These results have made possible a preliminary mapping of mutants within the acetate region.

Studies of recombination in the streptomycin region have indicated linkage among a group of markers including different levels of streptomycin resistance, dependence, and neamine resistance and dependence. However, there are complications in the data which have not yet been resolved [9, 28].

The detailed pattern of segregation of the acetate markers has been examined in a number of crosses, both with and without UV irradiation. Some of the results are summarized in Fig. 6. In general, no striking quantitative differences were observed in the rate of segregation of pure parental types from hybrid cells in the progeny of biparental zygotes recovered with and without UV. In Fig. 6 these results are compared as well with those obtained in cross 2, a similar cross in which the female parent is streptomycin-sensitive rather than dependent and the male parent is the same as that used in cross 1.

The results summarized in Fig. 6 show clearly that starting with hybrid zoospores, segregation of pure parental types ($ac_1$ and $ac_2$) occurs at a constant average rate per doubling for at least eight doublings. Attempts were made to fit the experimental data to a series of theoretical curves computed for the random segregation of various numbers of copies of each parental gene. Models were considered assuming random distribution of the number of copies as well as those assuming constancy of number, but randomness of kind. The experimental curve shown in Fig. 6 falls between the theoretical curves for one-copy and for two-copy models, and does not fit well to any random model we have considered.

Models involving random segregation have also been excluded by examination of individual hybrid cells and their vegetative progeny.

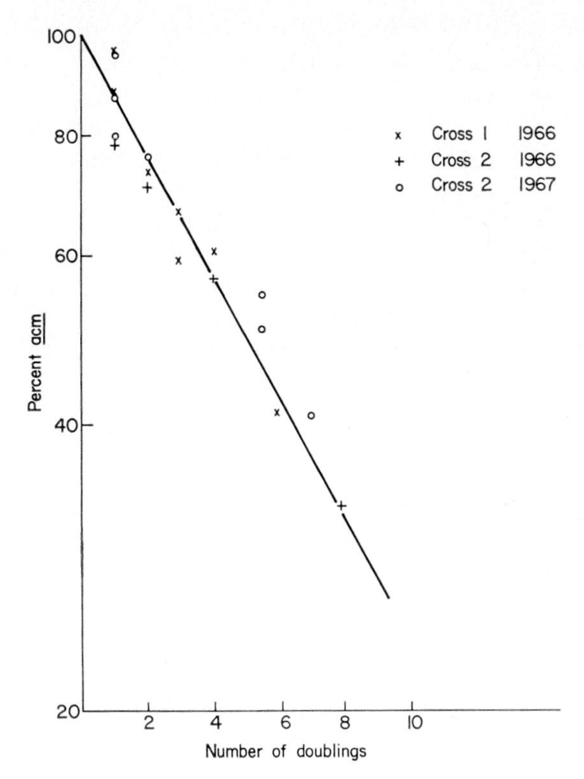


FIG. 6. Segregation of *acm* hybrids in zoospore clones. Population of zoospores were grown in liquid or on agar. Aliquots were sampled after known numbers of doublings and classified as to genotype: hybrid (*acm*) or pure type ($ac_1$ or $ac_2$).

Random segregation would result in the presence of daughter hybrid cells with unequal numbers of copies of the two parental types. This prediction is easy to test experimentally. One simply selects hybrid cells recovered after known numbers of doublings and allows each to form a clone which can then be tested for the relative frequency of $ac_1$ and $ac_2$. Extensive tests have demonstrated the remarkable uniformity of hybrid cells; with rare exceptions, they regularly produce equal numbers of the two parental types.

This result effectively excludes all random models involving two or more copies of each parental allele, as well as some one-copy models. All random one-copy models are excluded by the observed segregation rate, which is far too low. Thus we are led to conclude that some nonrandom process is involved in segregation.

A more detailed examination of the segregation pattern occurring in individual hybrid cells has provided further evidence of oriented segregation as well as certain complications. However, it seems evident that the observed segregation patterns of cytoplasmic genes reflect the replication and distribution of cytoplasmic DNA's just as Mendel's factors delineated the behavior of chromosomes at meiosis. The apparent complexities we find may result from the fact that hybrid cells contain twice the usual content of cytoplasmic genes, and that this extra burden is being reduced by segregation following partial replication. Such a hypothesis might be tested by noting the effect upon genetic segregation of induced changes in the replication of cytoplasmic DNA's, and experiments of this kind are in progress.

This paper has made no attempt to review the exponentially rising number of reports relevant to the topic under discussion. Rather, we have focused on work now in progress with a particular genetic system designed to locate and identify cytoplasmic genes by inducing them to mutate, and to map the mutants as a means of studying their physical organization.

Two clusters of genes have been investigated, and each cluster is now being mapped, as are other cytoplasmic genes outside these clusters. The clusters appear unlinked, but they may be far apart on the same linkage group, a matter which should be resolved soon. It seems likely that each linkage group comprises a replicon [12, 14] in which orderly replication and distribution to daughter cells is determined by attachment to a specific site on an organelle membrane.

## CONCLUDING REMARKS

No consideration of cytoplasmic genetic systems is complete without a few speculative remarks on their function and value to the organism. Since new information is now appearing so rapidly in this field, the following list is offered in the spirit of trial and error.

1. *Mutational stability.* Cytoplasmic genes exhibit great mutational stability toward potent chromosomal mutagens. We may surmise that this stability is as important to the organism from an evolutionary standpoint as is the mutability of chromosomal genes. Since both systems are made of DNA, mutability must be regulated indirectly, perhaps by the efficiency of DNA repair. Whatever the mechanism, the chromosomal and cytoplasmic genetic systems would have to be separated in space to assure the maintenance of differential mutability.

2. *Ribosomal differentiation.* Chloroplasts and mitochondria contain ribosomes which differ from the cytoplasmic ones present in the same organism, not only in physical properties, but also in response to various inhibitors. Perhaps this difference reflects an important regulatory de-

vice which requires that organelle ribosomes be compartmentalized during their synthesis as well as during their functional activity. Evidence of independent synthesis is provided in recent reports that some organelle RNA's are coded for by the corresponding organelle DNA.

3. *Membrane proteins.* The structural proteins of mitochondrial and other cytoplasmic membranes have special physical properties conferring insolubility under physiological conditions. These properties may require that the proteins be synthesized *in situ.* If so, it may be necessary that the transcription of messenger RNA as well as its translation occurs at the site of membrane formation.

4. *Regulatory flexibility.* In *Chlamydomonas* cytoplasmic DNA's are replicated at a different time than chromosomal DNA's, both in the vegetative cell cycle and in meiosis; this is probably a general feature of other cell types as well. This separation in time of replication suggests the existence of independent regulatory mechanisms. There is also some evidence that the number of copies of cytoplasmic DNA's may vary to some extent independently of the cell cycle and in response to environmental signals. If organelle biogenesis is to some extent independent of the cell cycle, it would probably be necessary to maintain an independent genetic system as well. However, the fact that many chromosomal genes also contribute structural components and regulatory control in organelle formation indicates that the question of regulatory flexibility is in fact a very complex matter and may well represent the focus of future studies of organelle biogenesis.

## References


1. Beisson, J., and Sonneborn, T. M., *Proc. Natl. Acad. Sci. U.S.* **53,** 275 (1965).
2. Caspari, E., *Advan. Genet.* **2,** 2 (1948).
3. Correns, C., *Z. Induktive Abstammungs-Vererbungsehre* **1,** 291 (1908).
4. Correns, C., *in* "Handbuch der Vererbungswissenschaften" IIH. (F. V. Wettstein, ed.) Borntraeger, Berlin, 1937.
5. Edgar, R. S., and Wood, W. B., *Proc. Natl. Acad. Sci. U.S.* **55,** 498 (1966).
6. Ephrussi, B., and Hottinguer, H., *Nature* **166,** 956 (1950).
7. Epstein, R. H., Bollé, A., Steinberg, C. M., Kellenberger, E., Boy de la Tour, E., Chevalley, R., Edgar, R. S., Susman, M., Denhardt, G., and Leilausis, A., *Cold Spring Harbor Symp. Quant. Biol.* **28,** 375 (1963).
8. Gillham, N. W., *Genetics* **52,** 529 (1965).
9. Gillham, N. W., *Proc. Natl. Acad. Sci. U.S.* **54,** 1560 (1965).
10. Gorini, L., Jacoby, G. A., and Breckenridge, L., *Cold Spring Harbor Symp. Quant. Biol.* **31,** 657 (1966).
11. Hayashi, T., and Szent-Györgyi, A. G., eds., "Molecular Architecture in Cell Physiology." Prentice Hall, Englewood Cliffs, New Jersey, 1966.
12. Jacob, F., Brenner, S., and Cuzin, G., *Cold Spring Harbor Symp. Quant. Biol.* **28,** 329 (1963).

13. Jinks, J. L., "Extrachromosomal Inheritance." Prentice Hall, Englewood Cliffs, New Jersey, 1965.
14. Lark, K. G., *Bacteriol. Rev.* **30,** 3 (1966).
15. Lerman, L., *J. Cellular Comp. Physiol.* **64,** Suppl. 1, 1 (1964).
16. Meves, F., *Arch. Mikroskop. Anat.* **72** (1908).
17. Novick, A., and Szilard, L., *Proc. Natl. Acad. Sci. U.S.* **35,** 591 (1949).
18. Randall, J., and Disbrey, C., *Proc. Roy. Soc.* (*London*) **B162,** 473 (1965).
19. Rhoades, M. M., *in* "Encyclopedia of Plant Physiology" (W. Ruhland, ed.), Vol. 1, p. 19. Springer, Berlin, 1955.
20. Sager, R., *Proc. Natl. Acad. Sci. U.S.* **40,** 356 (1954).
21. Sager, R., *Proc. Natl. Acad. Sci. U.S.* **48,** 2018 (1962).
22. Sager, R., *New Engl. J. Med.* **271,** 352 (1964).
23. Sager, R., *Proc. Roy. Soc.* (*London*) **B164,** 290 (1966).
24. Sager, R., and Hamilton, M. G., *Science* **157,** 709 (1967).
25. Sager, R., and Ramanis, Z., *Proc. Natl. Acad. Sci. U.S.* **50,** 260 (1963).
26. Sager, R., and Ramanis, Z., *Proc. Natl. Acad. Sci. U.S.* **53,** 1053 (1965).
27. Sager, R., and Ramanis, Z., *Proc. Natl. Acad. Sci. U.S.* **58,** 931 (1967).
28. Sager, R., and Ramanis, Z., Unpublished observations, 1967.
29. Sager, R., and Toback, F. G., Unpublished observations, 1966.
30. Setlow, R. B., *in* "Regulation of Nucleic Acid and Protein Biosynthesis" (V. V. Koningsberger and L. Bosch, eds.), Biochim. Biophys. Acta Library Vol. **10,** pp. 51–62. Elsevier, Amsterdam, 1967.
31. Smith-Sonneborn, J., and Plaut, W., *J. Cell Sci.* **2,** 225 (1967).
32. Srb, A. M., *Symp. Soc. Exptl. Biol.* **17,** 175 (1963).
33. Sueoka, N., Chiang, K. S., and Kates, J. R., *J. Mol. Biol.* **25,** 47 (1967).
34. Vogel, H. J., Lampen, J. O., and Bryson, V., eds., "Organizational Biosynthesis." Academic Press, New York, 1967.
35. Wilkie, D., "The Cytoplasm in Heredity." Methuen, London, 1964.
36. Wolstenholme, G. E. W., and O'Connor, C. M., eds., "Principles of Biomolecular Organization," Ciba Found. Symp. Little, Brown, Boston, Massachusetts, 1966.
37. Woodward, D. D., and Munkres, K. D., *Proc. Natl. Acad. Sci. U.S.* **55,** 872 (1966).
38. Woodward, D. D., and Munkres, K. D., *in* "Organizational Biosynthesis" (H. J. Vogel, J. O. Lampen, and V. Bryson, eds.). Academic Press, New York, 1967.

# AUTHOR INDEX

Numbers in parentheses are reference numbers and indicate that an author's work is referred to although his name is not cited in the text. Numbers in italics show the page on which the complete reference is listed.